PRAISE FOR THE FIRST EDITION

". . . a wonderfully useful description of how science works. The examples are informative and effective. The exercises are imaginative, both thoughtful and thought-provoking. And the components of scientific method are clearly presented with enough detail to see not only how they work but both their strengths and limitations."

Peter Kosso in *Science & Education*

"More often than not students acquire content knowledge about science, deprived from any explicit reflection about the methods, the reasoning and the uncertainties that characterize it. . . . But this is not how science is done. If there are recipes, they are open to creativity and they vary enormously. *Recipes for Science* excellently shows this and provides very useful materials for explicit reflection about the nature of science."

Kostas Kampourakis, *University of Geneva, Switzerland*

T0321022

Recipes for Science

Scientific literacy is an essential aspect of an undergraduate education. *Recipes for Science* responds to this need by providing an accessible introduction to the nature of science and scientific methods appropriate for any beginning college student. The book is adaptable to a wide variety of different courses, such as introductions to scientific reasoning, methods courses in scientific disciplines, science education, and philosophy of science.

Special features of *Recipes for Science* include contemporary and historical case studies from many fields of physical, life, and social sciences; visual aids to clarify and illustrate ideas; text boxes to explore related topics; plenty of exercises to support student recall and application of concepts; suggestions for further readings at the end of each chapter; a glossary with helpful definitions of key terms; and a companion website with course syllabi, internet resources, PowerPoint presentations, lecture notes, additional exercises, and original short videos on key topics.

KEY UPDATES TO THE SECOND EDITION

- 13 short chapters of uniform length that make it easier to adapt to a college semester
- Case studies and examples featuring new research and important historical research across many fields of science
- Added discussion of timely topics, including large research collaborations, trust and distrust of science, machine learning and other technology-driven advances, diversity in science, and connections to indigenous knowledge
- Streamlined and simplified discussion of some topics, such as experimentation and statistical hypothesis-testing
- Exercises that are clearly aligned with learning goals and sorted into types: Recall, Apply, and Think
- Additional online exercises and a series of original videos on key topics
- Exercise solutions available on an instructor-only section of the website

Angela Potochnik is Professor of Philosophy and Founding Director of the Center for Public Engagement with Science at the University of Cincinnati, USA.

Matteo Colombo is Associate Professor in the Department of Philosophy at Tilburg University in The Netherlands.

Cory Wright is Professor of Philosophy at California State University Long Beach, USA.

Recipes for Science

An Introduction to Scientific Methods and Reasoning

Second Edition

Angela Potochnik,
Matteo Colombo, and
Cory Wright

Routledge
Taylor & Francis Group

NEW YORK AND LONDON

Designed cover image: "Nautilus" by Rafael Araujo—www.rafael-araujo.com.
© Rafael Araujo, used with permission.

Second edition published 2024
by Routledge
605 Third Avenue, New York, NY 10158

and by Routledge
4 Park Square, Milton Park, Abingdon, Oxon, OX14 4RN

Routledge is an imprint of the Taylor & Francis Group, an informa business

First edition published by Routledge 2019

ISBN: 978-1-032-29097-3 (hbk)
ISBN: 978-1-032-29096-6 (pbk)
ISBN: 978-1-003-30000-7 (ebk)

DOI: 10.4324/9781003300007

Typeset in Berling
by Apex CoVantage, LLC

Access the Companion Website: http://www.routledge.com/cw/potochnik

For all the excellent teachers from whom we've learned our love of science

Contents

Introduction

SCIENCE AND YOUR EVERYDAY LIFE

A highly contagious and potentially fatal virus spreads through respiratory droplets but is easily preventable through a simple vaccination. Sound familiar? Since you've gone through a pandemic—one of the events in your lifetime that will make history—we know what you're thinking: Covid-19!

But no, the virus we're talking about is canine distemper. This is one of the most serious diseases a dog can get. It can also affect foxes, pandas, skunks, and raccoons. Like rabies, distemper is nothing you or your furry friend want to mess with. It can lead to some gnarly symptoms, including dehydration, vomiting, difficult breathing, and odd fits of snapping or chomping. As the condition worsens, infected dogs can suffer grand mal seizures followed by death.

Dog owners find the idea of canine distemper distressing. Most want effective prevention and treatment for their pets and, if they hear of canine distemper, follow expert veterinary advice. What's tragically ironic is that some of the same dog owners who do not hesitate to vaccinate their puppies against distemper are skeptical towards vaccines for humans, discounting the advice of medical experts.

Human papilloma virus (HPV) is widespread worldwide; it is one of the world's most frequently sexually transmitted diseases. Among other effects, HPV substantially increases the risk of various forms of cancer. Thankfully, there's a vaccine for HPV. The vaccine first became available in 2006, after thorough testing for safety and efficacy. The World Health Organization (WHO) recommends HPV vaccines as part of routine vaccinations in all countries. And yet, the global HPV vaccination rate in 2018 was estimated to be only 12.2%. This low rate is a combination of limited availability and affordability in low- to middle-income nations and parents' resistance to their children receiving the HPV vaccine, especially in high-income nations.

At some point in their lives, most people will need to decide about whether to be vaccinated for some disease, whether to have their children undergo routine vaccinations, and even whether to have their dog vaccinated. And, when it comes to vaccines for humans, sometimes vaccine skeptics have louder voices than doctors and other experts. So, it can be difficult to get a clear understanding of vaccines' safety, effectiveness, and necessity for public health. What can help is a sound understanding of the medical research about vaccines and a deeper understanding of what makes this and other scientific research trustworthy.

DOI: 10.4324/9781003300007-1

FIGURE 0.1 Routine vaccinations are medically essential for puppies and children

All vaccines approved for use in humans or animals have undergone thorough testing for safety and efficacy. Most vaccines do have potential side effects, such as achiness for a day or two, but serious side effects for any approved vaccine are very rare. And vaccine skeptics' claims about substantive risks of vaccinations have been thoroughly debunked.

But you don't need to just take our word for it, or the word of any individual scientist. Science's credibility and its trustworthiness as a source of knowledge result from its methods and reasoning processes and how these are carried out by scientists and endorsed by scientific institutions. This is what makes established scientific knowledge trustworthy. Learning more about these methods and reasoning processes can give you a sense for why scientific research results in trustworthy knowledge, and it can also prepare you to evaluate specific reports of scientific research findings.

Understanding how science works is important because scientific findings, and the public's reactions to them, dramatically shape our world. Science regularly influences your life, whether you are looking for it to or not. Consider the Covid-19 pandemic, the internet and the rise of AI, and climate change policy. The ability to understand and assess scientific reasoning enables you to make educated decisions related to your health, medical care, lifestyle, and more. It also enables you to critically evaluate reports of scientific findings and the credentials of experts in order to decide not just what to believe, but—more importantly—why.

Someone with a sophisticated understanding of science is also well positioned to make judgments about science more globally. Is the level of public funding for scientific research sufficient? Should we worry if private corporations fund science? Are there important topics that aren't yet adequately targeted in scientific research? Answers to these questions require thinking about the status of the scientific enterprise as a whole, how it should relate to society, and whether and how funding sources matter.

RECIPES FOR SCIENCE

Scientists are, of course, the main practitioners of science. Other researchers have as their primary focus understanding what science—and scientists—are up to. These researchers investigate what science is and how it works, its challenges and limitations, and its relationship with society. These topics are what this book is all about. Several disciplines investigate science in this way; primary among these are history, sociology, and philosophy. Historians of science research how science has changed over time. Sociologists research social and cultural influences on science. This book draws from the history and sociology of science, but its main approach is philosophical. This is because we, its authors, are philosophers of science.

If you haven't studied philosophy of science, it may sound obscure. But ***philosophy of science*** is just the investigation of science, focused especially on questions of what science should be like in order to be a trustworthy route to knowledge and to achieve the other ends we want it to have, such as usefulness to society. This book is written from a philosophical perspective, but it does not dwell on philosophers' different theories and debates about science. Instead, we aim to combine insights about science from philosophy with concrete examples to illuminate science's aims, methods, and patterns of reasoning, without getting bogged down in controversies, technical terminology, or too many details.

The title of this book, *Recipes for Science*, is meant to evoke two ideas. First, culinary recipes have many variations: in what dish they are used to create, which version of the dish, and details of implementation. Different recipes are used to make bread and lasagna, bread recipes may call for leavening with yeast or baking soda, and ingredient measurements may be in weight or volume. Science is also like this. Scientific research can have many different aims and proceed using a variety of different methods.

On the other hand, any culinary recipe employs intentional combinations of ingredients using time-tested methods to lead to a specific outcome. And different recipes for one type of food tend to have some elements in common, even if many of their features vary. So, for example, breads usually incorporate grain of some kind as a major ingredient, most use a leavening agent, and they are cooked—usually but not always by baking in the oven. There are resemblances among different breads and the recipes used to make them, even if there's no simple definition of bread and no one recipe required to make bread. Science is like this as well. Even as it proceeds in different ways, and even as there's no one overarching set of instructions or mechanical procedures that guarantees good science, there are certain generalizations that can be made about how good science is conducted.

FIGURE 0.2 Recipes for bread have many variations, but both the recipes and the resulting breads tend to have some features in common with one another. Scientific methods are like this: endless variation but with themes in both methods and results.

This book aims to facilitate a clear understanding of the main elements of science, and the importance of those elements, even as it illustrates the tremendous variety of projects that count as science.

The first part of the book, Chapters 1–5, addresses the nature of science and its key methods. Chapter 1 surveys what is distinctive about science's aims, methods, and institutional structure and suggests those features as a checklist for distinguishing science from non-science and fake science. Chapter 2 introduces how science is shaped by values, varies in its aims, and consists in a variety of recipes with a few commonalities. Chapter 3 examines the nature of scientific experiments and the roles experimentation plays in science. Chapter 4 catalogues varieties of non-experimental studies and discuss their advantages and disadvantages. Chapter 5 focuses on scientific models: how they are constructed and used, and the main varieties in which they come.

The second part of the book, Chapters 6–10, focuses on important patterns of reasoning in science. Chapter 6 examines patterns of deductive reasoning and their use in scientific hypothesis-testing. Chapter 7 addresses the importance of and challenges with inductive reasoning and the scientific significance of abductive reasoning, also known as inference to the best explanation. Chapter 8 introduces basic concepts and interpretations of probabilistic reasoning. Chapter 9 surveys graphical and numerical representation using descriptive statistics and misuses of statistical reasoning, while Chapter 10 discusses how inferential statistics is used in estimation and hypothesis-testing.

The final part of the book, Chapters 11–13, addresses the ultimate aims of science and its relationship to society. Chapter 11 addresses how scientific methods are used to acquire causal knowledge. Chapter 12 examines how scientific findings yield theoretical knowledge and understanding of the world and discusses change in scientific theories and how science makes progress. Finally, Chapter 13 discusses the complicated relationship science bears to society, analyzes how values influence science, and surveys the changes and challenges facing science in the 21st century.

INTENDED AUDIENCES AND HOW TO USE THE BOOK

This book is not just for students of philosophy or science majors. Indeed, the primary audience we had in mind in developing this book is an undergraduate student in a general education course, who may not take any additional science courses in college. We asked ourselves, what would that student most benefit from knowing about how science works? What episodes from historical and current science would that student be interested to read about and contemplate?

We expect this book will also be useful for some more specialized or more advanced courses. These include science education courses, especially those that focus on the nature of science and scientific reasoning; introductory philosophy of science courses, especially if supplemented with primary readings that address some of the major philosophical controversies about science; and introductory science courses, especially methods courses, when supplemented with material specific to the particular scientific field of study.

This textbook was designed to be usable in its entirety in a standard semester, with approximately one chapter covered per week. But, to be useful in a wide range of course levels and disciplines and for different teaching goals, we have also designed the book to be modular: each chapter can be used independently from the others. Instructors (or independent readers) can thus choose to use only the chapters that suit their needs. Each section may rely on earlier sections in the same chapter, but not later sections in the same chapter or other chapters. Instructors may choose not to assign later sections in some chapters that seem overly specialized or too difficult for their teaching needs.

Box 0.1 Citation and using sources

Scientific integrity requires the proper acknowledgement of the sources of ideas or words researchers use from someone else. There are standard citation practices for ideas and words from published material such as articles, books, and encyclopedia entries. But even using a tool like a large language model (LLM) for generating and editing text should be transparently acknowledged in a paper. When you directly use others' words, that should be indicated with quotation marks and precise citation, including page number where available. A similar point applies to paraphrases. In this textbook, we have compiled our sources for each chapter in a bibliography at the back of the book. Failing to cite one's sources in academic writing or using parts of a text written by someone else or the ideas of others—whether intentional or not—is plagiarism. But scientific integrity not only requires proper acknowledgement of sources; it also requires relying on the best available sources. Sources must be credible, up-to-date, and pertinent. There's no general standard for where to find high-quality sources. Depending on the purpose of a piece of writing, peer-reviewed academic research articles and books may be the only acceptable sources, while, in other instances, popular content from newspapers and magazines might also be fine. Internet sources might be okay too, but because anyone can publish a blog or pontificate on social media, it's essential to carefully evaluate the credibility, reputation, and timeliness of a source on the internet and whether it comes from a genuine expert on the topic.

Each section within every chapter ends with a list of exercises to solidify understanding and challenge students to apply what they have learned. We encourage instructors to make use of these exercises for in-class group or individual activities, homework, or exam questions. Exercises are divided into three categories: *Recall* exercises can be completed just by consulting the section. *Think* exercises provide opportunities to consider implications of or questions about ideas from the section. In *Apply* exercises, students are asked to apply ideas from the section to new examples or circumstances.

A list of further readings at the end of each chapter provides inroads into more in-depth investigations of individual topics covered. At the end of the book, there is a glossary of technical terms and other specialized vocabulary that students can consult as needed. Terms defined in the glossary are indicated in the main text with bold and italics when they are first defined, as *philosophy of science* was earlier in this introduction.

Finally, there is a website to accompany the textbook. The website includes example syllabi for different kinds of courses utilizing this text, additional exercises, an answer key for some exercises, some image files related to the text, and links to content that will enrich students' experience with the topics covered in this book. The website also includes a series of brief videos, each engaging with one of the core topics in this book.

EXERCISES

0.1 **Recall:** List three ways in which science is relevant to your life (and others' lives). Watch Video 1

0.2 **Think:** What do you expect to learn from this textbook and the course you're reading it for?

0.3 **Think:** What most concerns you about this textbook and the course you're reading it for?

0.4 **Think:** What do you think is most valuable about learning about science and scientific reasoning? Give reasons to support your response.

0.5 **Apply:** Describe your relationship to science. For example, have you taken many courses in science or read about science on your own? If so, on what topics, and did you enjoy the experience? Do you know any scientists? Do you think there are reasons to distrust or dislike science? If so, what are the reasons?

FURTHER READINGS

For an overview of immunization and vaccination, see World Health Organization. *Vaccines and immunization*. www.who.int/health-topics/vaccines-and-immunization

For a concise overview of how vaccination has contributed to global health, see Greenwood, B. (2014). The contribution of vaccination to global health: Past, present and future. *Philosophical Transactions of the Royal Society B, 369*(1645), 20130433.

A general reference on topics in philosophy, including philosophy of science, is the *Stanford encyclopedia of philosophy*. https://plato.stanford.edu

CHAPTER 1

The nature of science

1.1 SCIENTIFIC KNOWLEDGE OF CLIMATE CHANGE

After reading this section, you should be able to:

- Explain why climate change is a serious practical concern
- Describe how scientific research supports the finding of human-caused climate change and why public opinion lags behind the research
- List three indicators that scientific knowledge is trustworthy

A serious practical concern

In November 2023, the 28th United Nations Climate Change conference was held in the United Arab Emirates. This event was the 28th Conference of the Parties (COP28) to the United Nations Framework Convention on Climate Change (UNFCCC). COP28 was also the 18th meeting of the parties to the Kyoto Protocol, which in 1997 extended and expanded the nations supporting the UNFCCC, and the fifth meeting of the parties to the Paris Agreement, signed by 196 countries in 2015. As this book was printed, COP29 was in planning for November 2024, and annual international meetings in this series are intended to continue.

International conferences about the challenge of climate change have been occurring for 30 years, with increasing consensus about the measures needed to counteract rising global temperatures. A primary goal is to limit the rise in global mean temperature—the average of all land and ocean surface temperatures—to 1.5° Celsius or below compared to preindustrial levels. This temperature change would be minor if it were a single-day temperature in one place. But a 1.5° Celsius increase in global mean temperature is a major change with radical consequences.

Think of this temperature increase like a fever. The human body maintains a relatively constant temperature in the range of 36.5°–37.5° Celsius (97.7°–99.5° Fahrenheit). When your body temperature increases 1.5° Celsius, which is nearly 3° Fahrenheit, you have a fever. If your body were suddenly that much warmer on average, day in and day out for months and years, it would be a serious medical emergency.

An average global temperature increase of 1.5° Celsius would be similarly devastating for Earth. But why? First, because it changes the Earth's climate. As a result of

DOI: 10.4324/9781003300007-2

climate change, mountain glaciers are shrinking, and ice sheets are melting in the Arctic, Greenland, and Antarctica. These changes lead to sea levels rising, thereby flooding coastal areas. Precipitation patterns across seasons also become more unstable, leading to more droughts, heat waves, flooding from storms, and wildfires—even shifting the growth timing of plants and crops that makes them more vulnerable to loss.

Second, the changing climate has downstream effects. These effects threaten to push some animal and plant species to extinction, and even collapse ecosystems. Also threatened are human social conditions. Drinking water is scarcer and droughts more frequent or severe; crop yields are expected to decrease. Coastal cities and island nations are at risk of serious floods and devastating hurricanes. All of this affects global health, poverty, hunger, and national security. The World Bank estimates that 200 million people will be forced to migrate between 2020 and 2050 due to the impacts of climate change. Ultimately, global warming will make the Earth less hospitable for all creatures, including humans, and a more unjust place in virtue of who will suffer and how this suffering will be managed. International climate change work therefore includes not just efforts to mitigate climate change but also, increasingly, attention to how to adapt human societies to a changed climate.

Greenhouse gases work like a blanket. As incoming radiation from the Sun permeates our atmosphere, some hits the Earth and is reflected back out to space. But greenhouse gases, such as methane (CH_4), carbon dioxide (CO_2), and water vapor, trap some of the heat in the atmosphere. This trapped heat warms the planet's surface, making it hospitable to life. Higher concentrations of greenhouse gases lead to a warmer planet; lower concentrations lead to a cooler planet.

Changing atmospheric concentrations of greenhouse gases are a major factor in Earth's climate. Other factors include variations in the Earth's orbit, the motion of tectonic plates, the impact of meteorites, and volcanism on the Earth's surface. So, our climate has never been static: it has been fluctuating for billions of years. What's special about the current climate changes, then? Why is this time different?

What's different is that human activities have led to extreme changes. Since the beginning of the Industrial Revolution, human activities that burn fossil fuels like coal and oil have resulted in carbon dioxide being released into the atmosphere at unprecedented rates. Since carbon dioxide is a greenhouse gas, this increases the heat retention of our atmosphere. Other human activities—primarily agricultural activities such as raising livestock—release methane, which is another greenhouse gas even more potent than carbon dioxide in trapping heat. These human-caused greenhouse gas emissions are so extreme that they've led to a historically unprecedented increase in the Earth's global mean temperature.

This discovery isn't recent. Scientists have known since the 18th century that burning carbon-based fossil fuels releases carbon into the atmosphere. Systematic research on the relationship between carbon dioxide emissions and climate change began in the 19th century, when the American engineer Marsden Manson discovered how the heat-trapping power of the atmosphere varies with only slight changes in its makeup. A few years later, the Swedish physicist and chemist Svante August Arrhenius completed calculations showing that changes in carbon dioxide also function as a "throttle"

FIGURE 1.1 Swedish chemist and physicist Arrhenius in his lab in 1905

on other greenhouse gases like water vapor. He calculated that there would be an arctic temperature increase of approximately 8° C (46.4° F) from atmospheric carbon levels two to three times their known value at the time. In 1908, Arrhenius predicted, "the slight percentage of carbonic acid in the atmosphere may, by the advances of industry, be changed to a noticeable degree in the course of a few centuries."

Just before the outbreak of World War II, the British steam engineer Guy Callendar presented a paper to the Royal Meteorological Society, in which he pointed out that the atmospheric concentration of carbon dioxide had significantly increased between 1900 and 1935, based on temperature measurements at 200 meteorological stations. In 1939, Callendar concluded:

> As man is now changing the composition of the atmosphere at a rate which must be very exceptional on the geological time scale, it is natural to seek for the probable effects of such a change. From the best laboratory observations it appears that the principal result of increasing atmospheric carbon dioxide . . . would be a gradual increase in the mean temperature of the colder regions of the earth.

This prescient recognition of the role of human activity on atmospheric temperatures had to wait several decades to become widely accepted.

In 1958, the geochemist Charles David Keeling installed four infrared gas analyzers at the Mauna Loa Observatory in Hawai'i. Measurement collection has occurred continuously since 1958, recording an ever-increasing atmospheric CO_2 concentration. The graph showing these measurements is known as the Keeling Curve (see Figure 1.2a).

Keeling's measurements provided evidence of rapidly increasing carbon dioxide levels in the atmosphere, and a 1979 report by the National Research Council—an American nonprofit organization devoted to scientific research—connected this increase to rise in average temperature. This report predicted that doubling atmospheric CO_2 concentration from 300 to 600 ppm would result in an average warming of 2.0°–3.5° C. (*Parts per million* refers to a unit for measuring small concentrations of a substance.) We haven't yet reached the ominous level of 600 ppm, but we're now long past safe levels of CO_2 in the atmosphere, estimated to be about 350 ppm.

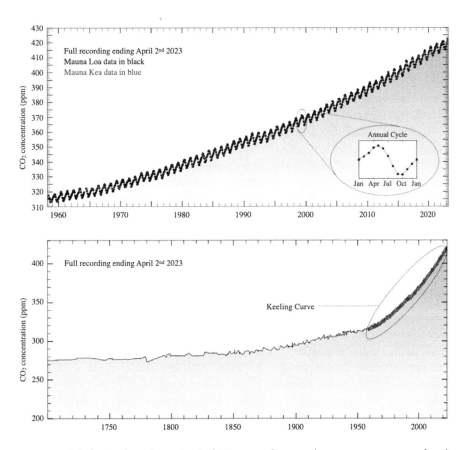

FIGURE 1.2 (a) The Keeling Curve (top); (b) Estimated atmospheric concentrations of carbon dioxide since 1700 (bottom)

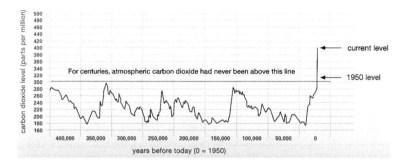

FIGURE 1.3 Unprecedented increases in atmospheric carbon dioxide after the Industrial Revolution

For several decades, climate scientists have tracked changing atmospheric carbon dioxide levels with increasing precision. Ice cores taken from various locations in the Antarctic have enabled scientists to extrapolate historic CO_2 levels for comparison to recent levels (see Figure 1.2(b)). A group of 78 scientists gathered data from several "climate proxies," including tree rings, pollen, corals, glacier ice, lake and marine sediments, and historical documents. These data provided multiple types of evidence supporting the conclusion that, at the end of the 20th century, atmospheric levels of CO_2 and global mean temperature are higher than at any point in the previous 2,000 years.

For the past 800,000 years, atmospheric carbon dioxide hadn't been over 300 ppm (see Figure 1.3). Since the Industrial Revolution began about 250 years ago, the concentration has spiked to 420 ppm. This is nearly 50% more than levels had reached in 800 millennia—reached in only a quarter of one millennium of human-caused change. (See www.co2.earth for an updated estimate; unfortunately, this number is still climbing steadily.) The last time CO_2 levels were this high, humans did not yet exist. In 2022, global mean temperature was 1.06° C (1.90° F) warmer than the pre-industrial period (1880–1900) and has been going up 0.18° C each decade. At this rate, 1.5° C will be surpassed before 2040.

Trustworthy scientific knowledge

Centuries of scientific research—including convergence of many types of evidence and broad consensus among scientists with relevant expertise—support the conclusion that we face an unprecedented climate crisis caused by human activity, sometimes called *anthropogenic climate change*. Like other scientific knowledge, it wasn't initially obvious and still might not be obvious to those without relevant expertise. Scientists had to try out various techniques and gather different kinds of data to discover this conclusion is true.

Fundamentally, science aims to produce knowledge—in particular, **scientific knowledge**: explanatory knowledge of why or how the world is the way it is. And it's the best approach we humans have developed for answering questions about the world around—and within—us. As we will see throughout the book, the scientific establishment has developed countless techniques to acquire knowledge. (You've already read about some of these at work in the climate science just described.) Understanding how scientists acquire new knowledge can give people greater reason to trust scientific knowledge, even if they themselves don't have a full understanding of the evidence, methods, and reasoning leading to it.

First, relevant expertise is important for scientific knowledge to be trustworthy. You should trust climate scientists to do climate science in the same way you trust your mechanic with your car or your favorite restaurant with your dinner. The types of expertise required for these positions takes years, even decades, to develop. But the expertise doesn't neatly transfer from one domain to another: don't trust the average climate scientist to fix your car or make you a delicious meal. Similarly, politicians and policymakers know things about political and legislative matters, but they should not be looked to as authorities on the science of climate change—regardless of whether they accept its existence. To do so would be an ***appeal to irrelevant authority***: appealing to the views of an individual who has no expertise in a field as evidence for some claim.

Second, consensus among the relevant experts is an important indicator that the findings are settled scientific knowledge. There is striking agreement among climate scientists about the existence of anthropogenic climate change. Reputable scientists and scientific societies, including the national science academies of the world and the Intergovernmental Panel on Climate Change (IPCC), agree that human-caused, or anthropogenic, climate change is occurring. This includes virtually all climatologists. In 2004, the historian of science Naomi Oreskes analyzed 928 abstracts on climate change published in scientific journals from 1993 to 2003; none expressed disagreement with the consensus position that anthropogenic climate change is occurring. In 2010, a group of researchers studied the views of the 200 climate scientists with the most extensive and productive publication records; more than 97% affirmed the existence of anthropogenic climate change as described by the IPCC.

Third, the convergence of different sources and types of evidence provides solid grounding for scientific knowledge. Well-established theories in physics explain how heat radiation works. Physical chemistry shows how carbon dioxide and other greenhouse gases in the atmosphere traps heat. Climatologists have developed extensive sources of evidence that support the same conclusions about climate change and its relationship to greenhouse gases and human activities. As we described earlier, some of this knowledge goes back centuries, and a range of techniques have been used to amass ever-more relevant evidence. Since the 1950s, scientific models and computer simulations have helped scientists to make testable predictions about what would happen to the global climate in response to different changes in human activities, and evidence has confirmed those predictions.

And yet, despite decisive scientific evidence supporting a consensus among scientists, public concern for climate change lags behind the research. According to surveys from the Pew Research Center from 2013, 44% surveyed across 23 countries did not view climate change to be a major threat; by 2018, that number had dropped to 33% across the same countries. Whether people are concerned about climate change largely

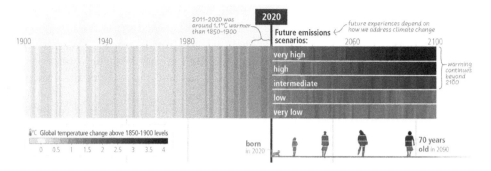

FIGURE 1.4 Today's decisions determine the extent of climate change by 2100

depends on understanding its human causes and level of education. In some countries like the United States, however, being better educated doesn't predict climate change concern as well as political views do.

People who don't know much about a given topic can experience an ***illusion of understanding***, in which they lack genuine understanding and so fail to appreciate the depth of their ignorance about that topic. For climate change, this means that people without advanced education in science—a demographic that includes most politicians—tend to have unwarranted confidence in their ability to assess the scientific findings. This includes those who are concerned about climate change as well as those who deny its existence. Worse, illusions of understanding have become easier to sustain in today's society. Finding information merely through internet searches (so-called *Google-knowing*) has diminished genuine understanding, and we have limited opportunities for truly public discourse. Our online and in-person conversations tend to happen with people who have beliefs similar to our own.

Improving public climate literacy can support public engagement about global warming. If one knows that Earth is warming up and genuinely understands why, this can lead to changed behavior—for instance, petitioning one's government to support more energy-efficient practices. More generally, understanding the processes that support trustworthy scientific knowledge—including relevant expertise, broad consensus, and extensive convergent evidence—is vitally important to assessing whether something qualifies as legitimate scientific knowledge and how to go about finding out.

EXERCISES

1.1 **Recall:** How do scientists know that human activities are radically altering Earth's climate? Why are these changes a serious concern?

1.2 **Think:** Do all scientists, in virtue of being scientists, have the expertise to make pronouncements about global warming? Or only just those scientists who specialize in the climate sciences? Give reasons to support your answer.

1.3 **Think:** Describe how a nonspecialist can know whether to trust someone claiming scientific expertise, listing at least three kinds of evidence of their expertise.

1.4 **Recall:** Identify three indicators that scientific knowledge is trustworthy described in this section. For each, briefly say why it is important.

1.5 **Apply:** Think of a scientific finding that strikes you as surprising or possibly wrong. Do a little internet research (trying to focus on reputable sources). Assess how that scientific finding fares according to the three indicators that scientific knowledge is trustworthy. Based on this, do you think the finding qualifies as settled knowledge at this stage?

1.6 **Think:** List three reasons why public concern about anthropogenic climate change lags behind scientific research. Given that lag, how should climate scientists affect environmental policy in the government? Should they merely collect evidence and produce knowledge, leaving policy decisions to public officials? Do they have any obligations to more actively engage with the public?

1.2 SUBJECT MATTER AND METHODS

After reading this section, you should be able to:

- Describe what it means for science to provide natural explanations of natural phenomena
- Define *empirical investigation* and *evidentialism* and state their importance for science
- Indicate the differences between falsificationism, falsifiability, and openness to falsification, and state which are essential to the nature of science

The subject matter of science

We just described some of the abundant scientific evidence for anthropogenic climate change that has amassed over centuries. But we also noted how other experts, including political leaders, have roles to play in public conversations and policy decisions bearing on climate change mitigation and adaptation. What, if anything, is the difference between these forms of expertise?

This question relates to the ***nature of science***, that is, the orientation, values, and methods that are specific to science and allow it to generate knowledge in the ways that it does. To begin, notice that things are more complicated than just saying science is in the business of generating knowledge. Scientific projects can be directed at a wide range of goals. Some scientific research aims at knowledge for its own sake; this is sometimes called ***basic research***. For example, scientists investigate the conditions under which rainbows form to learn more about the behavior of light. Such knowledge may have applications, but basic research is not primarily focused on identifying or developing applications. Instead, basic research often aims for ***explanatory knowledge***: sufficiently justified truths about how things work and why they are the way they are. We know so much about our world, such as how greenhouse gases influence the Earth's climate, how rainbows form, and how unemployment relates to inflation, because of discoveries and theories generated from basic research.

Yet, science also plays an important role in satisfying practical goals. Many life-changing innovations have come about through computer science. The biological and pharmaceutical sciences have vastly improved medical care and public health. Skyscrapers and airplanes wouldn't be possible without a lot of physics. The contrast with basic research is **applied research**. Scientific research is applied when it makes use of scientific knowledge to develop some tangible output, such as techniques, software, drugs, and new materials. Often, a core motivation for applied research is generating products for profit; successful research can result in patentable intellectual property.

Basic and applied scientific research can operate synergistically. Scientists aiming at the production of knowledge for its own sake often rely on materials and techniques created by scientists doing applied research, while scientists doing applied research often exploit pure scientific knowledge in order to develop new products. Still, basic and applied research are often carried out by different scientists, often using very different techniques, and sometimes in entirely different fields of science and types of institutions. For example, when Kathleen Montagu and Arvid Carlsson discovered the neurotransmitter dopamine in the human brain in 1957, this was basic research conducted in a hospital laboratory. In contrast, scientists employed by pharmaceutical companies to develop and improve dopamine-related treatments for Parkinson's disease are doing applied research.

Beyond this distinction between basic and applied research, there is also tremendous variety in topics among the various fields of science. Investigations range from sub-atomic particles like quarks, to DNA, to emotions, consciousness, and mental maladies, to languages, societies, and economic phenomena, and much else besides. It can seem as if there is a science of absolutely everything! Professional sports are a good example; some scientists devote their research to learning how to improve athletic performance. Other topics of scientific research are more abstract. String theory, for example, is highly theoretical physics that posits one-dimensional entities called *strings* as the basic building block of our universe.

Despite this variety, it's possible to give a unifying description of the sort of explanatory knowledge sought in science: science provides natural explanations of natural phenomena. This thesis is sometimes called **naturalism**.

Natural phenomena are objects, events, or processes that are sufficiently uniform to be susceptible to systematic study. Disease epidemics, lunar eclipses, and droughts are all natural phenomena. Inflation, poverty, and unemployment are all phenomena in human societies, but they also count as natural phenomena under this definition. The word *phenomenon* (plural *phenomena*) comes from Greek and means "that which appears or is seen." So, phenomena include all observable occurrences, that is, occurrences detectable with the use of our senses, including the use of our senses aided by technological devices like telescopes that extend their reach. Natural phenomena need to be somewhat uniform to enable systematic study. Occurring in a regular way is needed for scientists' observations across different times and places to be used to generate knowledge.

Natural explanations of natural phenomena invoke features of the world—that is, other natural phenomena—to account for these observable occurrences. If there's an epidemic in France or increased employment in Colombia, you might wonder how that came to be. A natural explanation of the epidemic might specify a contagion and a mechanism of transmission, for example, while a natural explanation for the increase in employment might specify private investments in industry and legislative choices made by political parties. These are both natural explanations of natural phenomena.

Science is always naturalistic in what it investigates and how it explains. The meaning of the term *natural* in the context of the naturalism of science can be better understood by contrasting it with *supernatural*. Supernatural entities and occurrences may not be governed by discernible regularities, may not be observable at all or not observable by other people, or are just supposed to transcend the range of physical human experience. Because supernatural entities or occurrences are not natural phenomena, science won't be able to deliver knowledge about them: they would be beyond its explanatory reach. Nor does science appeal to supernatural entities or occurrences to explain natural phenomena. "A miracle caused her to recover from disease" couldn't possibly be a scientific explanation, even though recovering from a disease is a natural phenomenon.

The methods of science

Science's goal of providing natural explanations of natural phenomena isn't all that is distinctive about the nature of science. Also significant are science's methods. One important ingredient of these methods is closely related to the idea of naturalism: science involves **empirical investigation**, which means using one's senses to inform one's beliefs about the world. What scientists see, hear, smell, touch, and so forth can all be used as empirical evidence for or against some attempted natural explanation.

The method of empirical investigation isn't special to science. We all use our senses in everyday life to learn about the world around us, beginning as infants. You know it's a clear day because you can see and feel the sun shining through the window. But science has fine-tuned and adapted this method to generate certain kinds of knowledge.

In science, empirical investigation is explicitly used to generate **evidence**. Science is thus based on **evidentialism**, the thesis that a belief's justification is determined by how well it is supported by evidence. For any scientific claim—particularly, any natural explanation of a natural phenomenon—it must be possible to state why that claim should be believed. This evidence ultimately traces back to empirical observations, but empirical evidence often confirms scientific claims only indirectly. We don't directly see human activity increasing atmospheric CO_2, for example; rather, scientists made predictions about changes to the atmosphere, historical trends in global temperature, and more, based on this conjecture; and then they tested those predictions with empirical evidence. So, evidentialism is important to science, but how evidence is gathered and used is not always straightforward.

Box 1.1 Is science always empirical?

Scientists typically use empirical evidence as the basis for knowledge. However, in fields like mathematics, computer science, and economics, some claims are not based on empirical evidence, or at least not directly. The mathematical claim that $\text{Log}_2\,(1/2) = -1$ is not an empirical claim, in the sense that it's not based on observation. Something similar—perhaps surprisingly—applies to physics too, where it can be very hard to obtain empirical evidence that bears on some phenomena. String theory, for example, is the idea that the fundamental objects in the world are extended, one-dimensional objects called *strings*. Empirical evidence of these strings cannot be provided by present-day instruments, but string theory has been developed to account for features of fundamental physics that are well confirmed empirically, and string theorists work to find nonempirical evidence that bear on this theory. Sources of nonempirical evidence include the simplicity and unifying and explanatory power of a theory, plus the logical relationships between the theory and other claims well confirmed by the empirical evidence or believed to be self-evidently true. Nonempirical evidence in favor of string theory includes that it can account for well-corroborated claims in fundamental physics, that it has been productively applied to a range of scientific problems like black holes and nuclear physics, and that it fits with both quantum mechanics and Einstein's theories of gravity. So, not all evidence is empirical evidence, and not all scientific research is based on empirical evidence.

Evidentialism in science leads to continual, self-corrective investigation in which ideas are fine-tuned or extended in light of new evidence. Significant empirical evidence is needed before some scientific claim, like the claim that human activity is causing global warming, is broadly accepted as settled scientific knowledge. There are many scientific findings that are so well supported by evidence that they seem entirely certain. We know atmospheric CO_2 is more than 50% higher now than at any other time in human history, and we know that the last four decades are the warmest on record. We know anthropogenic climate change is occurring. We also know that the Earth orbits around the Sun, that water molecules are composed of two hydrogen atoms and one oxygen atom, and so much more. Still, in principle, scientific claims are never taken to be absolutely beyond any doubt. And very occasionally, continuing investigation even leads widely held or long-established ideas to be significantly revised.

Karl Popper was a philosopher of science in the early 20th century who took this in-principle revisability of science to be especially important. Popper developed a principle called *falsificationism*, which names the thesis that scientific reasoning proceeds by attempting to disprove claims rather than to prove them right—that is, by advancing bold and risky conjectures, and then trying to falsify or refute them. This criterion for science has been very influential among scientists, but it is controversial. One problem is that the relationship between empirical evidence and a scientific theory can be complicated, so that it is sometimes hard to say when the evidence would disprove a theory. A second problem is that incessantly trying to prove central claims false would limit scientific progress. It seems scientists do accept theories and hypotheses that are well supported by evidence, moving on to downstream questions based on those theories and hypotheses.

Two other aspects of falsificationism do seem more plausible. First, any scientific claim should in principle be *falsifiable*. A claim is falsifiable when it is possible to describe what kind of evidence would, if found, show the claim to be false. This property is required for scientific claims to be subject to empirical evidence; without it, a claim would be unscientific. Notice that true claims can be falsifiable—you can describe what kind of evidence would, if found, disprove a true claim. It's just that, because the claim is true, you will never actually find such evidence. Even for false claims, scientists may never be in the right circumstances to obtain falsifying evidence. Putting forward falsifiable claims enables science to be based on empirical evidence and to reject ideas when the evidence warrants doing so.

Second, science requires honesty when empirical evidence does indicate a claim is false. When scientists discover apparently falsifying evidence, they should begin to doubt the ideas under investigation. In general, we humans try hard to hold on to our existing beliefs, even when those beliefs are challenged. Scientists are no different. But science's evidentialism requires scientists to doubt any scientific claims—even claims they had thought were really promising—when empirical evidence suggests they may be wrong. We might call this *openness to falsification*: any claim should be abandoned when the preponderance of evidence indicates that it's false.

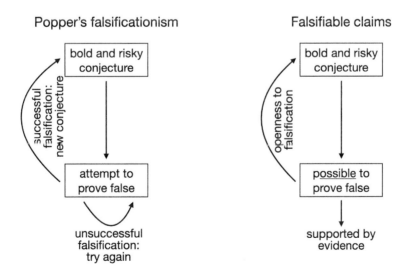

FIGURE 1.5 Schematic flowchart of Popper's falsificationism compared to falsifiable claims and openness to falsification

EXERCISES

1.7 **Recall:** Define *natural phenomena* and *natural explanations*, and describe the importance of each to science's ability to generate trustworthy knowledge.

1.8 **Apply:** Describe one real example of basic research and one real example of applied research. For each, describe your reasoning in considering it basic or applied research.

1.9 **Recall:** Describe what it means for science to provide natural explanations of natural phenomena. What are the limitations to the kinds of knowledge science can produce due to this requirement?

1.10 **Recall:** Define *empirical investigation* and *evidentialism*. Describe how they are different from each other and how each is important to science.

1.11 **Think:** Evaluate how and why the subject matter and methods of science are each relevant to the nature of science.

1.12 **Recall:** Define *falsificationism*, *falsifiability*, and *openness to falsification*, making sure you are clear about how each is different from the others. For each, say whether it is essential to science and why.

1.3 THE INSTITUTION OF SCIENCE

After reading this section, you should be able to:

- Define *confirmation bias* and give examples of how it works
- Describe how social structure is important to the nature of science
- Describe how social norms for individual scientists and the scientific community are important to the trustworthiness of science

Flaws in human reasoning

Empirical investigation is a basic aspect of human existence and so not special to science. Why, then, is science needed to give us knowledge about the world, beyond just our ordinary human powers of observation? Just as we humans are predisposed to investigate our world using our senses from our first days of infancy, we are also predisposed to some serious flaws in how we gather evidence and how we reason. Science is the best route to knowledge about the world in part because it incorporates ways to protect against those flaws in reasoning.

It is normal for people to favor some ideas over others. We can then use our experiences in the world, investigation of existing knowledge, and critical thinking to ensure that the ideas we favor are, in fact, good ideas. The problem is, we also seek out and interpret information in ways that fit with our favored ideas, and we avoid information that challenges those ideas. This is a well-established feature of human reasoning called *confirmation bias*: the tendency to look for, interpret, and recall evidence in ways that confirm and do not challenge our existing beliefs.

Imagine someone has just brought her friends to a restaurant she's selected. When she asks her friends if they like the restaurant, she may say, "It's good, isn't it?" Framing the question in this way promotes agreement with the judgment she already has of the restaurant—it's a way of looking for confirming evidence. Similarly, someone who's skeptical about climate change may perform an internet search for the phrase *evidence against climate change* to learn more, or they may focus on what critics say and ignore what climate scientists say. Someone concerned about genetically modified crops is more likely to make time to read an article entitled "Dangers of Genetic Modification" than an article titled "Genetic Modification Boosts Soy and Corn Performance." These are ways of seeking evidence that confirms one's existing ideas rather than challenging those ideas. We are also prone to interpreting evidence as supporting our existing ideas. In one study, people who were in favor of and opposed to the death penalty both read the same discussion of the death penalty. People on each side of the issue interpreted the discussion entirely differently; each side thought it supported their own view.

Confirmation bias can involve looking only for evidence that supports your existing beliefs, cherry-picking which research to believe and which to ignore, holding evidence against your views to a higher standard than evidence in favor of your views, and more easily remembering supporting evidence than contrary evidence.

We all do this; it doesn't matter what views about the world we have, what political views we have, whether we've graduated from college, or whether we have been trained as scientists. In fact, some evidence suggests that confirmation bias worsens with increased education. Although everyone is prone to confirmation bias, the effect tends to be stronger for politically or emotionally charged issues, such as vaccinations, climate change, and health.

Scientists' expectations or desires about the results of scientific research can lead to incorrect findings. One way in which this can happen is through the **observer-expectancy effect**, when a scientist's expectations lead them (perhaps, unconsciously) to influence the behavior of experimental subjects. A famous example of this involved Clever Hans, a horse who was thought to have sophisticated abilities including performing arithmetic calculations. Hans's owner, Wilhelm von Osten, was a mathematics teacher, horse trainer, and phrenologist. (Phrenology is the now-discredited study of the shape of the skull as an indicator of personality and mental abilities.) Hans was trained to recognize numerals from 1 to 9 and to tap his hooves to indicate which ones he recognized. Eventually, van Osten had Hans tapping out correct answers to questions like: what's the number of 4s in 16?

In 1891, van Osten traveled around Germany to exhibit his amazing horse. There was such fanfare that the famous psychologist Carl Stumpf appointed a special commission to provide critical scrutiny. In 1904, the commission concluded that Hans's abilities were legitimate. The horse was able to answer questions from simple arithmetic to square roots, fractions and decimals, units of time, musical scales, and the value of coins. Hans could even respond accurately when van Osten wasn't present.

FIGURE 1.6 Clever Hans and Wilhelm von Osten

The commission was wrong. Stumpf's pupil, Oskar Pfungst, demonstrated that Clever Hans was not performing sophisticated mental calculations. Pfungst used blinders to vary whether Hans could see the questioner, and he varied who played the role of questioner. Hans produced the correct answer even when van Osten himself did not ask the questions, but Hans's performance fell apart when the questioner did not know the answer or when the horse was asked the question from behind a screen. When the visibility of spectators and questioners was masked, Hans's ability to produce correct answers fell dramatically from 89% to 6%. Further observations confirmed that Hans was being unwittingly cued by his human audience. Questioners' body language and facial expressions became taut as his tapping approached the correct answer, and then more relaxed upon the final tap; this change prompted Hans to stop tapping.

Science as a social enterprise

Like van Osten and all the other people who asked questions of Clever Hans, our expectations can affect how matters play out, even when we don't intend this to happen. This possibility makes it hard for people—including scientists—to reason their way to the right answers. For this reason, one element of science's great success in generating knowledge about our world is its institutional features that protect against or counteract the basic flaws in human reasoning.

Scientific research requires communities of many people working together. Teams of scientists work together on research projects; it is common for research publications to have multiple authors. Scientists also regularly make use of techniques, data, or ideas developed by other scientists. And all new scientific research is based in part on the findings of previous scientists. Such collaboration is essential to the development and refinement of scientific knowledge: no one scientist can produce scientific knowledge on their own.

We have seen how science is based on empirical investigation. And yet, empirical evidence bearing on scientific claims often doesn't come directly from an individual scientist's own observations. Instead, an important source of evidence is other scientists' reports of their observations as detailed in research publications. Keeling and his collaborators first measured the increasing concentration of atmospheric CO_2 depicted in the Keeling curve, but many more climate scientists later made use of those data in their own research. Scientific collaboration thus greatly amplifies the reach of empirical investigation.

Collaboration and competition among scientists also help detect and correct flaws in human reasoning, giving rise to the self-corrective process of refining scientific knowledge. Collaboration among scientists creates opportunities for people with other viewpoints to analyze the evidence and ideas from their own perspectives and methodologies. While new research projects are based on the findings of previous scientists, they are also opportunities to refine or challenge those earlier findings. Competition among scientists—to make a discovery before anyone else, to get their research projects funded, and to show that an idea is better supported by the evidence than an opposing idea—also spurs reexamination of ideas that other scientists might take for granted. Collaboration and competition in science should combine to increase the trustworthiness of scientific knowledge. If a large and diverse group of scientists agree about some finding, we should be more confident that it is legitimate.

This raises another point about the importance of science as a social enterprise. To adequately protect against individual flaws in reasoning, scientific communities need to be diverse in order to provide satisfactory interpretations of the available evidence, as well as to formulate and test a variety of ideas, including perspectives from different nationalities, races and ethnicities, gender identities, cultures, and more. This kind of diversity benefits science by guarding against any individual biases and personal values.

Social norms of science

Because the institutional structure of science is essential to its ability to generate knowledge, science has important social norms—rules or guidelines that scientific activities should adhere to and against which they are evaluated. One set of norms applies to the behavior of scientists. Scientists are obligated to have scientific integrity, which involves expectations of honesty and avoiding improper influence by others. Norms of scientific integrity are so important that their violation is severely punished by the scientific community, such as with bans from publishing in scientific journals or even loss of one's job as a scientist.

Examples of scientific dishonesty include plagiarism and fabricating data. ***Plagiarism*** is the fraudulent theft of someone else's ideas, scientific results, or words, which are subsequently presented as one's own work without giving proper credit. Fabricating data occurs when, rather than collecting empirical evidence, scientists create records of observations they didn't actually make in order to use them as evidence to support a desired conclusion.

In 2011, a Dutch social psychologist, Diederik Stapel, published a widely read study in *Science*, one of the most prestigious scientific journals, presenting evidence that trash-filled environments lead people to be more racist. But rather than collecting actual data, Stapel just made it up. When this was discovered, his reputation immediately collapsed. All his other publications were scrutinized, and approximately 60 other papers were retracted for data fabrication. Other scientists have also been forced out of science after their ethics violations were discovered, such as the Seoul National stem-cell researcher Hwang Woo-suk and the Harvard evolutionary biologist Marc Hauser. Some science journalists have helped increase awareness of issues like plagiarism and data fabrication by running blogs such as *Retraction Watch*.

Box 1.2 Merton's social norms of science

Social norms are informal rules that govern behavior in groups and societies. American sociologist of science Robert Merton specified four social norms that govern scientists' attitudes and behaviors towards each other and their research, thereby enhancing the moral integrity of scientific communities and supporting the expansion of scientific knowledge.

1. *Communism*: scientific findings and methods are common goods owned by all and should be shared freely.
2. *Universalism*: scientific work should be evaluated based on impersonal criteria like coherence with other bodies of knowledge and empirical confirmation. In other words, scientific work should be independent of the socio-political or personal status of the scientists involved.
3. *Disinterestedness*: scientific work should not be aimed at personal gain.
4. *Skepticism*: scientific work should be scrutinized critically and transparently by relevant scientific communities before being accepted.

These four norms relate to how we are discussing collaboration and competition and social norms of science in this section.

Scientists also are expected to avoid *conflicts of interest*: financial or personal gains that have the potential to inappropriately influence scientific research, results, or publication. Conflicts of interest, especially when research is funded by organizations with a financial stake in the findings, can result in researchers intentionally or unintentionally altering what research they conduct, their findings, or what they report in publications. Thus, scientists are obligated to disclose any potential conflicts of interests they may have. The existence of potential conflicts of interest does not necessarily lead to bias, but transparency about them allows others to evaluate the possibility of improper influence.

Here's an important example. Clair Patterson, a geochemist at Cal Tech in California, led the campaign to remove lead from gasoline in the 1960s and 1970s. Leaded gasoline contained lead tetraethyl, which is extremely toxic to human and non-human animals alike. Because the campaign against leaded gasoline threatened their profits, the fossil fuel industry—particularly the Ethyl Corporation—fought bitterly against Patterson's research. Among their tactics was to pay another scientist, Robert Kehoe, to attest to the safety of leaded gasoline. Eventually, his dishonesty was revealed, and honest science carried the day. Lead was removed from gasoline, but only after generations of people in many countries around the world suffered from elevated lead levels in their blood, which leads to brain damage, chronic illness, birth defects, and increased death rates. Urban areas around the world still have elevated levels of lead in their soil from this period.

Another important form of protection against flaws in reasoning involves social norms and incentives governing the scientific community as a whole. One such norm is trust. Scientists' trust in one another is the glue of scientific communities. For example, collaborative projects on climate change involve scientists with a range of different expertise, including climatologists, ecologists, physicists, statisticians, and economists. None of these scientists alone possesses comprehensive expertise to collect, analyze, and interpret the full range of evidence that bears on our understanding of anthropogenic climate change. These scientists must rely on each other and must trust one another's scientific work.

Scientists also are expected to critically evaluate one another's work by deciding whether results warrant publication, evaluating the strengths and weaknesses of research findings, and choosing whether and how to respond to published findings. One important form of critical evaluation is attempting to replicate others' research. In *replication*, an experiment or study is performed again (often by different scientists) to determine whether the same findings obtain. If successful, the replicated results further confirm the ideas under investigation. If the results are not replicated, this raises doubts about the original work, such as the possibility that something unexpected was instead responsible for the finding. This is one way to think about what happened with continued evaluation of Clever Hans's apparent math skills.

EXERCISES

1.13 Recall: Describe three types of influence of confirmation bias, and define *observer-expectancy effect.*

1.14 Recall: Describe how social structure is important to science, listing at least three ways in which it's important that are discussed in this section.

1.15 Think: What does it mean to say scientific knowledge is produced by scientific communities instead of individuals? In light of your answer, explain why scientific communities need to be diverse across a range of characteristics.

1.16 Recall: Describe how social norms for individual scientists and for the scientific community are both important to the trustworthiness of science.

1.17 Recall: Describe three kinds of scientific fraud or scientific misconduct, giving an example of each.

1.18 Think: How should trust and criticism be balanced in scientific communities, and why is this important to science? How should trust and skepticism of the public toward scientific findings be balanced, and why is this important for the public's relationship to science?

1.4 DEFINING SCIENCE

After reading this section, you should be able to:

- Describe why it is difficult to define science and distinguish it from other pursuits
- Define *pseudoscience* and give examples
- Analyze whether a claim or topic of research counts as scientific using the checklist for science

Pseudoscience and the tricky work of defining science

Science is unrivaled in its ability to generate knowledge about our world. It has earned authority and legitimacy from centuries of successes and improvements that go beyond the expertise of any individual scientist or investigation. Many people and organizations are eager to lay claim to scientific legitimacy, and it's sometimes difficult to discern whether they are entitled to it.

This is not a new problem. Karl Popper, the 20th-century philosopher encountered in the previous section, argued that some investigations thought to be scientific were instead **pseudoscience**, which means false, fake, or bogus science. Such nonscientific activities are designed to look enough like science to deceive people into thinking they have scientific legitimacy. A standard example of pseudoscience is astrology (not to be confused with astronomy, which is the scientific field that studies celestial objects in space). Astrology is commonly associated with horoscopes, which use zodiac signs to make predictions about future events, relationships, destiny, and the like. Tests of astrological ideas have generated lots of empirical evidence against them, and advocates of astrology rarely engage in systematic attempts to empirically test their claims—claims that haven't changed much since astrology peaked in popularity centuries ago. And yet, even though astrology is bunk, it is still promoted as a legitimate source of knowledge. Massive numbers of astrologers, clairvoyants, psychics, and other charlatans take in billions of dollars every year for their consultations.

If astrology is pursued purely for entertainment, without pretense of generating knowledge or misleading anyone into thinking it's doing so, then perhaps there's no grounds for complaint. The central problem with pseudoscience is the deceptive attempt to appear scientific and, thus, to have the ability to generate scientific knowledge when it doesn't.

In many cases, the specific intent of the advocates of pseudoscientific theories is to appeal to science's self-correcting nature to call into doubt scientific findings supported by enough evidence to be considered established scientific knowledge. Anti-vaccination advocacy is like this. One popular anti-vaccination argument is that childhood vaccines increase the risk of autism. Extensive testing has demonstrated clearly and conclusively that there is no causal connection between vaccination regimes and the incidence of disorders like autism. This conclusion is scientific: it is based on evidence, is open to falsification, and would be rejected if sufficient evidence against it were found. But existing research is so extensive and compelling that the possibility of newfound disconfirming evidence is virtually nonexistent. Nonetheless, propaganda outlets and anti-vaccination groups peddle misinformation, trying to induce doubt by misconstruing the relevant research and with stories of children who were diagnosed with autism after vaccination. (This does regularly happen, for the simple reason that vaccination regimes and many symptoms of autism both tend to emerge in the same stage of early childhood.)

Another example of pseudoscience is creationism and ***intelligent design***, which are attempts to explain the characteristics of living organisms by appeal to supernatural events, inspired by religious teachings. For close to a century now, "creation science" and later intelligent design were effectively advocated in the United States as alternative scientific theories to evolutionary theory, including publication of glossy textbooks and even a creationism-based alternative natural history museum. The basic thought behind both creationism and intelligent design is that living organisms are so complex that they couldn't possibly have come about by evolution. This idea is not supported by evidence; it is actually debunked by the evidence, which, at the same time, clearly indicates the workings and effects of biological evolution. Notice that this does not imply that evolutionary theory has proven that there is no supernatural involvement; that would be beyond the purview of science. Rather, it's just that the natural explanation of evolution successfully accounts for the natural phenomenon of complex lifeforms.

Healthcare is a common target for pseudoscience. Besides anti-vaccination advocacy, another example is conversion therapy, which is intervention intended to change a person's sexual orientation. Conversion therapy pretends to be like psychological therapy and is still practiced in some circles, but it has been thoroughly shown to be ineffective and psychologically harmful. Other instances of pseudoscience might be less clear-cut. Naturopathy is an approach to healthcare that emphasizes thinking about conditions of the whole body and looking to natural, folk, or indigenous remedies for health concerns. This approach might have some value when medical research is focused on precisely targeted medical intervention and when pharmaceutical drugs (but not herbal supplements) are subject to rigorous testing and regulation. Further, naturopathy training programs and licensure exist in some places. Nonetheless, some approaches endorsed in naturopathic medicine have been disproven with evidence. It can be difficult to judge whether naturopathy should be dismissed entirely as pseudoscience or might be rendered more legitimately science-based with continued development or integration into mainstream medical practices.

Here's another example of pseudoscience that comes from inside science. Naomi Oreskes and Erik Conway's book, *Merchants of Doubt*, which also inspired a film of the same name, details how one group of well-respected scientists in the United States provided legal testimony and spurred research that misled the public and enabled corporations to dodge responsibility for the health and environmental catastrophes of cigarettes, acid rain, climate change, and the toxin DDT. Apparently inspired by their political views, these scientists misused the authority of science to delay acceptance of established scientific knowledge that was inconvenient for powerful corporations.

As these examples reveal, discerning science from pseudoscience can be essential for health and safety, but doing so can be very difficult. Where is the line between harmless entertainment and pernicious fake knowledge, or between a new, underexplored alternative idea and a cynical attempt to inspire doubt in well-established scientific knowledge? It seems we cannot just rely on whatever individual scientists tell us to believe. And some features of the nature of science described in the previous section might be shared by varieties of pseudoscience.

A checklist for science

We have discussed many distinctive features of the nature of science. These include aiming to generate knowledge, naturalism, empirical investigation, evidentialism, falsifiability and openness to falsification, and characteristic institutional structures. Some people have advocated one or another of these as the best way to define science, as with Popper's falsificationism. Others have suggested these different features are together the hallmark features of science. We think that is the most promising approach. So, we define **science** as the inclusive social project of developing natural explanations for natural phenomena. These explanations are tested against empirical evidence and must be subject to open critique, refinement, and rejection.

The characterization of science developed here can provide a kind of checklist for assessing to what extent some activity qualifies as scientific, as pictured in Table 1.1.

Consider how this characterization of science relates to our earlier example of climate change. First, science aims to generate knowledge. Climate science aims to

TABLE 1.1 Checklist of hallmark features of science

✓ Aims to generate knowledge (*knowledge-oriented*)

✓ Provides natural explanations of natural phenomena (*naturalism*)

✓ Advances claims that can be tested against observational evidence (*empiricism*)

✓ Updates claims based on available evidence (*evidentialism*)

✓ Abandons any idea that has been thoroughly refuted (*openness to falsification*)

✓ Involves the broader scientific community (*social and institutional structure*)

generate knowledge of the extent and ways in which human activities are transforming Earth's climate and of the impacts this transformation will have on weather systems, ecosystems, and human societies. Because science is naturalistic, it is limited to natural explanations of natural phenomena. The warming of the Earth's climate is a natural phenomenon, subject to empirical investigation. The proposed natural explanation for this phenomenon is that human activities have generated unprecedented levels of greenhouse gases and the warming effect of those gases.

All scientific claims must be testable, or falsifiable, with the use of empirical evidence, and claims must be supported by significant evidence to be accepted (or disconfirmed by sufficient evidence to be discarded). The claims that the concentration of atmospheric greenhouse gases has dramatically increased since the Industrial Revolution and that the last four decades are the warmest on record (for example) are both testable. We can describe the kinds of evidence that would lead us to reject these claims, but scientists have not found that evidence despite extensive investigation. These claims have not been falsified; they are accepted by the scientific community only because there is strong evidence in their favor.

As new evidence becomes available, scientific claims are corroborated, revised, corrected, or rejected through the collaborative work of researchers embedded in the social and institutional structures of science. Climate change research involves numerous scientists utilizing techniques from different fields of science, and our understanding of climate change and predictions of its effects are constantly fine-tuned. The basic idea of anthropogenic climate change has persisted through all of this—indeed has become more broadly held—because no challenges to the idea or to the research supporting it have been successful. Multiple studies published in peer-review scientific journals independently confirm that glacier retreat and climate-warming trends over the past century are due to human activities, and most of the leading scientific organizations worldwide endorse this conclusion.

Here's an obvious contrast with science: jazz. Jazz artists do not collect measurements or other similar forms of evidence to test hypotheses about the value of a piece of work, and disagreements about the value of, say, Ella Fitzgerald's *Over the Rainbow* cannot be settled by running experiments, conducting empirical studies, or developing models. Unlike scientists, jazz musicians do not aim to find natural explanations of features of the natural world with their practices.

Now consider astrology, a canonical example of pseudoscience introduced previously. The primary claims made in astrology, such as horoscope predictions, are not designed to be falsifiable; in fact, many are specifically designed to be unfalsifiable. They are vague in ways that allow many different interpretations, and so, for any interpretation that is wrong, another can be offered in its place. Further, the systems of horoscopes used by astrologists are inconsistent with well-understood basic theories of biology, physics, and psychology. This violates the expectation of the collaborative exchange of ideas among scientists. Astrology is not science.

Astrology may be a harmless fad, with negative consequences largely confined to misspent leisure time and money. Other pseudoscientific projects are much more dangerous. Denials of anthropogenic climate change—despite overwhelming evidence—have contributed to a lack of political will to address the climate crisis, a failure that is beginning to lead to catastrophic consequences. The campaign of denialism described in *Merchants of Doubt* involved well-established scientists introducing doubt and distraction about topics beyond their scientific expertise to influence political outcomes. Their denial of climate change was not designed to be falsifiable: no amount of evidence would change their mind. Some climate change deniers have even rejected the idea that science is a trustworthy source of knowledge in order to hold fast to their rejection of climate change.

Watch Video 2

Box 1.3 Evaluating scientific expertise

Imagine you are asked to vote on a policy about banning cannabis. The potential ban appeals to research showing that cannabis causes schizophrenia. Suppose you do not know much about cannabinoids and their psychiatric effects. Where would you search for relevant, trustworthy information? Without expertise in the relevant science, it can be difficult to evaluate scientific research. You might find on social media two alleged experts who disagree about the causal claim. How should you decide who is the most credible? The most straightforward way to evaluate scientific claims is to assess the quality of the arguments presented by the experts. But this can be difficult, as scientific information can be technical and hard to understand for non-experts. For this reason, it is also important to consider the credentials and reputation of the alleged experts, including the relevance of their qualifications and their accomplishments in their field, and look out for any possible sources of conflict of interest or bias. And, because science is a collaborative enterprise, try to learn what the consensus or near-consensus is in the relevant area of research. This is more important than what any individual scientist thinks. So, you should also beware the maverick scientist who claims to have refuted the consensus in the field!

Science's limitations

While science is our best route to knowledge about the world around us and to developing innovations based on that knowledge, it is also important to recognize what it doesn't do.

Scientists try to gain knowledge, that is, to develop natural explanations of natural phenomena. The list of the phenomena investigated in science is long; in principle, it includes everything in our universe. But there are some important limitations to the scope of science. Science doesn't replace or limit nonscientific intellectual pursuits, like literature, music, and painting—or politics for that matter. Basing our scientific knowledge about climate change on fluctuating political agendas would be a mistake. But, when it comes to addressing climate change with policy interventions, debating which steps are politically feasible and desirable is fair game for politicians. Of course, knowledge from climate science and other scientific fields such as economics, sociology, and psychology should be considered in these deliberations.

Scientific knowledge differs from theological doctrine and religious practice, too. Unlike religious practitioners, scientists attempt to explain things without appeal to supernatural entities or influences, such as deities or miracles, or to literary allegories or culturally significant myths. Furthermore, faith has a central place in many religions, while it should have none in science. Of course, one can be religious in myriad ways, and many people—scientists included—are both religious and believers in scientific knowledge. People disagree about the role religion should play in our society, but whatever role that might be, science is not designed to occupy it.

Scientism is a derogatory term for an excessive belief in science as a solution to every possible problem—including philosophical problems about the meaning of life and our place in the universe. Like pseudoscience, scientism expresses a kind of intellectual arrogance, where one gives excessive deference to science as the sole source of knowledge we might acquire and the only way to find correct answers to any question of human concern we might ask. In public debates, symptoms of scientism include generic slogans like "because science says so" and "science doesn't care what you believe," which are ironically designed to halt discussion rather than to promote it. Also included are quick dismissals of other humanistic endeavors and disciplines like history and philosophy as being "anti-science." We think it is important to distinguish the thought that science is a uniquely trustworthy source of a certain kind of knowledge from ideas that might sound similar, such as that professional science is the only way to have knowledge of any kind or should be the basis of one's entire worldview.

EXERCISES

1.19 Recall: Define the term *pseudoscience* and give two examples of pseudoscience discussed in the section. For each, describe why it counts as pseudoscience.

1.20 Apply: Choose one example of pseudoscience discussed in this section and evaluate it using the checklist of science. Describe how it is similar to science and how it is different. Can you identify features of the example you've chosen that seem to be intended to appear more like science than they are?

1.21 Think: What's distinctive about science, in comparison to activities like literature, music, and art, as a source of knowledge about the world? Do you think there are any important differences between scientific and artistic ways of gaining knowledge? Support your answers with justification.

1.22 Think: Why must science be limited to the study of natural phenomena? Why must it give only natural explanations? Can you think of any scientific projects that don't seem to satisfy these requirements? If so, describe one such project, making clear why you think it might not be naturalistic. If not, describe a nonscientific project that seems to be non-naturalistic and say why.

1.23 Apply: Search the internet (news websites, magazines, blogs, etc.) for a story about a finding purporting to be based on science, and answer the following questions about it. Include a link to your source when submitting your response. (Alternatively, your instructor may provide you with a story to analyze.)

Answer the following questions about the story:
a. What is the source? Is the person making the claims someone with genuine expertise in what they're claiming?
b. Does it seem like there's any conflict of interest? Why or why not?
c. Does the claim involve vague or ambiguous language?
d. Do the claims fit with other well-confirmed scientific theories?
e. What is the evidence cited in support of the claim?
f. Does this describe good science? Why or why not?

1.24 Recall: Define *scientism* and describe why it is a problem. Give an example of legitimate reasoning leading to knowledge that occurs outside of professional science.

FURTHER READING

For more on political influence used to cast doubt on climate change research and other scientific findings, see Oreskes, N., & Conway, E. (2010). *Merchants of doubt.* Bloomsbury.

For an accessible online resource about the nature of science and scientific processes, see the website *Understanding science.* https://undsci.berkeley.edu

For more on how social norms and social structures influence scientific inquiry, see Merton, R. K. (1942). Science and technology in a democratic order. *Journal of Legal and Political Sociology, 1,* 115–126. Reprinted with the title "The normative structure of science" in Merton, R. K. (1973). *The sociology of science: Theoretical and empirical investigations.* University of Chicago Press.

Also see on social norms and structures in science Boyer-Kassem, T., Mayo-Wilson, C., & Weisberg, M. (Eds.). (2018). *Scientific collaboration and collective knowledge.* Oxford University Press.

For more on the demarcation between science and pseudoscience, see Pigliucci, M., & Boudry, M. (Eds.). (2013). *Philosophy of pseudoscience: Reconsidering the demarcation problem.* University of Chicago Press.

CHAPTER 2

How science pursues its aims

2.1 PUBLIC HEALTH AND HOW VALUES SHAPE SCIENCE

After reading this section, you should be able to:

- Define *biological sex* and *gender* and say how each might be relevant to health outcomes
- Describe the germ theory of disease and social determinates of health, saying what is important about each focus in health research
- Describe how social values influence scientific research aims

Covid-19 and social determinates of disease

Covid-19 was declared to be a global pandemic by the World Health Organization in March of 2020. So early in the pandemic, little was known about the illness. Scientists very quickly identified the virus responsible, dubbed SARS-CoV-2. But how it was transmitted, what influenced people's vulnerability to the illness, and how the pandemic would progress remained a mystery for some time.

Studies based on early reports out of Wuhan, China, concluded that men were more than twice as likely as women to die from Covid-19. Other studies corroborated that men were much more likely to be admitted to intensive care units and to die from the illness. Early investigations of this phenomenon tended to target biological sex as the relevant factor, focusing on immune system differences between males and females due to hormonal balance and genes on the X and Y chromosomes. One study proposed that a "sex-based approach" to treating Covid-19 should be developed, with different medical support for male and female patients.

Biological sex includes the categories male, female, and intersex (perhaps with multiple varieties), which in humans is determined by combinations of X and Y chromosomes, hormones, reproductive organs, and other physical traits. The distinct but related term **gender** includes behaviors, social roles, appearances, and identities of individuals traditionally associated with the expression of masculinity, femininity, or non-binary features. A broader range of gender identities are acknowledged today, and some challenge the very idea of gender categories. Here, what's relevant are the behaviors and social roles related to genders traditionally associated with biological sex categories—that is, being a man or being a woman.

DOI: 10.4324/9781003300007-3

Biological sex and gender are distinct variables in health research, but which is relevant can be difficult to discern in some cases. Although biological sex is physiological and gender is behavioral and social, gender can still have physiological effects, including health effects. For instance, if men are more likely to continue to work outside the home during a pandemic stay-at-home order due to gender differences across occupations (a social role) or more likely to resist health recommendations like masking (a behavioral tendency), then this could result in increased susceptibility to Covid-19. If men are more likely to neglect preventive healthcare and treatment of health conditions (a behavioral tendency), then this could result in worse outcomes when they contract Covid-19. If it's unknown whether sex or gender differences are responsible for some health disparity, researchers might just refer to "sex/gender."

In light of the emerging focus on different Covid outcomes for men and women, a group of scientists and science scholars at Harvard University's GenderSci Lab began to comprehensively track sex/gender disparities in Covid-19 outcomes across the United States. These researchers found that the sex/gender differences in Covid outcomes changed over time, across different states, and in different social contexts such as urban and rural areas. This pattern suggested it was not biological differences between sexes but social/behavioral differences between men and women that were primarily responsible for the observed variation in Covid outcomes. This is because typical behaviors and social roles vary across locations and incidental circumstances in a way that the sex differences between males and females (chromosomes, hormones, and reproductive organs) do not. On average and in ways that vary across time and place, men and women tend to have different careers, different likelihoods of smoking cigarettes, different relationships to preventive care and health precautions, and more—all of which can influence one's susceptibility to Covid-19 and its severity if one does contract it.

And so, what had initially appeared to be a physiological difference in susceptibility to Covid-19 related to biological sex was revealed by further research to be instead—at least in large part—a social difference in exposure and vulnerability related to gender.

This difference is important for researchers to understand. In health research, biological factors like genes and hormones associated with biological sex tend to receive more attention than social factors like lifestyle choices and healthcare access. This can lead healthcare to focus too much on biological factors and neglect social factors. Recall, for example, the study mentioned earlier that suggested the need for different medical treatments for Covid-19 in men and women due to presumed immune system differences. That study did not consider whether social factors might instead be responsible for the sex/gender disparity in Covid-19. This focus on biological sex in health research and healthcare and the neglect of social factors related to gender can in turn reinforce stereotypes about differences between men and women tracing back to biological sex differences. It's hard to learn about the significance of social factors related to gender if they aren't investigated in research.

Note that similar points apply to health impacts of race. Health research has tended to emphasize physiological differences related on average to race, such as prevalence of certain genes or genetic conditions or susceptibility to conditions like heart disease or osteoporosis. But race is a social category, associated with social factors that have

(a)

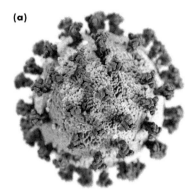

(b)

FIGURE 2.1 (a) Early in the Covid-19 pandemic, media coverage often featured artistic renderings of the viral particles SARS-CoV-2, reflecting a germ-centric orientation to the pandemic. (b) In summer 2020, the extensive Covid-19 outbreaks in meat processing plants began to receive special attention. Occupation is one of many social determinates of health.

clear implications for health, such as average socioeconomic status, access to health-care, exposure to racism and other sources of stress, and exposure to environmental hazards (like proximity of a neighborhood to polluting industries). As with gender and biological sex, if health research does not attend to race as a social factor, it risks essentializing race as only a biological factor. In the Covid-19 pandemic, racial and ethnic minorities in the United States suffered higher rates of infection and death from Covid-19 due in large part to social factors like living conditions, type of work, and access to healthcare.

There is a broader point behind these stories about Covid-19's variable impact by gender and race. Historically, health research and health campaigns largely focused on social influences on health: better sanitation, less crowded housing, safer work conditions, and neighborhood improvements. In the late 19th century, microbes were discovered: bacteria, viruses, and fungi so tiny they are invisible were found to be the source of many diseases. The ***germ theory of disease***, the theory that such microbes, or "germs," cause illnesses, won out over the miasma theory, which held that illnesses are caused by bad air from rotting matter and other bad sources. This led health research and campaigns to narrow their focus to preventing the transmission of germs and infection from germs, as with antibiotics and vaccination.

Yet there was also something right about the earlier health focus on living conditions. As we've illustrated with Covid-19, various social factors are relevant to disease susceptibility and severity: education, income, housing conditions and exposure to pollution, pervasive stress from racism, access to healthy food and activities, and more. These are collectively referred to as ***social determinates of health***. Over the past century or so, a focus on interventions related to germs, like avoiding and slowing disease transmission and developing vaccines and antibiotics, has predominated over research and interventions related to the social determinates of health.

Social values and the aims of health research

The health sciences include a variety of fields, two of which are medicine and public health. Medical research and practice focus on preventative and treatment options for individuals with, or at risk of, particular health conditions. Medicine primarily draws on the biological sciences, as reflected in the common term *biomedical science*. In contrast, public health research and practice focus on the overall health of populations, primarily in disease prevention, drawing from research in social sciences, psychology, and environmental science.

The field of public health is where social determinates of health are most likely to be investigated. But even public health has been shaped by the germ theory of disease to focus especially on disease transmission, to the extent that social determinates of health have been pushed to the side. Social determinates of health, like housing conditions, exposure to pollution, access to healthy food, and more, can be less obvious in their importance and can be challenging to address. But, for many diseases, addressing social determinates of health is crucial to supporting a healthier population. This is especially so for so-called lifestyle diseases that aren't caused by pathogens directly but rather things like diet, exercise, smoking, and drug and alcohol use. Examples of lifestyle diseases are heart disease, stroke, type II diabetes, and some forms of cancer. In many nations, lifestyle diseases are among the most common causes of death.

Focusing too much on the role of germs in diseases—a medical view of disease—can make for a sicker population by leading to the neglect and misunderstanding of the social determinates of communicable diseases, like Covid-19, and to the relative neglect of lifestyle diseases. In some articles in the popular magazine *The Atlantic*, science writer Ed Yong has suggested that the Covid-19 pandemic, including its inequities along race

TABLE 2.1 The contrasting features of medicine versus public health

	Medicine	*Public Health*
Target	Individuals	Whole populations
Aim	Prevent and treat particular health conditions	Improve overall community health, including disease prevention
Relevant Sciences	Biological sciences, pharmacology, bioinformatics/ bioengineering	Sociology, environmental sciences

and gender lines, is an opportunity to refocus public health on the social determinates of disease.

Covid-19 is a good illustration of how scientific research could focus on many different but related topics. Scientific research on Covid-19 ranges from the SARS-CoV-2 virus's structure and impact on specific cells; to the epidemiology of how quickly the disease would spread, how it would impact the economy, and how spread would change with different public health actions, like stay-at-home orders; to the influences of age, class, gender, and race, as described earlier; and more. And, of course, Covid-19 is just one threat to public health. There are also many other health topics in need of study, in general and even in how the pandemic influenced them. For the pandemic, these include mental health and educational impacts of stay-at-home orders and school closures; the effects of postponed care on health conditions like heart disease, cancer, and diabetes; how social distancing influenced the transmission of seasonal illnesses like the flu; and more. And this only scratches the surface of the range of potential health research topics in general, of course.

So, which of these many research topics should be prioritized in health research? And should the focus be on medical treatment, preventative and population health, or something else? The answers to these questions are influenced by social values of the communities that influence science. **Social values** are group priorities and moral ideas accepted in a community. Social values provide a shared background orientation that can influence decisions and activities, including in scientific research. But social values are not universally shared: different communities might have other social values or differently prioritize some of the same social values.

Here's an example related to our case study of how Covid-19 impacts different genders. Harvard University's GenderSci Lab, the group that tracked sex and gender disparities in Covid-19 outcomes, describes itself on its website as "dedicated to generating feminist concepts, methods and theories for scientific research on sex and gender." This group brings their shared feminist values to their biomedical research, including bringing a critical perspective to research that may emphasize sex differences more than the data warrant. This perspective leads to research that can shed light on how gender influences health outcomes.

As this example shows, social values influence which research topics are priorities in health research. Other examples include whether attention should focus more on medical options for difficult-to-treat diseases or on broad health outcomes in the overall population, including those with less access to medical care, and whether attention should focus more on theoretical knowledge of some disease, be it Covid-19 or cancer, or prevention of the most cases possible. Different orientations on these and other questions will shift which research topics are prioritized and what the focus is in researching any given topic. This helps make sense of Ed Yong's suggestion that attention to social injustice and health inequity arising out of Covid-19 may motivate additional investment in public health research and a greater focus on the social determinates of health within the field of public health.

The influence of social values on research topics and aims is not special to health science but is general across science. Incredibly important questions to ask to understand science, then, are these: What is the specific research topic and aim? And what are the reasons for prioritizing those topics and aims over others?

In Chapter 1, we suggested that science is ultimately structured to pursue knowledge about our world: natural explanations of natural phenomena. Here, we've seen how that general aim can vary widely in the specifics, and social values help drive which specific aims and topics are prioritized over others. The remainder of this chapter is devoted to an overview of the methods science employs to pursue these aims and how the institution of science developed to support them.

EXERCISES

2.1 **Recall:** Define *biological sex* and *gender* and describe how each was thought to Watch Video 3 be relevant to Covid-19 susceptibility and severity.

2.2 **Think:** Consider how sex/gender has both physiological elements (biological sex) as well as behavioral and social elements (gender). (a) Why is it important to consider these as distinct variables in health research? (b) Why can these influences be difficult to untangle? (c) Why do you think biological sex tends to be focused on more in health outcomes than gender is?

2.3 **Recall:** Define *germ theory of disease* and *social determinates of disease*, then (a) indicate how each relates to medicine and to public health, and (b) describe what is correct about each.

2.4 **Apply:** List at least six social factors that might be relevant to health outcomes. For each, describe the potential relevance and whether it could be changed to improve public health in general and/or the health of specific populations.

2.5 **Think:** What are some risks of health research focusing only on medicine (neglecting public health)? What are some risks of health research focusing only on public health (neglecting medicine)?

2.6 **Think:** Define *social values*, give at least three examples, and say in your own words how social values can influence research aims in science.

2.2 VARIETY OF SCIENTIFIC AIMS AND METHODS

After reading this section, you should be able to:

- Identify why there is not a single, unified scientific method
- Explain how different scientific aims and circumstances influence scientific methods
- Characterize the roles of the following in science: experiments and observational studies, modeling, scientific arguments, statistical reasoning, and theorizing

No simple scientific method

In some science class along the way, you probably learned about the scientific method. But, interpreted literally, the idea that science always uses the scientific method is a myth. There is no fixed series of steps that scientific research follows. There is immense variability in aims and in methods for getting there, and there is considerable room for creativity along the way.

In Chapter 1, we developed a checklist approach to defining science. That checklist focused a lot on methods common in science, like empirical investigation, evidence gathering, openness to criticism, and collaboration. Scientific methods are central to its ability to produce trustworthy knowledge about our world. But the reason we defined science with a checklist is because of all the variety scientific projects can have, and that includes variety in methods.

Some of the most important scientific breakthroughs had decidedly unscientific-seeming origins. For example, there was no real method by which 19th-century German chemist August Kekulé discovered that the benzene molecule was structured like a ring; allegedly, he just had a daydream of a snake biting its tail. (However, this daydream came after Kekulé had been studying chemistry and the nature of carbon-carbon bonds for years.) Similarly, the idea that natural selection is the mechanism of evolutionary change occurred to the British naturalist Alfred Russel Wallace during a feverish attack of malaria while traveling in Indonesia in 1858—or so he wrote in his autobiography.

We also saw in section 2.1 how scientific research can have a number of different aims—generating knowledge, supporting effective action, making predictions, designing and building new products and technologies—and can focus on a number of different topics. For example, we encountered the different health conditions medical science and public health may prioritize, as well as whether the focus is on theoretical under-standing, medical treatment, prevention efforts, or something else.

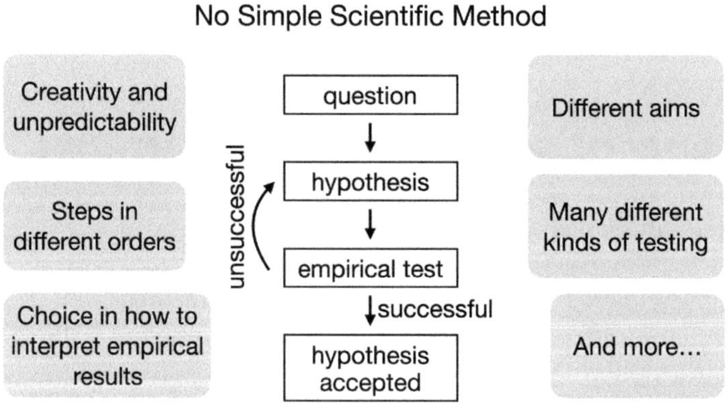

No Simple Scientific Method

FIGURE 2.2 Scientific methods are sometimes represented very simply, but there are many variations in science's aims and methods

So, there is no clear method supporting at least some important scientific discoveries, and there's lots of variation in the aims of scientific research. You might think that once an aim is decided on and then a hypothesis is formulated, the scientific method kicks into gear in how the hypothesis is tested. But there are also many differences in how and the degree to which scientific claims are tested by empirical evidence. Sometimes empirical investigation is exploratory and open-ended, without clear ideas in mind to test. Sometimes getting direct empirical evidence for or against an idea isn't possible, and scientists must be creative in how they find evidence or what they count as evidence. Sometimes the research doesn't involve collecting new empirical evidence at all but combining or reinterpreting existing evidence. These and other complications are reasons to say that there is no single thing we can call "the" scientific method.

Box 2.1 Descriptive versus normative claims in science

Descriptive claims are claims about how things are. Examples of descriptive claims are that this textbook has three authors and that the Nile River flows over 6,600 kilometers. Descriptive claims can be evaluated for their truth or falsity. "Rabat is the capital of Australia" is a false descriptive claim. In contrast, *normative claims* are claims about how things should be. For example, people should read more; scientists should be sincere; there ought to be peace. These claims can be evaluated for the robustness of the rules that they specify. Descriptive and normative considerations are both part of science. Certain theories in economics, for example, make claims about rational decision-making, which express how agents should make decisions. If people do not make decisions that way, it does not follow that the theory is false—though economists, depending on their goals, might revise their normative theories to make them more descriptively accurate. Just as science involves both normative and descriptive claims, both kinds of claims can be made about science. One can simply attempt to characterize what science is—that is, how scientists in fact develop theories and test claims. *Metascience*, using scientific techniques to study science itself, is a discipline that pursues this route. Or, one can attempt to say how science should work, that is, what features science should have for it to succeed at generating knowledge. Philosophy of science sometimes pursues this route. This book does both.

Different methods for different aims and circumstances

Let's add more detail to the idea of a variety of scientific aims introduced in section 2.1. First off, that variety of aims regards what natural phenomenon is studied—such as in the case of health sciences, Covid-19, heart disease, and environmental pollution, to name just a few options. This variety of aims also regards what specific aspects of the phenomena are investigated. An illustration of this is

TABLE 2.2 The variety of scientific aims

Types of Variety	Examples
Phenomenon studied	Covid-19, heart disease, or environmental pollution
Focal aspect of the phenomenon	Structure and biological action of SARS-CoV-2, epidemiological models of disease spread, or how virus affects different genders and races
Goal of the research	Specific theoretical knowledge, connections to knowledge of other phenomena, medical treatment, prediction, or public policy guidance

all the different aspects of Covid-19 that were investigated by different scientists, such as the structure and biological action of the SARS-CoV-2 virus, epidemiological models of disease spread, and how the virus affected people of different genders and races differently.

The variety of scientific aims also extends to the specific research goals when targeting some aspect of some phenomenon. Is the focus of studying the structure and biological action of SARS-CoV-2, for example, to develop theoretical knowledge of this virus in particular, or to explore its similarities to other coronaviruses, or to spur vaccine discovery, or to predict how the virus will likely evolve, or to determine what kinds of public policy would mitigate its spread? All of these are reasonable goals for scientific research into this aspect of the phenomenon, but each can lead to research with different features.

Besides having a variety of different aims, scientific research is also carried out in a variety of different circumstances. Some phenomena you can directly influence, while others you can just watch from a distance. Some phenomena you can see with your own eyes, and others you can merely detect distant evidence of. Some phenomena change quickly, and others are very slow. Sometimes scientists have specialized equipment to support exactly what they're trying to do, and at other times they need to make do with old equipment or no specialized equipment at all. Perhaps the desired equipment hasn't yet been invented.

These differences in the aims and circumstances of scientific research give rise to differences in scientific methods. A good bit of resourcefulness is needed to find the empirical evidence to support developing natural explanations of natural phenomena. The specific aims and circumstances influence what methods will be useful to that project. Most of the main topics of this book can be thought of as exploring one or another aspect of how scientific methods vary in response to aims and circumstances.

As we've already suggested, one major type of variability in methods regards the manner and extent of empirical investigation. Recall from Chapter 1 that *empirical investigation* is inquiry that grounds the justification for beliefs about the world in sensory information and observations. More empirical evidence that directly bears on ideas under investigation is always better. What varies is, first, how empirical evidence can be gathered and, second, how directly that evidence bears on the ideas being investigated. Experiments are a highly valuable way of conducting empirical investigation. In Chapter 3, we will survey the ways in which experiments are conducted, identifying their core features and also recognizing how different experiments can be from one another.

Sometimes it just isn't possible to conduct an experiment. When this is the case, there are various ways to conduct observational studies and use other methods to gather empirical evidence bearing on the ideas under investigation. We survey several of those methods in Chapter 4. One valuable set of methods to indirect empirical investigation is scientific modeling; this is the topic of Chapter 5.

We said just a few paragraphs ago that the aims and circumstances of scientific investigation influence the extent to which empirical evidence bears directly on the ideas under investigation. Oftentimes, whether an idea is true can't be directly determined using empirical investigation. Instead, scientists need to deploy arguments, or reasoning, that use empirical evidence to support conclusions about the ideas they are interested in. This reasoning can follow different patterns. Sometimes it's what we call deductive reasoning, as when an empirical test provides grounds for definitively rejecting a hypothesis; we examine that pattern of reasoning in Chapter 6. Other times, reasoning patterns in scientific arguments involve drawing general conclusions from a limited set of evidence or, in other ways, reasoning beyond what the evidence guarantees. These forms of reasoning, called inductive and abductive, are discussed in Chapter 7.

A variety of mathematical tools are also put to use in science, though whether and how mathematics is relevant is yet another feature of science that varies. Particularly widespread and important uses of mathematical tools involve reasoning with probabilities, which we discuss in Chapter 8, and using statistics to describe phenomena (Chapter 9) and to make inferences and predictions (Chapter 10). Statistical tools can be used to uncover patterns in what might otherwise seem merely like random variation.

One important aim of scientific research is uncovering causal relationships. Chapter 11 explores how methods encountered in earlier chapters can be used in causal reasoning. Chapter 12 examines how all of the various methods of empirical investigation surveyed throughout the book can be used to develop scientific theories and explanation, important forms of scientific knowledge. Finally, Chapter 13 explores more fully how social values influence scientific practices and surveys salient features of science in the 21st century and new challenges it faces.

By the time we've worked through all the topics of this book, we will have seen the great extent of variation in science's methods and in how these methods relate to the different aims and circumstances of scientific research. But first, let's take a quick look at some of the commonalities across these methods.

Watch Video 4

Box 2.2 How to read a scientific article

Scientific knowledge is typically communicated in articles published in professional journals written by experts, for experts. Thus, scientific articles tend to use specialized jargon, formalism, images, equations, and tables that most people will find hard to understand. For non-experts reading scientific articles, one helpful method involves a set of five questions for structured reading:

1. What problem did the authors try to address? You will typically find the answer to this question in the article's Abstract and/or Introduction.
2. What's the point of addressing that problem? To answer this question, focus on the Abstract, Introduction, and Discussion sections.
3. What did the authors do to address their question or problem? The Methods and Results sections should provide you with an answer; but do not get lost in the details—just focus on the independent and dependent variables the researchers manipulated and measured and the general kind of methods they employed.
4. What did the authors find? Focus on the key finding presented in the Results section. This is where the specific research question and rationale are addressed.
5. How are the results interpreted? Read the Discussion section to answer this question, considering whether the authors' interpretation is warranted by what they did and what they set out to address, and try to figure out if an alternative interpretation of the results may be more warranted.

EXERCISES

2.7 Recall: What are some reasons to think there's no single, unified scientific method?

2.8 Think: Consider Figure 2.2. Describe the idea that there's no unified scientific method, and then evaluate this idea. Try to raise at least three considerations in favor of the idea as well as three considerations opposed to the idea.

2.9 Recall: Describe how scientific methods are influenced by (a) the specific aims of investigation and (b) the circumstances of investigation.

2.10 Think: Look at this book's table of contents, and perhaps flip through some of the chapters to come. Write out at least three questions about scientific methods and reasoning that you want to find answers to from reading this book.

2.11 Apply: Go to www.science.org, the website of a prestigious scientific journal, *Science*. Choose an article featured on the website, read the title and abstract, and look at all the section headings. Alternatively, your instructor may provide you with an article to analyze.

(a) Write the title of the article; (b) summarize the main point(s) of the article in 1–2 sentences; and (c) describe how well the structure of the article matches the article structure described in Box 2.2. Finally, (d) identify which of the tools summarized in this section (experiment, observational study, model, argumentation, statistics, explanation, theorizing) were relevant to the research depicted in the article, and say how each was relevant.

2.3 RECIPES FOR SCIENCE

After reading this section, you should be able to:

* Describe what the metaphor of recipes and ingredients is intended to mean here
* Define each of the three ingredients found in most recipes for science, and describe why each is a challenge
* Describe at least three ways in which each of the ingredients of science—hypotheses, expectations, and observations—can vary

Methods in science

In this chapter, we have surveyed how scientific aims and methods vary. Science proceeds in myriad ways, and there's not a simple, unified scientific method. And yet, as the title of this book suggests, even as scientific methods vary, science does follow some familiar recipes.

Consider culinary recipes like you find in a cookbook. These recipes have some standard components, like the name and origin of the dish, the ingredients along with their quantities and proportions, cooking times, and the necessary equipment to make the dish. These recipes also vary: in their ingredients, the equipment and processes they involve, their difficulty, how long they take, and—of course—what will result from following the recipe. Furthermore, simply following the steps doesn't guarantee a delicious dish. Many cooks have margin notes in their cookbooks adapting recipes to their circumstances and tastes. And some recipes are just bad recipes!

Like culinary recipes, recipes for science have multiple components, involve a wide array of techniques and instruments, vary from one to another, accomplish different tasks, and are improved by others. As with cooking, there is no single set of mechanical instructions and step-by-step procedures that guarantees good science. Just like great cooking, good science is a highly variable and creative process. It also can be messy.

Still, just as culinary recipes tend to have some common features, recipes for science tend to use some common patterns. To start, most involve something like these three ingredients, in one form or another: hypotheses, expectations, and observations. What is sometimes thought of as "the" scientific method describes one way these ingredients can come together: scientists may formulate hypotheses about the world, described in Chapter 1 as bold and risky conjectures, and then use those hypotheses to generate specific expectations regarding their experiences. If their observations conform to those expectations, their hypotheses are confirmed. If not, they return to the drawing board.

These three ingredients—hypotheses, expectations, observations—can be combined in different orders, and they can be combined again and again in different patterns. And each ingredient can vary in its features. For example, sometimes scientists investigate a specific hypothesis, while at other times research is more exploratory and open-ended. Sometimes hypotheses have obvious empirical implications, and at other times scientists need to use statistics to develop their expectations. Sometimes scientists design experiments to test their expectations, and at other times they develop models. Further, some scientific research isn't described well by this trio of hypothesis, expectations, observations, such as highly theoretical research carried out without making observations.

Nonetheless, these three ingredients are integral to the production of scientific knowledge. They are the basic ingredients that, with tremendous variation, occur in the many successful recipes for science we survey in this book.

Hypotheses

Empirical investigation is how we learn about our world. Scientists make observations to try to figure out what's out there, why things are the way they are, and how things change. But simple observations can't accomplish these tasks by themselves; scientists also need theoretical claims. **Theoretical claims** are claims made about entities, properties, or occurrences that are not directly observable. As an example, consider a claim about all of something of some kind, like the claim that all salt dissolves in water. You might have seen plenty of salt dissolve in water, but you will never be able to witness all of the salt that exists dissolving in water. Because you can't directly observe that all salt dissolves in water, this is a theoretical claim. We have plenty of evidence that this is true, but the claim is theoretical because it goes beyond what we can directly observe.

Theoretical claims investigated in science are called hypotheses. A **hypothesis** is a conjectural statement based on limited data—a guess about what some aspect of the world is like, which is not (yet) backed by sufficient, or perhaps any, evidence. Scientists do not yet know whether any given hypothesis is true or false; when there is sufficient evidence in favor of some hypothesis, it graduates from that category. Theoretical claims that we have sufficient evidence to conclude are true become scientific knowledge.

Formulating a hypothesis requires some imagination: if you could observe something we can't—if you could witness the beginning of life on Earth, or see all the salt in the world—what would you find? Sometimes scientists may formulate a hypothesis before any observations have been made, just with the use of their imagination. But often, initial observations, other hypotheses, or background knowledge about related phenomena help inspire new hypotheses. Before scientists knew about the properties of potassium chloride, they'd seen that table salt—sodium chloride—dissolves in water. This informed their expectations for potassium chloride, a similar compound. Scientists' hypotheses about the first lifeforms were shaped by what they know about organisms, existing and extinct, and how the Earth has changed over geologic time.

Scientists can have different levels of confidence in different hypotheses. If a hypothesis is informed by lots of experience with similar objects or significant background knowledge of related phenomena, scientists might be much more confident in it than if it were a random guess. But, by their very nature, hypotheses are guesses. This is why hypotheses must be tested.

Expectations

Learning whether a hypothesis is true is often more circuitous than just making direct observations. A second ingredient is usually needed to test hypotheses: this is developing expectations based on hypotheses. *Expectations* are conjectural claims about observable phenomena based on some hypothesis. These claims are conjectures since they go beyond what scientists have observed so far, but, unlike hypotheses, their truth or falsity can be discerned directly from making the right kind of observations. Expectations are claims about what scientists expect to observe if a given hypothesis is true.

Expectations do not regard just any possible observations, but observations that scientists anticipate being able to make. We could say what we would expect to experience if we were present for the beginning of life on Earth, but since we don't have a way to make those observations, such expectations are useless. Instead, expectations based on a hypothesis regarding the origin of life on Earth should be about what scientists expect to see today, in present lifeforms or in traces of past life.

Depending on the nature of a hypothesis, developing expectations based on the hypothesis can be straightforward or complicated. On one extreme, the hypothesis that all salt dissolves in water leads directly to an expectation: any sample of salt will dissolve when placed in water. But even then, the expectation needs to be fine-tuned. Should salt dissolve when placed on a chunk of ice (frozen water)? What if some salt is already dissolved in the water; should we still expect the sample of salt to dissolve? And expectations for present observations that bear on some hypothesis about the origins of life on Earth are, of course, much more complicated to develop.

No matter whether deriving expectations is relatively straightforward or very complicated, this is an important and nontrivial ingredient of scientific research. Expectations set scientists up to make observations that can provide evidence for or against the truth of a hypothesis. Deriving expectations thus serves as a bridge between theoretical claims (hypotheses) and observations (data).

Observations

All or nearly all science fundamentally depends on observations. It's not enough to think up interesting ideas about how the world might work; those interesting ideas must also be evaluated by how well they fit with our observations of the world. This is why both empirical investigation and evidentialism are on our checklist definition of science from Chapter 1. *Observations* include any information gained from your senses—not only what you see, but also what you hear, smell, touch, taste, and any other way you may be able to experience the world.

Your sensory experiences belong only to you. If we are on a hike together, we might both hear a rattling sound coming from behind a boulder. But each of us only has access to our own experience of the sound. Data are different. **Data** are public records produced by observation, sensory experience, or some measuring device. Observations are important because they are your only way to directly access the world. Data based on observations are important because they allow us to record and compare our observations.

Observation isn't passive. We can move our heads to see different things and relocate our bodies to different places where we can hear different things. We can also use observations from multiple senses together. If you're wondering about that rattling sound from behind the boulder, you can walk around to the other side to see whether there's a rattlesnake there. Besides changing our position and using multiple senses to enhance our observations, we can also change the world around us to create opportunities for different observations. Crushing a leaf lets you better smell whether it's sage or mint.

Humans have also found many ways to use tools to enhance our powers of observation. Light can be refracted with mirrors, prisms, and lenses to extend the reach of vision. We now can see not just through our eyes alone, but also through our eyes aided by telescopes, microscopes, and other devices. To help us hear beyond our ears' capabilities, we have developed microphones, stethoscopes, and more. These technological enhancements range from observational correctives like eyeglasses and simple sensory aids like microscopes to much more complex technology with highly specific purposes, like an fMRI machine that can show brain activity and the Large Hadron Collider, which uses superconducting magnets to cause streams of high-energy particles to collide in a detectable way. Such enhancements have allowed humans to generate what we might call **super-observational access**: using tools to enhance our powers of observation beyond what they ordinarily include.

Making observations, and collecting data as records of those observations, is at the heart of science's ability to generate knowledge of our world. But observations aren't always independent from the ideas about the world we already have. Changes in what we believe to be true can have significant impact on what we observe. For instance, when we observe the Sun at the horizon, what we seem to see is the Sun at one point on its path across the sky. Geocentricism, the historically dominant idea that the Earth is the center of the universe, organizes this and similar observations into an easily understood pattern, and those observations confirm geocentrism. But from the perspective of heliocentrism—the idea that the Earth and other planets revolve around the Sun—once your head is slightly turned to the side, the Sun at the horizon and the other planetary bodies that appear comprise a different observation. Figure 2.4 schematically diagrams this conceptual shift.

The switch in theoretical orientation to heliocentrism thus provides a different perspective on astronomical observations. It may also create a different perceptual experience: instead of the Sun moving below the horizon as it sets, your position on Earth rotates away from the Sun. New ideas can sometimes have a strong effect on what we think we see. Thus, observations are a crucial ingredient of science, but they aren't passive, aren't always the starting point, and aren't always decisive.

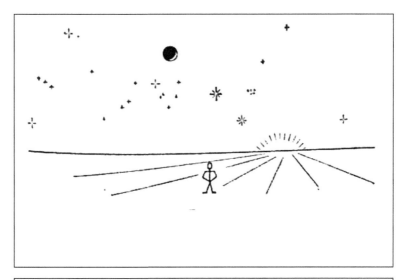

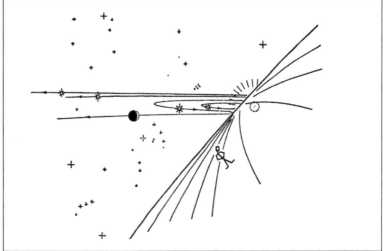

FIGURE 2.3 Schematic diagram of conceptual shift from geocentrism to heliocentrism

EXERCISES

2.12 Recall: This section develops the metaphor of recipes with common ingredients. Watch Video 5 What about science do recipes correspond to, and why is it a plural—*recipes* instead of *recipe*? What about science do the ingredients correspond to?

2.13 Think: Describe at least three aspects of science the metaphor of recipes with common ingredients is intended to highlight. Evaluate the metaphor: what do you think is useful about it, and what is a limitation or potentially misleading about it?

2.14 Recall: What is the difference between observations and data? What is important about observations in particular, and why? What is important about data in particular, and why?

2.15 Think: Hypothesis, expectations, and observations are all important ingredients for most science. Describe the importance of each, a typical way that the three ingredients work together, and what they accomplish together.

2.16 Think: Hypothesis, expectations, and observations are all important ingredients for most science. Describe a difficulty with each, or circumstances in which it can be difficult.

2.17 Apply: Go to www.science.org, the website of a prestigious scientific journal, *Science*. Choose an article featured on the website, read the title and abstract, look at all the section headings, then read more of the article as needed to complete the following steps. Alternatively, your instructor may provide you with an article to analyze.

(a) Characterize what you think the hypothesis under investigation is and, in 1–2 sentences, say why you think so. If you aren't sure what the hypothesis is or you don't think there is a hypothesis, give your reasoning.

(b) How explicitly did the researchers describe their expectations? See if you can distinguish the specific expectations, or expected observations, from the general hypothesis under investigation. Whatever your answer on this, give your reasoning.

(c) Describe the kinds of observations made in the research. What were the researchers' findings? If there weren't observations of any kind made as part of the research, describe what you think the point of the article is.

2.4 SCIENCE'S ORIGIN AND KNOWLEDGE ACROSS CULTURES

After reading this section, you should be able to:

* Describe how scientific methods for gaining knowledge about the world are similar to and different from strategies employed in daily life
* List innovations from the Islamic Golden Age and the Scientific Revolution and their significance for science
* Indicate three areas of investigation to which indigenous knowledge is particularly relevant and describe why it's relevant

Science and everyday reasoning

The ingredients of science's recipes discussed in section 2.3—hypotheses, expectations, and observation—comprise a distinctive and powerful combination that support science's distinctive ability to generate knowledge about our world. They are also to some extent common strategies employed by people in their everyday lives, as well as strategies harnessed historically by different cultures around the world, to some extent building incrementally toward the modern institution of science.

Alison Gopnik, a psychologist at the University of California at Berkeley, has conducted research exploring ways in which early childhood development involves empirical investigation, much like the research conducted by scientists. Babies and children conduct informal experiments: they test their ideas about the world around them by checking out whether what they observe matches up with what they expect to happen. Over time, children develop something akin to theories about things that are important to them, including about what other people around them think and believe. These theories grow more sophisticated as children develop, for example, by incorporating the recognition that someone with different experiences from oneself will have different ideas. Around four or five years old, a child begins to recognize, for example, that someone who did not witness a jar of cookies being moved to a new location will have a false belief about where the cookies are, even though the child knows the true location of the cookies.

Much like scientists, children conduct experiments and develop theories as attempts to better understand the world around them. This use of observation to develop theories about the world extends into adulthood, though adults tend to be more confident about what we will encounter and thus more fixed in our ideas. Perhaps, then, scientific research can be thought of in some respects like an extension of childhood curiosity and openness, cultivated into habits of investigation and openness to refutation about the topics a scientist investigates.

The basic pattern of making guesses, collecting observations, and then adjusting ideas in response is common across people of all ages, at least when the circumstances call for it. What is distinctive about science is, first, harnessing this basic pattern to systematically gain knowledge about our world, and, second, developing an institution around this project, including the important social norms introduced in Chapter 1. Next, we explore more about the historic development of science as an institution.

The development of science as an institution

Science's aim of producing knowledge traces back to the origins of the very word *science*. This word derives from the Latin words *scientia* and *scīre*, which pertain to knowledge. So, science, from its origins, has been about the pursuit of knowledge.

Most historians of science agree that cultural, social, and technological changes that unfolded in Europe between roughly 1550 and 1750 are very important to the development of the modern institution of science. This period is often referred to as *the Scientific Revolution*, beginning with the work of Nicolaus Copernicus, who put forward a heliocentric theory of the cosmos, and ending with Isaac Newton, who proposed universal laws of physics and a mechanical universe. The Scientific Revolution brought about fundamental transformations in our knowledge of the natural world and in how knowledge claims were thought to be justified. Many of the methods, ideas, and institutional structures developed during that period remain central to science.

But let's start our consideration of science's history even further back. Way before the Scientific Revolution, a variety of innovations across diverse civilizations—including ancient Egypt, Iran, India, China, Greece, and the pre-Columbian Americas—provided fertile grounds for proto-scientific activity. A variety of civilizations developed measurement systems that were essential for collecting data and refining ideas about various phenomena. For instance, many ancient civilizations made sophisticated catalogues of constellations and their observed movements in the night sky, which provided a detailed record of data against which later astronomical predictions and discoveries could be checked.

One important period in the development of science prior to the Scientific Revolution was the 500 years from the 8th through 13th centuries known as the **Islamic Golden Age**, during which time early science saw significant development from Central Asia to the Iberian Peninsula. Here is a sample of some of the scientifically important developments from that period.

The Hindu-Arabic numeral system, which greatly advanced the symbolic representation of numbers and calculation, was invented between the first and fourth centuries in India. Muḥammad ibn Mūsā al-Khwārizmī further developed this system in the eighth century and brought it to Arabic mathematics, and his work later introduced this numeral system to medieval Europe. Al-Khwārizmī also made significant contributions to algebra, geometry, and astronomy. Shortly after, Abū Bakr Muhammad ibn Zakariyyā al-Rāzī was responsible for many innovations in medicine, including advocating for experimental methods and developing classifications of contagious diseases. In the ninth and tenth centuries, Ibn al-Haytham conducted revolutionary work in optics and vision, including the discovery that vision occurs by eyes detecting light deflected by objects.

Arabic polymaths, including especially Ibn Sina, known also by the Latinized name Avicenna, ibn Aḥmad Al-Bīrūnī, and Ibn Rushd, or Averroes, preserved and developed theories about the natural world from the famous fourth-century BCE philosopher Aristotle. This was the basis of ideas about the natural world in 15th-century Europe, with ideas added from Christian, Jewish, and Islamic theology. Based on Aristotle's views, the universe was thought to be geocentric—the Earth at the center—and with two regions: terrestrial for Earth and celestial for the planets and stars. The celestial region was thought to contain transparent concentric spheres that rotate around the Earth. In the first century, Ptolemy supplemented this with an account of the apparent motions of the stars and planetary paths, including detailed models and tables that could be used to calculate the positions of the stars and planets. Geocentrism in 15th-century Europe blended observations of planetary bodies with religious ideas about humanity's place in the universe.

A longstanding problem with the geocentric view of the cosmos was the appearance of **retrograde motion**. In observations made over a series of nights, planets sometimes seem to stop in their orbit, reverse course back across the sky, then stop again, and reverse yet again to continue on their original way. Following Ptolemy, geocentrists explained retrograde motion by positing epicycles: that the planets were on mini-orbits that also follow the larger orbits. This successfully accounted for retrograde motion, but it wasn't as intuitive as other elements of geocentrism.

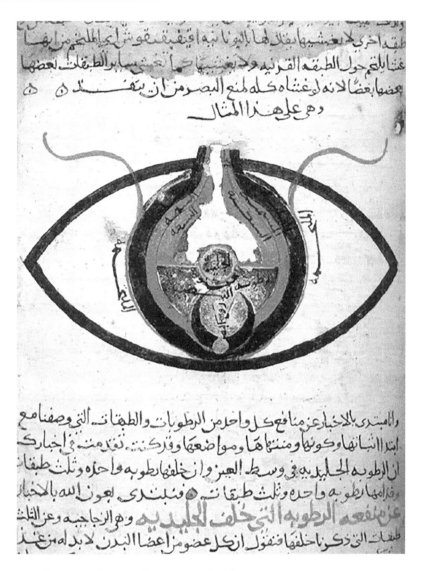

FIGURE 2.4 Diagram of an eye from a circa 1200 manuscript

In the 16th century, Nicolaus Copernicus presented a radical alternative conception of the cosmos as heliocentric, or centered on the Sun, and this provided an alternative explanation for retrograde motion. According to heliocentrism, retrograde motion of planets was due to Earth's changing position relative to other planets as these all revolved around the sun. Copernicus's proposed heliocentric conception of the cosmos was met with skepticism. It violated widely accepted beliefs and called for a fundamentally new physics of the heavens. Besides, the mathematics of Copernicus's system was just as complex as Ptolemy's epicycle solution to retrograde motion, and it did not

FIGURE 2.5 Appearance of retrograde motion

make predictions of planetary motion any more accurate. So, few astronomers were convinced by Copernicus's theory.

The situation changed with the research of Johannes Kepler and Galileo Galilei, each of whom developed and improved the Copernican heliocentric system. Kepler was a German mathematician and astronomer with interest also in astrology; he devised a set of laws that described the motions of planets around the Sun. Based on calculations of the orbits of Mars, he inferred that planets do not have circular orbits as proposed by Copernicus but ellipses instead. This simplified the Copernican theory and significantly improved the predictive accuracy of heliocentric models. Born in Italy, Galileo was instrumental in establishing Copernicus's heliocentric system and, more generally, in replacing Aristotelian mechanics of the separate terrestrial and celestial realms with a new, single physics. Galileo invented the telescope, which he used to observe the phases of Venus and to discover that Jupiter had moons orbiting it. This was a significant discovery for heliocentrism: if our Earth were the center of the cosmos around which all things orbit, then Jupiter's moons should be orbiting Earth instead.

In the Scientific Revolution, the rapid development of new ideas, methods, and tools resulted in the swift accumulation of knowledge. A similar process played out in the later development of the fields of chemistry, biology, psychology, and economics. But many of the pursuits that furthered scientific knowledge also included religious, theological, and philosophical ideas that we would not consider scientific nowadays. In the Islamic Golden Age and the Scientific Revolution, philosophy, theology, and science were not divided as they are now, and often the same ideas had significance for religious belief and views about the natural world. The Scientific Revolution was a decisive step toward the separation of scientific from nonscientific questions and thus toward science explicitly adopting naturalism.

Other central features of the nature of science were also established in the Islamic Golden Age and the Scientific Revolution, such as looking to sense experience and performing experiments to decide what's true, the systematic use of mathematics to study natural phenomena, and the institutionalization of investigation in formal organizations. Of course, these features have also continued to develop since those times. For example, the social organization of scientific activity was significantly transformed with the professionalization of science in Europe and North America beginning in the mid-19th century, which also established the French language, and later English, as the dominant language for scientific communication.

Institutional science and indigenous knowledge

This picture of scientific revolutions culminating in the development of contemporary science is helpful for demonstrating how science is a development shaped by various human societies, but it has the drawback of perhaps overemphasizing the extent to which contemporary institutional science is responsible for knowledge about our world.

For some investigations, such as astronomy and fundamental physics, specialized equipment and training are so essential that such knowledge is the special province of the institution of science. But it wasn't always so. As noted earlier, astronomical observation occurred across many historical civilizations, such as the Mayan in Central America, Polynesian, and Chinese civilizations. And, even today, scientific research about astronomy, as well as many other topics, sometimes incorporates observations or other contributions from people who aren't scientists but are just interested in astronomy. This is called *participatory research* (also *citizen science*), to be discussed more in Chapter 13.

Some types of scientific research aren't limited to only people with access to specialized equipment or training. And some types of scientific research relate as much or more to circumstances on the ground in some specific place as to general scientific knowledge. Examples of these kinds of research include investigations of biodiversity in different locations, effective land management techniques, and how specific communities can mitigate the most disastrous effects of climate change on them. For these kinds of scientific research, it's increasingly appreciated that locals have important expertise and the ability to contribute meaningfully to scientific research.

Beyond the value of localized expertise, it's also now recognized that traditional societies in many parts of the world have developed extensive stores of knowledge about the natural world around them, some of which are still maintained today. **Indigenous knowledge** refers to true claims based on observations, practices, and ideas developed about some geographic region by people native to the area. Indigenous knowledge is in part developed with the use of practices crucial to science—observation, systematic recordkeeping, and checking ideas against evidence—but typically outside the institution of science. There is increasing interest in the value of indigenous knowledge within the institution of science, especially about local environmental sustainability and resource use. This becomes only more valuable as scientific research increasingly turns to questions of sustainability and climate change adaptation, as these are shaped by local conditions and the subject of local knowledge.

Although the institution of science aspires to be fully inclusive and international, it inherits a history of exclusion and is still limited in whom it involves. Historically, in the 18th and 19th centuries, sea voyages of European nations played dual roles as both scientific expeditions and also commercial trips to expand colonization. Even today, there are fewer opportunities to become scientists in the Global South (Latin America, Asia, Africa, and Oceania), and scientists in the Global South face more professional challenges in their research. We will discuss diversity in science in greater depth in Chapter 13.

In the context of this chapter's discussion of how science pursues its aims, this discussion reveals that, though the development of professional science as an institution was very important for its success in uncovering knowledge, it's also increasingly appreciated that institutional science isn't the only place where relevant scientific knowledge is found. The scientific value of indigenous knowledge shows that scientific reasoning and scientific knowledge are not the purview of any single culture or institution.

EXERCISES

2.18 Recall: Describe two ways in which children's reasoning is like scientific reasoning and two ways in which scientific reasoning is distinct.

2.19 Recall: Choose one proto-scientific development from the Islamic Golden Age or the Scientific Revolution. Describe how that development constituted progress (a) in the subject matter of science and (b) in the methods of science.

2.20 Think: It was discovered in the 19th century that the planet Mercury was not following the orbit predicted by Newton's theory of gravity. When this happened, Newton's theory was not considered falsified. Instead, it was hypothesized that this anomaly was the result of another planet, named Vulcan, orbiting between Mercury and the Sun. Despite a systematic search, Vulcan was never found. The anomalies exhibited by Mercury's orbit could be explained only a century later by Albert Einstein's theory of general relativity.

a. Why do you think scientists initially refused to consider Newton's theory falsified?

b. Was this a failure of science? Should the scientists have given up Newton's theory sooner? Why or why not?

c. Does this mean Newton's theory of gravity was not falsifiable? Why or why not?

2.21 Recall: Define *indigenous knowledge*. List three areas of investigation to which indigenous knowledge is particularly relevant and, for each, describe why it's relevant.

2.22 Apply: Mythology and science are generally understood to be very different from one another. And yet early science had its origins in, and then grew out of, mythology, and both myths and scientific theories provide explanations of the natural and social phenomena observed in the world around us.

a. Look up two creation myths from different cultures and historical periods—that is, myths of how the world began and how people first came to inhabit it.

b. Identify similarities and differences across the two myths.

c. Describe similarities between the creation myths and scientific theories of human origin.

d. Describe differences between the creation myths and scientific theories of human origin.

FURTHER READINGS

For resources on gender and sex disparities in Covid-19, see the Harvard Gender-Sci Lab's Teaching Module. *Gender/sex in Covid-19.* www.genderscilab.org/gender-sex-in-covid19-teaching-module

For more on Covid-19 and the importance of social determinants of health, see Yong, E. (2021). How public health took part in its own downfall. *The Atlantic.* www.theatlantic.com/health/archive/2021/10/how-public-health-took-part-its-own-downfall/620457/

For a consideration of how values influence science, see Potochnik, A. (2020). Awareness of our biases is essential to good science. *Scientific American.* www.scientificamerican.com/article/awareness-of-our-biases-is-essential-to-good-science/

For more on the Scientific Revolution, see Shapin, S. (1996). *The scientific revolution.* University of Chicago Press.

On science in the Islamic Golden Age and other periods around the world, see the *History of science society introduction to the history of science in non-western traditions.* https://hssonline.org/page/teaching_nonwestern

For more on science's relationship to indigenous knowledge, see Nicholas, G. (2018). When scientists "discover" what indigenous people have known for centuries. *Smithsonian Magazine.*

CHAPTER 3

Scientific experiments

3.1 THE NATURE OF LIGHT AND EXPERIMENTAL DESIGN

After reading this section, you should be able to:

- Define *experiment* and provide examples
- Describe three ways in which existing scientific knowledge shapes experiments
- List five features of experimental design and identify them in an example experiment

Experimenting on light

In Chapter 2, we discussed how many of the successful recipes for science involve three ingredients: hypotheses, expectations, and observations. Scientific experiments provide scientists with a structured way to make observations and compare them to what we would expect to observe if the hypothesis under investigation is true.

An ***experiment*** is a type of empirical investigation where researchers perform an intervention that changes some feature of a system and observe the effects, with the aim of understanding how the system works or why a certain outcome occurs. Ideally, an experiment changes a system in such a way that the effects depend only on the intervention rather than on other possible factors. For example, by giving the plants on one windowsill in my apartment different types of fertilizer and observing what happens, I can know that the fertilizer is what's making a difference to the plants' growth. This is how experiments enable us to figure out what would happen to a system if some of its features were different—for example, what would happen to my plants if I gave them urine as fertilizer? Being able to figure out what would happen if something were different matters for policymaking and applied research—for example, a company might decide to invest in urine-based fertilizers, or governments may incentivize the use of urine-diverting toilets for recycling waste, if it turned out that urine was an excellent fertilizer.

To begin to explore the main aspects of experiments, let's consider the nature of light. Have you ever seen how a glass prism can make little rainbows appear around a room? When sunlight passes through a piece of glass, the white light is separated into different rainbow colors. Why does that happen? Where do those colors come from?

The nature of light and its relation to the color spectrum visible in rainbows have been studied for millennia. In Chapter 1, we mentioned Ibn al-Haytham (Latinized as

DOI: 10.4324/9781003300007-4

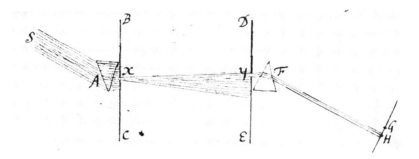

FIGURE 3.1 Newton's original illustration of his two-prism experiment. S represents sunlight; the light between the planes BC and DE is a color spectrum; these colors are recombined to form white light on the pane GH.

Alhazen), who during the Islamic Golden Age advanced the scientific understanding of vision, optics, and light. Through experiments using lenses and mirrors, al-Haytham showed that light travels in straight lines. From dissections, he began to explain how the eye works and began synthesizing the medical knowledge of previous scholars. In particular, al-Haytham demonstrated that light is not produced by the eye, as some theories had claimed, but instead that it enters the human eye from the outside. This research formed the basis of scientific knowledge about light.

In the 17th century, many scientists—or "natural philosophers," as scientists were called then—believed in what we might call the *modification hypothesis* of light. They thought that sunlight is essentially white, and that the spectrum of colors we see from a glass prism is caused by the impurities of the glass modifying the light. Isaac Newton was unconvinced. He hypothesized instead that white light is made of several colors, and that passing sunlight through a glass prism causes those colors to separate and become visible. If this hypothesis were true, then the modification hypothesis would be wrong: the impurities of the glass are not the cause of the rainbow colors we observe from prisms.

Newton performed several experiments to test his hypothesis. In one experiment, he darkened his room and bored a small hole in the window shutters so that only a thin beam of light could enter the room, casting a white circle of light on the wall. Then Newton placed a glass prism in the beam. A rainbow of colorful light appeared on his wall. This observation was consistent with the expectations of both the modification hypothesis and Newton's alternative hypothesis that white light is a mixture of colors. Both hypotheses lead us to expect that a beam of light travelling through a glass prism will produce different-colored bands. One thing was intriguing though: the shape of rainbow light was not a circle, as the white light had been, but oblong.

Then, Newton added a second prism to the path of the light. If the modification hypothesis was true—he reasoned—you would expect that the impurities contained in the two glass prisms would continue to modify the sunlight and just spread out the color spectrum further. But when the spectrum of colored light passed through the second prism, it recomposed back into white light! Because the modification hypothesis, if true, does not lead us to expect this observation, Newton concluded that the modification hypothesis was probably false. This result showed that white light could be composed of a rainbow spectrum: Newton's experiment had made this happen!

This finding also raised new questions. If light is a spectrum of colors, what's the nature of these colors? Are they particles of some sort? What makes them different colors? And why do the different colors spread out when they travel through the prism, making the light oblong instead of round? It would take more experiments performed by Newton and several other scientists to answer these questions and lead us to better understand the nature of light.

Experimental design

As Newton's experiments illustrate, experiments involve much more than simple observation. Experiments are strategically designed to enable the expectations drawn from a hypothesis to be tested.

An important feature of any experiment is its material aspects. Performing an experiment involves physical, concrete objects. While the experimental setup in Newton's experiments on light involved only cheap glass prisms, pencils, and notebooks, many present-day experiments require more expensive technologies, including specialized instruments, biological samples, chemical reagents, standardized questionnaires, computers, software for data analysis, and so on. All this apparatus costs money, which raises questions about how funding should be distributed among experimental research projects. For example, should we spend billions of dollars on a special superconducting collider for experiments in particle physics, or should funding agencies distribute money to prioritize several small-scale cheaper experiments in other fields?

Another important material aspect of experiments is what is being experimented upon. Experiments can focus on humans, non-human animals, or inanimate objects; these are the **subjects** of the experiment, also called experimental entities or participants. Human experimental participants are typically paid for participating in an experiment, while non-human animals for experimental testing often need to be bought, fed, and trained. Some inanimate objects on which certain interventions are performed are cheap to obtain or free, such as the rays of sunlight on which Newton experimented. But other objects studied in other experiments, such as rare metals, are hard to obtain and expensive. Experiments on humans vs. non-human animals vs. inanimate objects present different challenges and opportunities, which is discussed in section 3.3.

Box 3.1 Ethics and data management in experiments

Ethical evaluation is an integral part of contemporary scientific practices. If researchers want to run an experiment involving humans or non-human animals, typically they have to fill out an Ethical Clearance, Data Management and Protection form, where they describe the aim of the experiment, how it will be performed, what data will be collected, how the data will be analyzed, where the data will be stored, for how long, and who can access them. The researcher fills out and submits this form to the Ethics Committee of their institution. In the US, this is the Institutional Review Board (IRB). This committee includes other researchers and ethicists tasked with evaluating potential ethical and methodological issues with the proposed research. The Ethics Committee must give clearance before the research begins. The aim is to uphold scientific integrity and methodological soundness—especially when personal data are collected, such as sensitive demographic information, or when invasive interventions are performed. Before participating in an experiment, subjects must be informed of the research's potential risks and benefits, and they must explicitly consent to be part of the experiment and grant the researcher the right to collect and use their data for specific scientific purposes. Ethical clearance and data management and protection is a recent innovation. Many (in)famous experiments in medicine and social psychology, such as the Tuskegee Study in the 1930s, Harry Harlow's monkey studies in the 1960s, and Philip Zimbardo's Stanford Prison Experiment in 1971, did not receive ethical clearance and could not be performed nowadays.

Where an experiment occurs and over what period of time can also be important features of the experimental design. Newton's experiments on light took place in his room at Trinity College, Cambridge UK, around 1670. Many present-day experiments occur in laboratories located in universities and hospitals or take place in the field, that is, in settings like farms, subway stations, and forests. Newton's experiments on light had a short duration; other experiments can last years, such as present-day experiments in high-energy physics at the European Organization for Nuclear Research (CERN) in Geneva, Switzerland, where a very expensive and large superconducting collider called the Large Hadron Collider is used to accelerate and collide subatomic particles to study the nature of the fundamental constituents of matter and light.

Whether an experiment takes place in a laboratory or the field, their location and setup are always background conditions that can influence the data collected. **Background conditions** are the physical, technological, and social aspects of an experiment or study. The room at Trinity College where Newton performed his experiments had a certain ambient lighting, temperature, and humidity. The angle at which sunlight hit the room's windows varied by time of day and season. Prisms, the instruments Newton used, were not commonly thought of as scientific instruments in the 1660s and so were sold simply for their entertainment value. As a result, they were irregular in both size and composition. These features were all in the background of Newton's experiments. For his experiments to produce evidence against the modification hypothesis, Newton needed to show that none of these background factors were responsible for the results.

Experiments are designed to produce data. *Data* was defined in Chapter 2 as public records produced by observation or by some measuring device. Data provide evidence in favor of or against a hypothesis. Newton's data consisted of records in a notebook of the effects of passing a beam of sunlight through one or more glass prisms. For physicians, the results of blood tests and testimony about one's medical history can both count as data. Fossils, tracks, and recordings of the chemical features of mammoth tusks and of rocks in different locations all can count as data for a paleontologist. Climate scientists collect data from things like glaciers, oceans, and the atmosphere—for example, glaciers' mass balance, sea surface temperatures, and the atmospheric pressure at sea level.

Another feature of experimental design is who carries out the experiment. A scientist may conduct an experiment alone, but collaborative experiments are common in contemporary science. Most collaborative experiments involve scientists with different scientific expertise who rely on one another's expertise. Experiments at CERN, for example, are highly collaborative, run by hundreds of scientists and engineers from all over the world, each of whom brings some specific expertise to bear.

Even when an experiment is run by a single lab or an individual scientist, the broader scientific community shapes the experimental design. Communities of scientists, represented by scientific institutions and societies, determine protocols to be followed in experimental design and data analysis. These protocols include ethical guidelines for what sort of experiments are permissible, how experimental participants—humans or non-human animals—should be treated, and how data should be managed. As it happened, the Royal Society—the learned society for science of which Newton was a member—criticized his results, suggesting that the prisms' bubbles, veins, and other impurities caused the light to become colored as it passed through and that the modification hypothesis could account for the results of his experiments.

EXERCISES

3.1 Recall: Define *experiment* and give two examples of experiments.

3.2 Apply: For each of the two example experiments from Exercise 3.1, identify the hypothesis or hypotheses under investigation, the expectations for the experiment based on the hypothesis or hypotheses, and important features of the experimental design.

3.3 Recall: What is the modification hypothesis of light? Describe Newton's one-prism experiment and his two-prism experiment. For each, say what evidence it provided that challenged the modification hypothesis.

3.4 Recall: Describe three ways in which existing scientific knowledge shapes experiments. Illustrate each with an example from Newton's experiments on light.

3.5 Recall: List five features of experimental design. Identify each feature of experimental design for both Newton's prism experiments and experiments at CERN, the Large Hadron Collider.

3.6 **Apply:** List five features of experimental design. Find an example experiment described in a research article (or your instructor may provide you with an example). As best you can, identify each feature of the experimental design from what is said in the article. Then, describe how the experiment relies on existing scientific knowledge.

3.2 THE POWER OF EXPERIMENTS: INTERVENTION AND CONTROL

After reading this section, you should be able to:

* Define *intervention, independent variable, dependent variable,* and *extraneous variable* and indicate the role of each in a perfectly controlled experiment
* Indicate how defining expectations and collecting data can introduce confounding variables
* Describe two strategies of variable control: direct and indirect

The perfectly controlled experiment

Experiments have some ingenious features that make them a powerful way to gain scientific knowledge. To appreciate those features, it will be useful to start by describing an ideal experiment, even if real experiments usually deviate from this ideal. In an ideal experiment, experimenters perform an intervention that changes only one single feature of a system or situation while all other features remain the same. As a result of all the other features being unchanged, scientists know that any change in the system or situation is due only to the intervention.

That's the basic idea, but introducing some technical vocabulary will help make this precise. A **variable** is anything that can change, vary, or take on different values. For example, the number of books you read in one year, people's height, and the temperature in your hometown are all variables, since these are all things that can change over time or vary in different people or places. The **value of a variable** is just the particular state of the variable in some instance. For example, the value of the variable *number of books you read in 2024* might be 0, 12 or 56; the value of the variable *Matteo's height* is now 1.85 meters (6 feet); and your hometown temperature might have the value 62° Fahrenheit (16.67° Celsius) one summer evening and 92° Fahrenheit (33.33° Celsius) the next evening.

In experiments, there are three types of variables, distinguished by the roles the experimenters want them to play. An **independent variable** is a variable that's changed or observed at different values in order to investigate the effect of the change. A direct manipulation of the value of the independent variable is called an **intervention**. For example, the independent variable in Newton's experiment on light was the number of glass prisms through which the beam of sunlight passed; this variable had the value of 0, 1, or 2 in the experiments we described earlier.

A *dependent variable* is a variable researchers measure for changes after they intervene on the independent variable. The researchers anticipate how the value of the dependent variable depends on, or is affected by, the independent variable. When scientists perform an intervention in an experiment, they do so to investigate how that change affects one or more dependent variables. For example, Newton varied the number of prisms through which sunlight passed (independent variable) and then looked for changes in the color and shape of the light that had passed through the prisms (dependent variables).

The goal of an experiment is to isolate the relationship between an independent variable and one or more dependent variables. This requires controlling all extraneous variables. *Extraneous variables* are all other variables besides the independent variable that may influence the value of the dependent variable. In Newton's prism experiments, extraneous variables included bumps and impurities in different prisms, the time of the day, the angle at which a beam of light hits the prism, the ambient temperature, and much more. If you are experimenting with a new homemade fertilizer to find out how it affects the growth of your plants, then the amount of light, water, soil quality, ambient temperature, and more are all extraneous variables.

In an ideal experiment, recall, experimenters intervene on one feature of a system or situation (the independent variable) while all other features (extraneous variables) remain the same, so they know that any change (dependent variable) is due to the intervention. Keeping fixed the values of all extraneous variables is known as *controlling* the extraneous variables. More precisely, *variable control* is creating conditions such that no extraneous variable can change the value of a dependent variable during or as a result of an intervention on the independent variable. In a *perfectly controlled experiment*, any change in a dependent variable can only be due to the intervention on the independent variable. This isolates how the independent variable affects the dependent variable(s). In Newton's prism experiments, his goal had been to control the conditions of the light beam so carefully that any changes to the light (color, shape, etc.) could be due only to the prism(s).

Controlling variables

A perfectly controlled experiment is simple to describe, but it's difficult even to get close to this ideal in practice. Many extraneous variables can influence the dependent variable in a given experiment, and some may be hard to control or even to identify. When you investigate the effect of a homemade fertilizer on your houseplants' growth, you need to somehow ensure no other conditions—amount of sunlight, season, warmth in your home, humidity, soil quality, the prior health of the plants, different growth rates in different kinds of plants, etc.—are instead responsible for any changes in growth.

When extraneous variables are not controlled and affect the relationship between the independent and dependent variables, we call them *confounding variables*, or confounds. Confounding variables can undermine researchers' ability to draw conclusions about the influence of the independent variable. So the goal of variable control is to avoid any confounding variables.

TABLE 3.1 Types of variables and their use in experiments

Type of Variable	Use in Experiments
Independent variable	An intervention is performed upon the independent variable
Dependent variable	Dependent variable(s) are measured for changes after an intervention
Extraneous variable	Extraneous variables are controlled directly or indirectly so they do not vary during an experiment
Confounding variable	An extraneous variable that was not controlled and may have interfered with the relationship between independent and dependent variables

There are two basic strategies of variable control: direct and indirect. ***Direct variable control*** is when experimenters hold extraneous variables at constant values during an intervention. This is why Newton ran his experiments at the same time of day and in the same darkened lighting conditions. Keeping the values of those variables constant ensured that they didn't change the light's behavior. Newton also attempted to directly control the confounding variable of air bubbles and other impurities in the prisms by using higher-quality prisms. The carefully managed conditions in today's laboratories help scientists to directly control many variables. For example, there is a standard temperature and pressure used for laboratory experiments, abbreviated STP. In experiments conducted with the Large Hadron Collider at CERN, scientists use sophisticated technologies to keep many variables under direct control, such as the magnetic fields and temperature in the collider.

Some extraneous variables are difficult to hold constant or even to identify as potentially relevant. For these variables, indirect variable control is best. ***Indirect variable control*** is when experimenters allow extraneous variables to vary but ensure that variation is independent from the value of the independent variable. The key to indirect variable control is to study multiple systems, situations, or groups, all with the same variety of features and facing the same range of conditions other than the independent variable. Then, even though many variables vary, the value of the independent variable is the only systematic difference between them. In this case, any differences in the dependent variable between the systems, situations, or groups must be due to the difference in the independent variable. You might not be able to keep the humidity or temperature of your home exactly the same over days or weeks, and you certainly can't control the weather. But you can put two plants of the same kind on the same windowsill, give them the same amount of water (these are direct variable control), and let the other conditions the two plants encounter vary. Those conditions should affect both plants about the same.

But what if one of the two plants just happens to be heartier or quicker growing than the other? Indirect variable control often employs groups to also include variations across individuals. So, instead of studying two plants, perhaps you plant ten seeds from the same seed packet, five into each of two pots. You place both pots on the same windowsill and water both pots the same amount, at the same time. Perhaps you even switch which side of the windowsill each is on once a week. After three weeks, you thin the plants to the three tallest in each pot. Then, even though conditions for the plants are changing over time and you don't know if some of the plants will grow larger than others, the group of plants in each pot should experience about the same range of conditions.

With indirect variable control, the only systematic difference between groups is the value of the independent variable. One group, the **experimental group**, receives the intervention to the independent variable, or experiences the intended value of the independent variable. The other group, the **control group**, does not receive the intervention but experiences other value(s) of the independent variable. As you can probably guess by now, of the two pots of plants, one should be treated with your homemade fertilizer, while the other should not. What happens to that second pot of plants, your control group, depends on what you want to investigate. Do you want to know how your homemade fertilizer compares to no fertilizer at all, or to the fertilizer you used to buy from the store? What value you set the independent variable to in your control group determines what you will learn about the intervention under investigation.

One common strategy for forming experimental and control groups is **randomization**, which is the use of arbitrariness or some chance procedure like a lottery to assign experimental entities to experimental and control groups. Randomization is meant to ensure that the two groups are equal in all relevant characteristics except the intervention, since any differences among the experimental entities should vary randomly across groups. This is effectively what we were proposing when we described planting ten seeds from the same packet into two pots—seeds with different characteristics were equally likely to make it into each group. Many scientists, especially in the medical sciences, believe that randomized controlled trials (RCTs) are the gold standard of indirect variable control. But, as we will see later in this chapter and in the next chapter, there are other techniques for managing extraneous variables.

Random group assignment guarantees extraneous variables are not deliberately related to group assignment, and so not related to whether experimental entities experience the intervention. But random group assignment does not guarantee that all extraneous variables vary equally within the two groups. Randomization only ensures the similarity of the experimental and control groups in the limit—only when the experimental subjects are randomly divided in two groups an infinite number of times. In actual randomized experiments, the two groups may have systematic differences simply due to chance. Perhaps, just by chance, all the seeds that went into one pot were able to germinate sooner and grow larger. For this reason, for randomization to be an effective approach to indirect variable control, the groups must be large enough so that large chance differences across groups are very unlikely. Planting five seeds per group is better than just one seed, and planting ten seeds per group (with adequate space to grow) is even better.

Watch Video 6

Clarifying expectations and collecting data

To test a hypothesis with an experiment, clear expectations must be articulated for the outcome of the experiment. These expectations are predictions of the results of some intervention, assuming the hypothesis in question is true. Expectations should be clearly and precisely defined before running the experiment, in a way that makes them easily comparable to the data the experiment will produce. Yet, scientists' hypotheses can involve broad concepts and ideas that can be hard to know how to test.

Two techniques that scientists can use to generate precise expectations from hypotheses with broad concepts and ideas are operational definitions and cluster indicators. An ***operational definition*** is a specification of the conditions when some term applies, enabling measurement. In social science research, for example, *wealth* might be operationally defined as a household's combined material assets, as this would capture an important component of material wealth and offer a precise, measurable basis for comparison. But this operational definition also simplifies a complex concept. For instance, generational wealth not yet transferred into a household isn't considered. Because of this lack of nuance, economists often study wealth using a combination of indicators, such as yearly income, access to education and healthcare, and permanent housing. Such ***cluster indicators*** identify several markers of some variable in order to more precisely measure it while not oversimplifying it. Many terms can be defined in multiple ways without one being obviously best, but the choice between different operational definitions and cluster indicators is not arbitrary. This depends on the scientists' research goals and the details of the experimental setup, and the choice is important. Employing the wrong definition might introduce a confounding variable, such as generational wealth, or it might obscure impacts of the intervention. What if, in your fertilizer experiment, you simply measure plant height, but your fertilizer increases plant longevity or improves flowering?

Data collection is another opportunity for confounding variables. First, this often involves specialized ***instruments***, technological tools or other kinds of apparatus used in experiments, ranging from specialized equipment to surveys. These instruments are a possible source of error—and thus a potential confounding variable. Newton had to convince the Royal Society and other audiences that the data he collected using prisms was legitimate, as prisms' use in scientific inquiry was not yet broadly accepted. Questions about the reliability of instruments used in experiments still arise. The ***reliability*** of an instrument is the extent to which it accurately and consistently measures what it is supposed to measure.

Scientists must calibrate instruments to ensure they are reliable. ***Calibration*** is the comparison of the measurements of one instrument (for example, an electronic ear thermometer) with those of another (for example, a mercury thermometer) to check the instrument's accuracy and adjusting the instrument if needed. In experiments with human subjects, data collection often involves surveys or questionnaires, and these also need to be calibrated and assessed for reliability just as technical apparatus does. A poorly designed question can prime subjects to answer in a certain way, for example, or questions might be ambiguous, eliciting different kinds of responses from different people or unintentionally asking about more than one thing at once.

Calibration and assessment of reliability are ways of directly controlling extraneous variables related to data collection. Another set of extraneous variables that must be controlled during data collection is human expectations. We learned about the observer-expectancy effect in Chapter 1, how a scientist's expectations can lead them to unconsciously influence the behavior of experimental subjects. For example, it's well established that the expectation that a medicine will be effective can lead to improved health; this is called the *placebo effect*. Researchers' expectations and, for experiments involving human subjects, subjects' expectations thus need to be controlled or they may become confounding variables.

The strategies of direct and indirect variable control that we have talked about so far don't help with the extraneous variable of human expectations. To control for this, scientists rely on *blinding* (or masking), which is when researchers or subjects are temporarily kept unaware of group assignment, hypotheses under test, or other experiment details. Blinding aims to reduce the risk of biased observations by directly controlling the information researchers and/or experimental subjects have access to and, thus, indirectly controlling their expectations. In *double-blind* experiments, researchers and subjects are both unaware of which subjects are in the experimental and control groups.

EXERCISES

3.7   **Recall:** Define *intervention, independent variable, dependent variable,* and *extraneous variable* and indicate the role of each in a perfectly controlled experiment.

3.8   **Apply:** Review the discussion of Newton's prism experiments from section 3.1. Identify the two hypotheses under investigation, the independent variable, and the dependent variable(s). Describe the intervention and how Newton controlled extraneous variables. What were the expectations based on each hypothesis? What did Newton conclude on the basis of his experiments?

3.9   **Recall:** Define *extraneous variable* and *confounding variable.* What is the relationship between the two? Why are experiments designed to limit confounding variables?

3.10   **Recall:** Define *direct variable control* and *indirect variable control.* Give three examples of directly controlled variables and three examples of indirectly controlled variables in the fertilizer experiment sketched in this section.

3.11   **Think:** (a) What is the purpose of having an experimental group and a control group in an experiment? (b) How does division into two groups achieve this purpose, and why is random group assignment important? (c) What are the limitations of this strategy?

3.12   **Recall:** Describe how defining expectations and collecting data can introduce confounding variables and how each can be controlled.

3.3 LEARNING FROM EXPERIMENTS

After reading this section, you should be able to:

* Define *internal experimental validity, external experimental validity, ecological validity*, and *population validity* and indicate why each is important
* Analyze how sample selection, sample size, group number, and group assignment influence experimental design
* Describe the challenges that reliance on background knowledge and alternative hypotheses pose to testing hypotheses with experiments

Internal and external validity

Laboratories give researchers control over many aspects of an experiment. Depending on the kind of experiments performed, a lab's design features may include constant temperature, sterile environment, special equipment to produce unusual conditions, or, for experiments with human subjects, carefully selected lighting and furniture, soundproofing, and so forth. Those design features, and the direct variable control they provide, constitute one of the greatest advantages of the **laboratory experiment**. These features can enable scientists to discover regularities that are not easy to discern in the outside world.

The high degree of control enabled by laboratory conditions brings with it a high degree of internal experimental validity. **Internal experimental validity** is the extent to which researchers can infer accurate conclusions about the relationship between the independent and dependent variables. This amounts to the absence of confounding variables, achieved by direct or indirect control of extraneous variables. Another advantage of laboratory experiments is that the experimental setup and data analysis can follow standard procedures, which make it easier to assess and replicate an experimental finding.

However, lab research also has some disadvantages. Some phenomena are not easily investigated in a lab. Suppose you are investigating the effects of climate change on large marine mammals, especially the effects of elevated Arctic Ocean temperatures on the deep-diving behavior of narwhal whales. Narwhals—sometimes called unicorns of the sea because of their tusks—can dive as deep as 1.8 kilometers in Arctic waters. To directly investigate this phenomenon in a lab, you will need—for starters—a huge tank of freezing salt water nearly two kilometers deep. Investigating this in the lab is thus all but impossible.

The unusual conditions in a lab that make it easy to control variables also make the lab setting different from the outside world, and that has some disadvantages too. The artificiality of the experimental setting might mean that the results obtained in the lab do not generalize well to real-life settings outside the lab. This is problematic since it's ultimately the real-world phenomena that we want to know about. Laboratories thus facilitate high internal validity, but potentially at the cost of external validity. **External experimental validity** is the extent to which experimental results generalize from the experimental conditions to other conditions—especially to the phenomena the experiment is supposed to yield knowledge about.

One component of external validity is ecological validity. *Ecological validity* is the degree to which an experiment's circumstances are representative of real-world circumstances. Experimental settings or what subjects are asked to do can be artificial, unlike real-world circumstances, in ways that impact the phenomenon under investigation. One way to enhance the ecological validity of an experiment is to conduct it "in the field," that is, in participants' everyday environment outside of the laboratory; such experiments are called *field experiments*. Field experiments tend to have more external validity than lab experiments because they occur in circumstances similar to everyday circumstances. Their ecological validity is higher as a result.

A downside to field experiments is decreased internal validity. Less influence over the circumstances and the selection of experimental subjects is linked to decreased control over extraneous variables and sometimes a decreased ability to intervene in the desired way. The decreased influence on experimental design also makes it more difficult for other researchers to replicate the experiment. Researchers conducting field experiments may also be constrained in what they can be in a good position to observe or measure, the number of subjects they can involve, and how long they can run the experiment. Many field experiments require special permissions from subjects or from authorities that control access to areas like nature preserves. Uncontrollable events like inclement weather or warfare can also disrupt observation or limit the length of study that's feasible.

Watch Video 7

Samples and groups

Alongside ecological validity, the second component of external validity is *population validity*: the degree to which experimental entities are representative of the broader class of entities or population of interest. With a *representative sample*, the experimental entities studied do not vary in any systematic way from the general population. The more representative a sample is of the broad class or population, the more confident scientists can be of the experiment's external validity.

Here's an illustration of the importance of population validity. Many clinical trials testing the efficacy and side effects of drugs have been performed only on men, but the results are expected to generalize to women as well. This decreases the population validity of the results, since women and men differ in several medically relevant ways. There is thus relatively limited experimental knowledge about the effects of some drugs on women, and this may have serious consequences for health and medicine. Indeed, some prescription drugs have been withdrawn from the market after they were belatedly revealed to pose greater health risks for women than for men.

One way to increase population validity is *random sampling*: using a chance method for selecting a sample to investigate from the population (this is not the same as randomization, which relates to group assignment). In a random sample, every member of the population has an equal chance of being selected for participating in an experiment. Unfortunately, hardly any experimental sample of human subjects is randomly selected. Many experiments only involve "convenience samples" like college students or social media users. But neither college students nor social media users are representative of the entire population of any country. Indeed, the criticism has been made of psychological research that almost all has been conducted on "WEIRD" subjects (Western, educated, industrialized, rich, and democratic), which are not representative of people across the world.

Sample size is the number of individual sources of data in a study; often this is simply the number of experimental entities or subjects. In section 3.2, we mentioned that indirect variable control requires a large enough sample to ensure chance variation doesn't differently influence the experimental and control groups. A larger sample size also can improve the sample's representativeness, simply by including more variation present in the broader population. Of course, this won't help with variable values that are systematically excluded, like medical trials that enroll only men or psychological studies that enroll only college students in the United States. But, for variables that are included in the sample just by chance, a larger sample increases the values represented. In general, then, a larger sample size increases the success of indirect variable control, thus increasing an experiment's internal validity, and increases population validity, thus increasing an experiment's external validity.

But the advantages of a large sample size must be balanced against the practical disadvantages. Large samples are more difficult to assemble and are more difficult and expensive to manage in the experiment. There are also diminishing advantages to samples beyond a certain size that are not intentionally more representative. It would be a bigger improvement for a psychological study to enroll subjects of different ages and education levels than it would be for the study to enroll 5,000 US college students instead of 1,000 US college students.

Another choice in experimental design concerns how many groups to include in an experiment. So far, we have focused on experiments with two groups: an experimental group and a control group. More complicated experimental designs include multiple experimental groups, each of which experiences a different but related intervention. For example, returning to our imagined fertilizer experiment, we may want to compare our homemade fertilizer both to a commercial fertilizer and to no fertilizer at all. Including multiple experimental groups can be enlightening, and it can lead to surprising outcomes. But this also complicates experiments, making them more difficult to perform, and it makes it more difficult to get an adequately large sample size for each group. Having multiple experimental groups also can make analysis of the results more difficult. For these reasons, experiments are usually performed with as few experimental groups as researchers deem absolutely necessary to get the needed comparative data.

Background knowledge and alternative hypotheses

The design of experiments is influenced by existing scientific knowledge. Existing knowledge shapes what hypotheses are developed and what expectations scientists have based on a hypothesis. Due to past research by others, Newton's experiments presumed that light is a substance that travels from a light source. From Newton's hypothesis that white light is composed of colors, Newton developed the expectation that a rainbow of light could be recombined into white light—an expectation his two-prism experiment was developed to test. This expectation relies on the idea, part of background knowledge about light, that changes to light can be reversible. Existing scientific knowledge also helps guide scientists' interpretation of experimental results as evidence in favor or against some hypothesis. For example, the finding that a rainbow spectrum becomes oblong surprised Newton, since he knew from existing research that light typically travels in straight lines.

The reliance of experiments on existing knowledge is unavoidable, but it does create some challenges. What if any of that existing knowledge is wrong? Scientists have to rely on existing ideas to design properly functioning experimental instruments and to know what expectations to draw from their hypotheses. But they also have to rely on properly functioning instruments to test their theories with trustworthy experimental data and on the relationship between hypotheses and expectations to know what conclusions to draw about hypotheses on the basis of experimental data. Indeed, the existing knowledge that informs experimentation could turn out to be wrong.

But scientists have some resources to diminish the risk. First, researchers often have a good sense for what ideas are plausible enough to be trustworthy as the basis for experiments and which are new and risky. Second, over time, researchers refine not just their theories but also instruments and experimental designs. This can involve experimental work in replication or calibration, as discussed in section 3.4. Scientists can conduct an experiment or analyze data using different instruments or techniques to detect any variation depending on instruments or experimental design, called *triangulation*, and then use this to refine their theories, instruments, and experimental methods.

Another challenge in using experiments to definitively test hypotheses is *underdetermination*: the evidence may not be sufficient to determine which of multiple hypotheses is true. Consider again the experiment we imagined to test whether homemade fertilizer improves plant growth. Even if you clearly defined expectations if the hypothesis is true (plants receiving the homemade fertilizer grow faster than others), perfectly controlled extraneous variables with your experimental design, and collected data that matched your expectations, the truth of your hypothesis wouldn't be guaranteed. At the outset of the chapter, we imagined testing the hypothesis that urine is an effective fertilizer, and so let's suppose that's the fertilizer under investigation. You take a daily multivitamin: is that the only reason the urine was an effective fertilizer? We can't know without a new experiment. These kinds of questions are always possible, but scientists can't think of every possible hypothesis. It's always possible that some unimagined hypothesis that hasn't been tested accounts for the experimental results.

How should scientists proceed in the face of underdetermination? One response would be to suspend judgment about which hypotheses should be accepted. But this isn't an option when we need to build a bridge or design an effective drug. Instead, scientists must diligently attempt to consider alternative hypotheses that might account for the data, testing alternative hypotheses that seem potentially credible. Additionally, hypotheses that fit with the available experimental data are sometimes more or less appealing in other regards; for example, they may be simpler or fit better with other scientific knowledge. These considerations may push us toward one or another hypothesis when the data underdetermine which is true. Ultimately, underdetermination simply requires continuation of the spirit of openness to falsification that we identified in Chapter 1 as essential to science.

EXERCISES

3.13 Recall: Define *internal experimental validity*, *external experimental validity*, *population validity*, and *ecological validity*. For each, describe its specific importance and how it can be improved.

3.14 Recall: What are the main advantages and disadvantages of a laboratory experiment? How about a field experiment?

3.15 Apply: For each of the following hypotheses, indicate (a) whether you would study it in the lab or in the field, briefly describing why, (b) what steps you would take to control extraneous variables, and (c) whether you are more confident about the internal or external experimental validity of your study and briefly describe why.

a. Whether peacocks' colorful trains play a role in their mating success
b. Whether a new fertilizer improves plant growth
c. Whether watching a television show about sharing improves children's ability to share their toys

3.16 Recall: Describe how sample selection, sample size, group number, and group assignment are each important to experimental design. For each, describe the negative effect a wrong choice can have.

3.17 Apply: Suppose you want to test the hypothesis that baseball players who eat pizza every day hit more home runs. Let's suppose that to test this hypothesis, you want to divide the baseball players of some team into two groups that are balanced in all important background variables that can affect players' performance. The only difference you want between the two groups is that the members of one group eat pizza every day and the members of the other group do not.

Rank the following four strategies from best to worst for accomplishing this goal:

1. Sit in the clubhouse after a game. The first players who enter the clubhouse are assigned to the group of pizza eaters (the experimental group), while the following players are assigned to the control group.
2. Allocate players born in the first six months of the year to the experimental group and players born in the second six months of the year to the control group.
3. For each player in the team you toss a coin. If the coin lands on heads, then the player is in the experimental group; otherwise, the player is assigned to the control group.

4. Assign all players over 230 pounds to the experimental group and the rest of the players to the control group.

Justify each of your rankings by describing how well or poorly you expect that strategy will control the extraneous variables.

3.18 Recall: Describe the challenges that reliance on background knowledge and untested alternative hypotheses pose to testing hypotheses with experiments. How can scientists manage each challenge?

3.4 OTHER USES OF EXPERIMENT

After reading this section, you should be able to:

- Define *crucial experiment* and indicate three limitations of these experiments' decisiveness
- Describe the use of experiments in replication and calibration and give examples of each
- Indicate how exploratory experiments differ from experiments to test hypotheses

Crucial experiments and replication

Albert Einstein's theory of general relativity revolutionized our understanding of space and time. While Newton believed that space is a sort of stage on which events unfold, Einstein conceived of space and time as a single interwoven manifold, a fabric of sorts. For Newton, gravity was a force; Einstein instead explained gravity as the curvature of the space-time manifold. Just as marbles placed on a fabric sheet held in the air bend the sheet around them, massive objects like the Sun warp space-time in their vicinity. That's why other objects accelerate toward those massive objects. Einstein's theory of general relativity leads to testable hypotheses. One of these hypotheses is that light, just like any other form of matter, is affected by gravity. This hypothesis, in turn, generates the expectation that if a beam of starlight passes near the Sun it will bend toward the Sun.

This expectation was first tested on May 29, 1919, when a total solar eclipse blocked out the light of the Sun. A group of scientists led by Arthur Eddington took photographs of stars visible near the dimmed Sun. These scientists compared these to other photographs taken at night when the light of those same stars did not pass close to the Sun before reaching Earth. These data confirmed Einstein's prediction of the starlight's deflection. The Sun changed the path of nearby starlight as the theory of general relativity predicted, providing confirmation of the theory. When the press reported that a key prediction of Einstein's theory had been borne out by observation, Einstein became a famous public figure.

Eddington's test of Einstein's new theory was a ***crucial experiment***, an experiment that decisively adjudicates between two hypotheses. This kind of dramatic experimental result is exciting, but even crucial experiments do not yield scientific knowledge

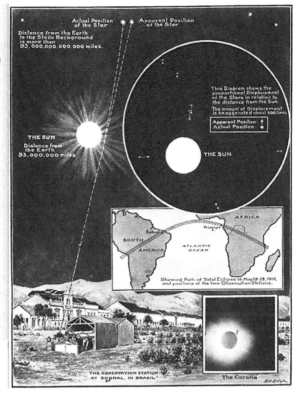

LIGHTS ALL ASKEW IN THE HEAVENS

Men of Science More or Less Agog Over Results of Eclipse Observations.

EINSTEIN THEORY TRIUMPHS

Stars Not Where They Seemed or Were Calculated to be, but Nobody Need Worry.

A BOOK FOR 12 WISE MEN

No More in All the World Could Comprehend It, Said Einstein When His Daring Publishers Accepted It.

FIGURE 3.2 Headlines from the *New York Times* (left) and the *Illustrated London News* (right) reporting on Eddington's observations during the 1919 eclipse, which confirmed Einstein's theory of general relativity

all on their own. As we have seen, even well-designed experiments face numerous potential sources of error: unidentified confounding variables, reliance on background knowledge that might be wrong and techniques or instruments that might not work as intended, and unidentified alternative hypotheses that might account for the data rather than the hypothesis under investigation. These potential sources of error cannot be eliminated with one single experiment, and so science has developed experimental techniques to minimize the risks of these errors.

One is replication. *Replication* is performing the original experiment again—often with some modification to its design—in order to check whether the result remains the same. If, for example, Newton's two-prism experiment is replicated by different people, using different prisms, in different places and at different times, and they also observe the spectrum of light recombining into white light, this additionally supports Newton's hypothesis that white light contains a spectrum of color. This shows that no oddities of Newton's equipment or circumstances is responsible for the result, increasing internal experimental validity. Replication attempts that vary features of the experimental design—the experimental subjects or subject selection, the types of instruments or surveys used, and so on—can increase external experimental validity.

The replicability of experimental outcomes is an important ingredient of science, so much so that a persistent failure to replicate experimental findings may undermine a scientific field's credibility. If some experimental result cannot be replicated—if different scientists follow similar experimental procedures but do not get the same result—then the original experimental result might be a fluke, or it might be due to some confounding variable in the experimental setup that the scientists haven't yet identified. For example, it has been suggested that the field of social psychology faces a crisis in replicability, where different research groups have tried but failed to replicate some classic experimental results. Thus, we shouldn't put too much stock in those findings, unless this failure in experimental replicability is resolved.

Box 3.2 The replication crisis

Nobel Prize winner Daniel Kahneman wrote an alarmed open letter to social psychologists in 2012. He pointed out that prominent results in priming research, which is the study of how subtle cues can unconsciously influence social behavior, failed to replicate. Kahneman claimed that this field had become "the poster child for doubts about the integrity of psychological research." This was the alarm bell that psychology was probably suffering from a replication crisis: original experimental findings widely cited by other researchers and taught to students as "facts" failed to replicate when similar experiments were performed again by other researchers. This was worrisome, as the trustworthiness of science partly depends on the idea that studies repeated under similar conditions will produce similar results. Further, it's not only research in social priming that suffers from problems in replicability. Important published results in the psychological, social, behavioral, and biomedical sciences have also failed to replicate. Researchers in these fields have started to discuss various potential reasons for the crisis, including sloppy experimental design and statistical analysis, publication bias, and outright fraud. And various remedies have been proposed too, such as reforms in statistical analysis, changed publication practices, more careful and powerful study designs, and—more generally—a call for *open science*, whereby all research practices and data are made freely available for anyone to scrutinize or use.

The importance of replication fits with the idea that science is essentially a collaborative, social venture. Gaining scientific knowledge via experimentation is generally more complicated and slower than a single dramatic experiment. This also means that scientific knowledge can go in unexpected directions. A surprising finding that upends something we thought we understood might be right around the corner.

Calibration

To persuade other members of the Royal Society that his hypothesis about light was true, Newton had to show that his prisms were reliable scientific instruments. For this reason, many of Newton's experiments aimed at testing how prisms with different shapes and composition affected the light spectrum produced. Supported by Newton's extensive data and theory of light, prisms became accepted scientific instruments.

As this illustrates, experiments also can be used to evaluate whether scientific instruments function as intended and to calibrate scientific instruments. Calibration, comparing the measurements of one instrument with those of another to check the instrument's accuracy and adjusting it if needed, was introduced in section 3.2. This is required not just for technical apparatus like prisms and thermometers but also for surveys or questionnaires used with human subjects. Any instrument for data collection must be calibrated using known measurements before it can be used in an experiment with uncertain results.

Brain imaging techniques are a nice illustration of using experiments to establish the function of an instrument and to calibrate it for data collection. fMRI machines track blood flow in the brain. They do not directly measure neural activity, but that is what the scientists employing these machines want to assess. Neuroscientists use data about blood flow to track neural activity because they know that greater neural activity requires more energy, which requires increased metabolism, which uses more oxygen, and oxygen is delivered by blood flow. The confidence that blood flow in a brain region is a reliable measure of neural activity is confirmed by experimental findings concerning brain metabolism and the relationship between different brain areas and functions. Experiments have been run over decades to additionally develop fMRI techniques and to improve their calibration as a tool to study neural activity.

Calibration requires establishing ***measurement standards***, or rules to regulate the use of quantity concepts and to create a meaningful scale to apply across instruments. Examples include the standard kilogram for weight or mass measurements and setting the freezing and boiling points of water at 0° C and 100° C, respectively, as temperature reference points. A standardized scale enables measurements to be compared over time and across instruments. This body of measurement data might then be used to construct more stable measurement scales and more accurate instruments.

Measurements can be based on the value of ***physical constants***, or quantities that are believed to be universal and unchanging over time, and these are also determined experimentally. From 1889 until 2019, the standard kilogram was defined by a physical measurement standard—a physical prototype with a mass that was, by stipulation, exactly one kilogram. In 2019, the kilogram was redefined on the basis of the Planck constant, a quantity in quantum physics.

However, no scientific instrument is perfectly precise. In 2017, scientists at the National Institute of Standards and Technology (NIST) used a Kibble balance—an instrument that uses electric current to produce extremely accurate measurements of mass—to determine the most precise value yet of the Planck constant. Even after more than 10,000 measurements with this specialized instrument, a small degree of uncertainty about the exact value of the Planck constant remained. Future experiments that further refine the measurement of the Planck constant will in turn change (very slightly) how the mass of a kilogram is defined. (The value of the Planck constant is about $6.62607015 \times 10^{-34}$ J/Hz if you were wondering.) Another physical constant measured experimentally is the speed of light in a vacuum. The measure of this constant has been refined from the 18th century through late 20th century, and it is used to define standard measures of both distance (the meter) and time (the second).

Exploratory experiments

In 1800, the British astronomer William Herschel used a telescope to observe sunspots, which are regions on the Sun that appear temporarily dark. Observing sunspots is hazardous for the eyes, and so he used colored glass filters to reduce the intensity of the rays. Herschel noticed that he could feel the Sun's heat coming through the filters, and different filters seemed to differ in temperature. Since the filters were made of the same material, Herschel wondered whether the different colors of the filters might actually be responsible for the temperature differences. Notice that this wasn't what Herschel had set out to investigate; sometimes experiments take us in unanticipated directions.

Herschel tested his hypothesis about a relationship between light's color and temperature by directing sunlight through a prism to spread the spectral colors, as Newton had. Then he measured each color—red, orange, yellow, green, blue, indigo, violet—with a mercury thermometer. He also measured the ambient temperature in the room in order to have a baseline temperature to compare with the temperature measurements of the light. This setup yielded data in the form of measured values of color and measured values of temperature, which could be used as evidence to evaluate the hypothesis that different colors of light differ also in temperature. The evidence confirmed this hypothesis: Herschel found that the temperatures increased incrementally from the "cool" colors like blue to the "warm" colors like orange.

Another of Herschel's observations in this experiment introduced a new question about light. Herschel also measured the temperature of the air just beyond the beam of red light, outside the edge of the spectrum created by sunlight through the prism, where no light was visible. His hypothesis was that this temperature would be the same as the ambient temperature in the room, since it was beyond the edge of the light spectrum. To his surprise, the temperature at that location was much warmer than the ambient room temperature, even higher than any of the temperature measurements for the light spectrum. How could that be?

Herschel's observation immediately led to a new hypothesis: invisible, hot light exists just beyond the red part of the visible spectrum. This hypothesis—anticipated by the French physicist Émilie du Châtelet almost 65 years earlier—explained the observation that the temperature continued to increase beyond the edge of red light. Later observations confirmed this hypothesis, and we now accept the existence of this hot, invisible light. It's called *infrared* light.

A further role of experiments, beyond hypothesis-testing, replication, and calibration, is exploratory. **Exploratory experiments** may not rely on existing theory and are

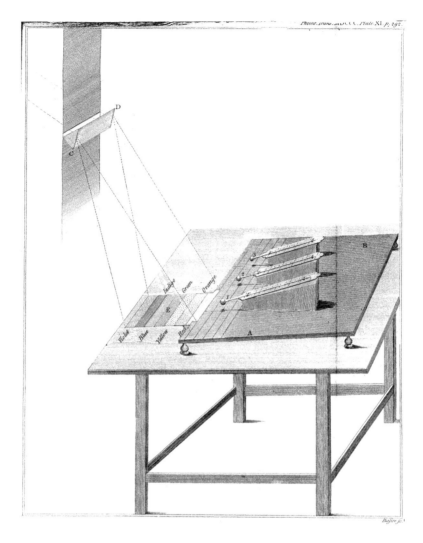

FIGURE 3.3 Herschel's experimental setup to test the relationship between color and temperature of light

not aimed to test a specific hypothesis; instead, they are used to gather data to suggest novel hypotheses or to assess whether a poorly understood phenomenon actually exists. Herschel's work on the relationship between heat and light did not rely on a particular theory or a hypothesis about that relationship. When, while investigating sunspots, he discovered that red light is warmer, Herschel surmised that the light spectrum is made of both heat and colors. This idea was on the right track, but it was not until James Maxwell's theory of electromagnetic radiation that Herschel's observations could be adequately explained and his work vindicated.

FIGURE 3.4 Portrait of Émilie du Châtelet, French physicist and mathematician in the 18th century

EXERCISES

3.19 Recall: Define *crucial experiment* and describe three limitations of crucial experiments' decisiveness.

3.20 Recall: Describe why replication is important and how replication can improve internal and external experimental validity.

3.21 Recall: Describe how calibration relies on measurement, and what role experiments play in measurement.

3.22 Think: Briefly describe three roles for experiments other than testing hypotheses, and give an example of each. Then discuss how each of these might relate indirectly to testing hypotheses.

3.23 Think: Describe how exploratory experiments differ from experiments to test hypotheses. Consider the important features of experimental design. What are some drawbacks to exploratory experiments, related to expectations, extraneous variables, and experimental design?

3.24 Apply: Before Ibn al-Haytham's work, some thought that vision involved light shining out of the eye, coming into contact with objects, and thereby making them visible. This was known as the emission theory of vision. Ibn al-Haytham set up the following experiment to test the emission theory. He stood in a dark room with a small hole in one wall. Outside of the room, he hung two lanterns at different heights. He found that the light from each lantern illuminated a different spot in the room. For each, there was a straight line between the lighted spot, the hole in the wall, and one of the lanterns. Covering a lantern caused the spot it illuminated to darken, and exposing the lantern caused the spot to reappear.
   a.   What is the independent variable and what is the dependent variable?
   b.   How did Ibn al-Haytham control extraneous variables?
   c.   How did the experimental result provide evidence against the emission theory?
   d.   Describe one way in which the emission theory might be adapted to account for the data (but still remain an emission theory of vision).
   e.   Describe a new hypothesis you can formulate based on the results of Ibn al-Haytham's experiment.

FURTHER READING

For an introduction to the philosophy of experiments with a focus on the natural sciences, see Hacking, I. (1983). *Representing and intervening: Introductory topics in the philosophy of natural science.* Cambridge University Press.

For more on the experimental approach in the social sciences with a focus on economics, see Guala, F. (2005). *The methodology of experimental economics.* Cambridge University Press.

For a case study on the role of instruments and measurements in experiments and studies, see Chang, H. (2004). *Inventing temperature: Measurement and scientific progress.* Oxford University Press.

For a recent perspective on replicability and the so-called replicability crisis, see Romero, F. (2019). Philosophy of science and the replicability crisis. *Philosophy Compass, 14*(11), 1–14.

CHAPTER 4

Non-experimental investigation

4.1 PALEONTOLOGY AND NON-EXPERIMENTAL STUDIES

After reading this section, you should be able to:

*   Describe how scientists have learned about woolly mammoth and mastodon life histories
*   Give examples of when an experiment is impossible, impractical, and unethical
*   List three varieties of empirical investigation that can be used when intervention is not possible

Prehistoric life histories

The woolly mammoth (*Mammuthus primigenius*) is an extinct species most closely related to today's Asian elephants. It's one of the best understood prehistoric animals, with research based on preserved bones and dung, depictions in prehistoric cave paintings, and even frozen carcasses found in Siberia and North America.

Research on woolly mammoths isn't based on conducting experiments. Paleontologists can't control the conditions encountered by woolly mammoths, of course, or recruit them into experimental and control groups. Indeed, they can't intervene on woolly mammoths at all, since the last mammoths died millennia before the invention of science. Scientists can directly study some features of mammoths—namely, physical and chemical features of their preserved carcasses, skeletons, teeth, and dung—but they are also interested in other mammoth features that can't be studied directly, like their diets, habitats, social interactions, and migration patterns. These features relate to woolly mammoth's *life history*: the traits and circumstances that affected survival and reproduction of members of the species.

The impossibility of experimentation doesn't mean paleontologists are at a dead end, unable to gain knowledge about woolly mammoth physical traits and life histories. Instead, they just have to get a little creative about where they find evidence and how they draw inferences.

For example, a 2021 article in the journal *Science* reported the results of chemically analyzing the tusk of a woolly mammoth that lived 17,000 years ago. The researchers compared the isotopes of two elements (oxygen and strontium) to geographic maps

DOI: 10.4324/9781003300007-5

FIGURE 4.1 Artistic depiction of woolly mammoths in their habitat

showing isotope variation in the environment. Because of how tusks grow over a mammoth's lifespan, this comparison enabled the researchers to identify where this particular woolly mammoth moved across its lifetime. It was found to have a very extensive geographic range, traveling very long distances across what is now Alaska.

The researchers could also see that the mammoth had different types of movements across its lifespan. In the first 16 years of its life, it moved repeatedly among the same territories, which is similar to how herds of elephants move today, suggesting the mammoth was moving with a herd. Then, in the middle of the mammoth's life, the isotopic variation increases, suggesting it wandered irregularly over long distances, which in turn suggests that it had left its herd, as mature males in elephant species also do. Finally, in the last year and a half of its life, the mammoth moved only within a very small area. Its tusk also showed an isotopic pattern associated with starvation, which is probably how it died.

In this study, researchers were able to use chemical analysis to determine where one mammoth lived and how this changed over time, and then to draw further conclusions about its health and even its social interactions—that is, whether it lived in a herd. These conclusions aren't just about this particular mammoth, but about other woolly mammoths as well, and potentially other similar species, especially that also lived during the Pleistocene era. Indeed, other scientists followed up on this study by performing similar chemical analyses on the tusk of an American mastodon (*Mammut americanum*) that lived about 13,000 years ago. Mastodons and woolly mammoths both lived in the Pleistocene era and were similar in appearance, but mastodons are less closely related to modern elephants.

Box 4.1 The roles of museums in science

Visiting a museum is a way to learn about history, art, and science in an enjoyable and leisurely way. But museums also play an important role in scientific research. To start, many have extensive collections of artifacts that are useful as objects of scientific research. These can include fossils, artifacts from human history, preserved plants and animals, and more. Usually, a given museum will only display small portions of its collection at a time; other items in the collection can be available for scientific research. Many science museums also have scientists on staff: curators often have advanced degrees in relevant fields of science and conduct their own scientific research. It's also common for science museums to have partnerships with nearby universities. Scientists at the university may access collections or collaborate with curators, and sometimes graduate students participate in research activities at the museum as well. There's a trend toward making this research activity visible to members of the public who visit the museum. Sometimes museums include discussion of ongoing related scientific research in their exhibits or have lab space in which scientists can conduct research while guests and visitors look on.

These scientists were able to identify how the mastodon's migration followed annual seasons, including adult springs and summers in an unusual location far from its typical range; this was inferred to be a mating ground. Life history research on both woolly mammoths and modern African elephants was also deployed to support conclusions from their data, for example, that the mastodon lived in herds, and also to draw conclusions from their research to a broader group of similar large mammals. Thus, both of these studies analyzed remnants from particular, long-extinct animals with reference to ecological data

28 Million Years Ago   7 Million Years Ago   2 Million Years Ago   Now

FIGURE 4.2 When mastodons, woolly mammoths, and African elephants lived and when they diverged evolutionarily

and knowledge about modern-day related animals in order to draw conclusions about long-ago happenings that we can never observe directly, let alone intervene upon.

When experiments aren't possible

In the paleontology research just described, the scientists couldn't directly or indirectly control variables, intervene on, or even directly observe the phenomena under investigation. Experiments just aren't possible. Instead, they brought together various forms of evidence to draw inferences about mammoth and mastodon life histories.

This is one reason experiments can't be performed: sometimes it's downright impossible to experiment on the phenomenon under investigation. The phenomenon may have occurred eons ago, like the life histories of Pleistocene mammals, or might occur far away, like star formation in nebulas, or it might be impossible to intervene on, like the Earth's orbit around the Sun.

Other times, experimental intervention might technically be possible but is impractical or inadvisable. We might wonder what would happen to the solar system if the Earth's moon exploded, but no one is rushing to develop the capacity to explode the moon in order to find out. And, though it's important to research how childhood trauma impacts health in adulthood, it's simply unethical to perform experiments to find out, as that would require randomly assigning children to the experimental condition of experiencing significant trauma.

In instances like these, scientists need to employ methods of empirical investigation that do not rely on experimentation. Some of these methods adhere closely to experimental methods, only deviating when necessary. For example, although scientists can't randomly assign children to an experimental group that experiences significant trauma and a control group that does not, they can compare the later health outcomes of children who experience trauma to those who do not, attempting to ensure other variables do not differ consistently between the experimental and control groups. This is an ***observational study***: data is collected and analyzed without performing interventions and sometimes without aiming to control extraneous variables. Observational studies are especially common in studying human populations, as in the health sciences, educational research, psychology, and economics, as it is unethical or impractical to intervene on many circumstances of human life.

Scientists have developed numerous approaches to observational studies. These differ from experiments in a variety of ways and can use creative methods to manage extraneous variables. Other methods of empirical investigation deviate more fully from experimental methods. We have seen that paleontology research into mammoth and mastodon life history utilizes several different forms of evidence to piece together a picture of what things must have been like. This is a common approach to empirical investigation in the ***historical sciences***, fields of science that investigate past events, such as archaeology, paleontology, and cosmology—the scientific study of the origin and development of the universe. Another category of empirical investigation relies on extensive data or simulation, such as computer simulations, as the basis for gaining knowledge when experiments aren't possible or feasible.

EXERCISES

4.1 **Recall:** Describe how scientists have learned about woolly mammoth and mastodon life histories. What forms of evidence have they used? What kinds of conclusions have they drawn?

4.2 **Think:** Describe at least two ways in which the mastodon research described earlier built on the woolly mammoth research also described.

4.3 **Recall:** What are three reasons that scientists can not always perform experiments? Give an example of each.

4.4 **Apply:** Give an example of potential scientific research in each of the following categories:
   a.   performing an intervention is impossible
   b.   performing an intervention is impractical
   c.   performing an intervention is unethical
   d.   direct observation is impossible

4.5 **Recall:** List three varieties of empirical investigation available when intervention is not possible; briefly describe each.

4.6 **Apply:** Imagine you would like to know what your friend's favorite restaurant is, but you can't ask her directly. List 10 ideas for how you might get evidence bearing on this question. Put a checkmark next to the 4–7 ideas that you think would be the easiest evidence to obtain. Then, put an asterisk (*) next to the 4–7 ideas that you think would generate the strongest evidence.

Write 2–3 sentences analyzing the potential kinds of evidence, the ease of collecting them, and their strength. Then, describe how you would approach your "study"—that is, which form(s) of evidence you would collect and how you would go about collecting it.

4.2 OBSERVATIONAL STUDIES

After reading this section, you should be able to:

• Describe and give an example of natural experiments, longitudinal studies, cohort studies, case studies, and phenomenological analysis
• Indicate three ways observational studies can manage extraneous variables
• Characterize the similarities and differences between each type of observational study and experiments

Natural experiments

Every now and then, nature gives rise to circumstances that are almost like an experiment. These so-called *natural experiments* occur when an intervention on an independent variable occurs naturally in real life, without scientists bringing it about. An

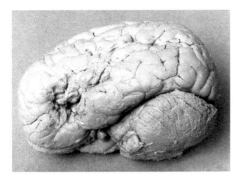 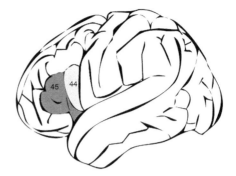

FIGURE 4.3 (a) The brain of patient Louis Leborgne, who lost the ability to speak (left); (b) Brodmann's brain regions 44 and 45, commonly known as "Broca's area" (right)

example is the case of Louis Leborgne, who lost the ability to speak when he was about 30 years old. He could utter only a single syllable, *tan*, which he usually repeated twice in succession, giving rise to his nickname "Tan Tan." Apart from his inability to speak, Leborgne exhibited no symptoms of physical or psychological trauma. He could understand other people, and his other mental functions were apparently intact. After Leborgne died at the age of 51 in a Paris hospital in 1861, the physician Paul Broca performed an autopsy. He found that Leborgne had a lesion in the frontal lobe of the left hemisphere of his brain, an area that came to be known as *Broca's area* and recognized as essential to speech production.

Broca's insight into brain activity was based on identifying accidental circumstances that altered speech production, which we can think of as the dependent variable, and then studying what was different in Leborgne's brain that might give rise to the dramatic change. The function of Leborgne's brain was not intervened upon by Broca; Leborgne just happened to suffer damage to a precise location in the brain such that Broca could later identify it as crucial for speech production. Broca conjectured that a specific area in the human brain is necessary for speech, from which he developed the expectation that an injury in Broca's area causes the loss of speech—an expectation confirmed with an autopsy after Leborgne's death. This inability to produce speech is now known as Broca's aphasia.

Natural experiments also occur when groups just happen to get sorted accidentally—without any scientific intervention—into something approximating experimental and control groups. Some natural or historical process separates them out, such that one group but not the other can be construed as receiving an experimental treatment or condition. For example, in the early 2000s, researchers investigated how having women village council leaders, known as Pradhan, in India might affect social services. This was a natural experiment relying on a 1993 Indian constitutional amendment that required one-third of Pradhan positions to go to women. The law was structured so that the change in leadership was randomly implemented across villages, mimicking a surgical intervention. Researchers collected data on 265 village councils in West Bengal and Rajasthan, including the minutes of village council meetings and Pradhan interviews. They also collected data about social services and infrastructure in each village and requests that had been submitted to the village council. The Pradhan's policy decisions and villagers' requests were unaffected by their interactions with the experimenters, since those requests and decisions were already made during data collection. The researchers found that women policymakers had important effects on social service policy decisions. Women Pradhan increasingly invested in the social goods that were more closely connected to women's concerns in a village: drinking water and roads in West Bengal and drinking water in Rajasthan. They invested less in public goods connected to men's concerns: education in West Bengal and roads in Rajasthan.

As this example illustrates, governmental policy can support natural experiments or at least indirectly control some variables in an observational study. Many studies examining the impact of the Covid-19 pandemic and pandemic policies like stay-at-home orders or remote school instruction relied on variation in policy to create quasi-experimental conditions.

For example, an observational study evaluating pandemic-related learning loss in elementary-school students compared the standardized test scores of Ohio third graders in school districts that began the 2021 school year in remote instruction, hybrid instruction, and in-person instruction. This study benefits from both direct and indirect variable control from law and policy. All Ohio third graders are required by law to take the same standardized test at the same time of year. But, school districts across the state varied in when they returned to in-person instruction—from the start of the 2021 school year in August to spring of 2022. That variation enabled researchers to determine whether additional time in remote instruction extended learning loss. And it did: all districts saw a decrease in student proficiency, but 11.2% fewer students who began the year with remote instruction met the state benchmark for advancing to fourth grade, whereas 6.5% fewer students with hybrid instruction and 5.3% fewer students with in-person instruction met the benchmark to advance.

Managing extraneous variables

As stressed in Chapter 3, experimental methods are designed to eliminate the possibility of confounding variables through direct and/or indirect variable control. The same principles apply in observational studies, though researchers must seek circumstances that naturally hold extraneous variables constant (direct control) or vary them randomly (indirect control), or they must use techniques to manage extraneous variables when direct and indirect control aren't possible. Circumstances sometimes perfectly mirror experimental conditions, giving rise to natural experiments, but it is common in observational studies for there to be extraneous variables that circumstances do not direct or indirectly control.

Scientists have developed various methods to manage uncontrolled extraneous variables in observational studies. For example, in the study on Covid-19 learning loss just mentioned, school districts were not randomly distributed across the conditions of remote, hybrid, and in-person instruction. So, researchers used the data to estimate the impact of other factors like unemployment and poverty level on learning loss, and then they calculated how that effect would impact remote versus hybrid versus in-person districts. They found that those extraneous variables did not account for the difference observed.

Another method sometimes used to help account for extraneous variables is to match subjects with different characteristics across the different conditions, and then only investigate outcomes for those subjects. For example, if this method had been employed in the Covid-19 study just discussed, researchers would identify individual third graders in remote, hybrid, and in-person instruction and attempt to match them with similar third graders across these instructional formats, attending to potential confounding variables like income level, race, prior academic achievement, and family unemployment or illness. This method isn't as successful at controlling variables as random group assignment, but it goes some way toward correcting for confounding variables (as long as researchers are already aware of them).

Other approaches to accounting for extraneous variables make use of the passage of time. In a ***longitudinal study***, observations are made of the same variables over time, in many cases over a long period of time. The study on Covid-19 learning loss we have been discussing is a longitudinal study: the researchers compared standardized test performance of all third graders in Ohio public schools in 2022 with the performance of third graders in Ohio public schools in prior years. Unlike all prior students, third graders in 2022 had experienced educational disruptions from Covid-19, and there's good reason to think Covid-19 and its downstream effects are the only major differences between these third graders and those of prior years.

Longitudinal studies can be carried out over many years. The Early Childhood Longitudinal Study was started by the US Department of Education in the late 1990s and has followed 20,000 American children, examining their development, performance at school, and early school experience. Researchers also conducted extensive interviews with the subjects' families. This study provides a lot of information about American children's development and family life, such as what is important about kindergarten experiences for later academic success, and whether offering Algebra I to eighth graders as an online course is effective. (It is.) Early Childhood Longitudinal Studies continue, with the latest following the kindergarten class of 2023–2024 through the fifth grade.

One type of longitudinal research is a ***cohort study***, where researchers select a group of subjects sharing some defining trait and study them over time, in comparison to another group of subjects that is as similar as possible except without this trait. Cohort studies can reveal changes over time in the characteristics of the group of subjects with the trait of interest. For example, subjects with Covid-19 may be grouped according to demographics like age and gender, or treatments they received, and then have their health outcomes assessed.

Case studies

One form of observational study that is quite different from a controlled experiment is a ***case study***, a detailed examination of a single individual, group, system, or situation in a real-life context. Case studies allow researchers to gain a first-hand understanding of a phenomenon as it occurs in its specific context. Often, various sources of data are used for case studies. Depending on the phenomenon under investigation, these may include observations of a person's daily routine, unstructured interviews with participants and informants, letters, e-mails, social media activity, legal or archival records, buildings, animal or plant behavior, and so forth.

A defining feature of case studies is their reliance on qualitative data. **Qualitative data** consist of information in non-numerical form, whereas **quantitative data** are in numerical form, which makes them easily comparable. For example, "the subject reports being very angry" and "the subject reports being not angry at all" are qualitative data, whereas "the subject reports their anger is a 3 on a scale of 1–5" and "the subject reports their anger is 1 on a scale of 1–5" are quantitative data. Case studies are often employed in the context of qualitative research in epidemiology, psychiatry, education, ethnography, archaeology, and other social sciences.

Case studies may focus on an instance of a common situation or condition, or they may focus on outliers, that is, individuals or situations that deviate from what's common. An example of a case study on a common situation is the research reported in the 2013 book *Paying for the Party* by education researcher Elizabeth Armstrong and sociologist Laura Hamilton. Armstrong and Hamilton conducted a five-year study focusing on the college experience of one group of women at a large, public university in the United States. Their research reveals ways in which social and academic life at such a university prioritizes a "party pathway" catering to affluent and well-connected students but disadvantages less affluent and first-generation students, to the detriment of their education and future careers.

A downside to case studies is that they offer no ways to control extraneous variables or to compare a variety of outcomes. Because the research focuses on only one individual, event, or group, results can be difficult to replicate and to generalize. Case studies are also vulnerable to bias due to the evaluation of qualitative data and no blinding.

Yet, in some fields, detailed first-person, qualitative reports of individual cases play important roles without necessarily aiming for reproducibility or generalizability. For example, clinical case reports often provide clinicians with accounts of surprising or novel conditions in an individual patient, which may generate contextually situated understanding of those conditions and new hypotheses about diagnosis and treatment. Case studies can also be valuable exploratory research, as a way to indicate promising foci for later experiments or more robust observational studies. For example, Armstrong and Hamilton's research may set the stage for a broader investigation into how a university Greek system or certain majors differently affect students with differing socioeconomic status.

Phenomenological analysis

Case studies sometimes reveal a first-person perspective on some phenomenon or situation, that is, how things feel or seem to the subject of the investigation. You may wonder whether such subjective accounts can be of scientific value; after all, science involves empirical investigation and comparison across individual perspectives. It's true that there is oftentimes scientific value in generalizing from one individual's perspective. But it's also true that first-person perspective on situations a person is experiencing can be essential to scientific knowledge. First-person accounts of lived experiences play various roles in fields such as psychiatry, health sciences, sociology, and anthropology.

First-person accounts sometimes are called *phenomenological* to emphasize that they are grounded in the methods of phenomenology, a philosophical tradition concerned with making sense of embodied subjective experience. In science, **phenomenological analysis** aims to describe and analyze what some experience is like for a particular individual. Phenomenological analysis is focused on subjective experience. It simply makes no sense to say that one's experience of, say, pain is incorrect if it varies from others, though we might contrast common experiences of pain with outlying experiences of pain.

Box 4.2 Phenomenological analysis

Phenomenology is a tradition in philosophy that originated in Europe in the early 20th century. It includes the work of philosophers like Edmund Husserl, Martin Heidegger, Edith Stein, Maurice Merleau-Ponty, Aron Gurwitsch, and others. The etymology of the word *phenomenology* comes from ancient Greek and means study of that which appears. What unifies strands of research within the phenomenological tradition is their focus on the structures of consciousness as experienced from the first-person perspective. One central structure of conscious experience is its intentionality—that is, being about or of something. Three methods in phenomenology to study the structures of conscious experience can be relevant to scientific fields like cognitive science, psychiatry, medicine, anthropology, and sociology. The first method consists in describing conscious experience as it is experienced, without influence by scientific or historical interpretation. Another method consists in interpreting conscious experiences by situating them in their social, material, and technological context. A third method consists in analyzing the essence of and conditions for different kinds of intentional states like perceptions, emotions, beliefs, and desires, pointing out their differences and similarities in relation to time, space, self, body, and interpersonal social relationships.

Phenomenological analysis has been used in psychology and neuroscience to distinguish actions of our body that we feel a sense of agency over from mere movements we are not responsible for. This distinction is salient when you are, say, pushed or jostled: you experience your body moving but do not experience yourself as having performed that movement. In light of this distinction, psychologists and neuroscientists have designed experiments to investigate whether there are different mechanisms responsible for these experiences. These experiments might help our understanding of symptoms of psychiatric conditions like the experience of *thought insertion* in schizophrenia, where an individual does not feel agency over some of their thoughts. In this example, phenomenological analysis helps clarify two distinct phenomena, which scientists can then probe with experiments and also compare with psychiatric patients' accounts of their own experiences.

Phenomenological analyses can also play a role in medicine, particularly in relation to what it is like to experience illness and how healthcare providers communicate with patients. For example, phenomenological insight can inform the development of methods for better understanding a patient's experience that go beyond verbal reports. Methods that tap into the experience of illness as an embodied phenomenon that affects one's experience can yield insight into changes in movements and behaviors. This can help bridge the gap between patients' experience of illness and healthcare providers' understanding of the illness and, accordingly, improve patients' trust in their doctors.

EXERCISES

4.7 **Recall:** Give a brief description and an example of natural experiments, longitudinal studies, and cohort studies, and then describe how each manages extraneous variables.

4.8 **Apply:** Look back at this section's description of the study of standardized test scores of Ohio third graders in school districts in 2022, after educational disruption from Covid-19. Describe how direct and indirect variable control were each used in the study (without researcher intervention), how the researchers managed the extraneous variables of unemployment and poverty level, and why being a longitudinal study was important for variable control.

4.9 **Recall:** Give a brief description and an example of case studies and phenomenological analysis.

4.10 **Think:** Why are qualitative data essential for case studies and phenomenological analysis?

4.11 **Apply:** For each of the following hypotheses, decide whether it's best investigated with an experiment or observational study and, if the latter, which variety of observational study (natural experiment, longitudinal study, cohort study, case study, or phenomenological analysis). Explain your reasoning, taking into account the hypothesis under investigation, the feasibility or ethics of experimentation, and other constraints.

a.   Whether rising interest rates make stock prices go up

b.   Whether men's resumes are judged more positively than women's, regardless of the resume content

c.   Whether Covid-19 patients admitted to a hospital sooner are less likely to die

d. Whether a patient who is fidgeting feels greater pain (regardless of her reported pain level)

e. Whether a new medicine is effective in decreasing symptoms of Parkinson's disease

4.12 Think: Describe the example of phenomenological analysis in psychology and neuroscience or the example from medicine. In the example you've described, what do you think is added by considering a person's subjective experience?

4.3 IMAGINATION AND COMPUTATION

After reading this section, you should be able to:

* Describe and give an example of thought experiments and computer simulations
* Define *big data* and *machine learning* and describe how these are used in scientific research
* Analyze the advantages and disadvantages of each of these non-experimental approaches

Thought experiments

The power behind experimentation is intervention and variable control. Observational studies employ a variety of different techniques to approximate this combination of intervention and variable control. Another way to proceed when direct or indirect variable control needed to run an experiment isn't possible is to perform an intervention on something else more under our control. An opportunity for this is provided by our rich imaginations.

Thought experiments involve an imagined intervention on an imagined system to learn about the role of the independent variable in the real world. Thought experiments may supplement empirical investigation or, in some cases, can replace it entirely. In the right conditions, your imagination can be a reliable guide to learn about reality. Here's a simple example. Someone who does not often drink in bars stays out late one night in a bar with their friends. They awake in the morning feeling physiologically and psychologically poor. They might ask themselves how they would feel if, instead, they'd gone home earlier from the bar. They know enough about the circumstances and their own propensities to be able to infer that they'd have had fewer drinks, slept a few more hours, and now would be less dehydrated and better rested. They can't go back in time to actually perform this intervention, and so they use this informal thought experiment to arm themselves with knowledge about how staying out late drinking in bars rather than going home to sleep (the independent variable) affects their physiological and psychological health the next morning (the dependent variable).

In science, thought experiments can be used to test a hypothesis, to show that nature does not conform to one's previously held expectations, or to suggest ways in which expectations can be revised. In 16th- to 17th-century Italy, Galileo Galilei used many thought experiments in his investigations of physics and astronomy. In one instance, he wished to investigate an idea shared by many "natural philosophers" of his time that objects with different weights fall at different speeds. Galileo asked his readers to assume that this was true: that heavier objects fall faster than lighter objects. He then imagined two objects, one light and one heavy, connected to each other by a string and dropped from the top of a tower. If the heavier object fell faster, then the string would pull taut. But, Galileo reasoned, both objects together are heavier than the heavy object on its own. So, the two objects together should fall faster than either object alone. The objects cannot simultaneously fall both faster and slower, and so the idea that was the starting point for this thought experiment could not be right. Galileo thus conjectured that the speed of a falling body is not dependent on its weight.

Newton also used thought experiments to help show how his theory of gravitation worked. He had readers imagine a cannon at the top of an extremely tall mountain and then asked: what would happen if somebody loaded the cannon with gunpowder, and then fired? Plausibly, Newton reasoned, the cannonball would follow a curve,

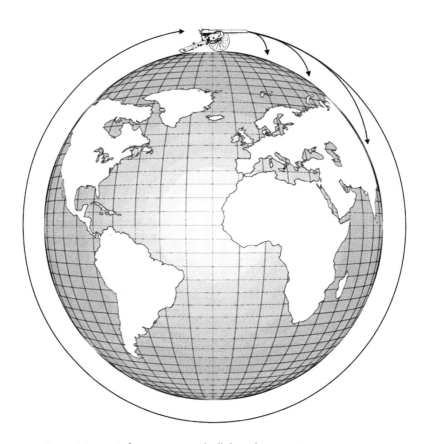

FIGURE 4.4 Isaac Newton's famous cannonball thought experiment

falling faster and faster because of gravity's force, and would hit the Earth at some distance from the mountain. But what if one used more gunpowder? The velocity of the cannonball would be greater, and it would travel farther before falling back to Earth following a curve trajectory. But if one used vastly more gunpowder, then, Newton suggested, the cannonball would travel so fast that it will fall all the way around the Earth, never landing. The cannonball would be in orbit, going around again and again just like the Moon! If the cannonball went even faster, then it would escape Earth's gravity heading out in space. Newton's theory of gravitation provided the resources to arrive at these same conclusions through mathematical calculations. Yet, imagining this situation gives a satisfying, intuitive sense for how an object like the Moon can stay in orbit by remaining in constant free fall.

In the 20th century, thought experiments were central to the development of both transformative theories in physics: the theory of relativity and quantum mechanics. An example that's received some popular attention is the thought experiment about Schrödinger's cat. Quantum mechanics treats subatomic particles mathematically as if they are not in discrete locations but probabilistically spread over possible locations, and there's a longstanding and unresolved debate about how to interpret this mathematical characterization. When a technological apparatus is used to detect an electron, it is always in a single, discrete location. Are subatomic particles like electrons really not in discrete locations until they are measured? If so, what changes when they are measured?

Erwin Schrödinger, Nobel Prize-winning physicist, asked us to imagine a cat, a readily observable living entity, in an apparatus that entangled its state with a subatomic particle such that the cat is killed if some quantum event occurs and not otherwise. Yet, quantum mechanics gives us just probabilities for whether the event occurs. If this is truly the state of the particle, then the cat is not definitively alive or definitively dead but in a probabilistic state between these. This result is inconsistent with our experience of reality, of the possible ways cats could be, and is meant to show that an answer is needed for how the mathematics of quantum mechanics should be interpreted for

FIGURE 4.5 Schrödinger's cat in a probabilistic state of being alive and/or dead

everyday experiences of reality. Note that this thought experiment was not intended as a challenge to quantum mechanics, which has significant evidence in its favor, but to demonstrate the need to offer an interpretation for how that theory relates to our everyday observations.

Just like experiments, thought experiments may suffer from poor experimental design or from scientists inferring unjustified conclusions from them. An additional, very significant limitation is that experimental results are always limited by scientists' powers of imagination. The world can surprise us, including in experimental results, in ways that thinking through the implications of ideas generally cannot. One possible response to the challenge of Schrödinger's cat, for example, is just to point out that our imagination may be ill-equipped to comprehend what we have never directly experienced, like the behavior of subatomic particles.

Computer simulations

As thought experiments can be used to imagine fictive interventions, computers can be used to simulate interventions. *Computer simulations* are computer programs developed from data to mimic a phenomenon. Examples of simulated phenomena might include ecological effects of global climate change, likely shapes of galaxies after two collide, and the progression of a pandemic under different circumstances.

Computer simulations play key roles in many fields of contemporary science, including climate science, astronomy, and epidemiology (the fields of the computer simulations used as examples in the previous paragraph). Computer simulations may be used in combination with experiments or observational studies, or they may be used when experiment or observational studies are not possible. Regardless, a significant amount of background knowledge about the phenomenon is required in order to develop computer programs that can adequately simulate the behavior of a phenomenon. This limitation is similar to the limitations of thought experiments: computer simulations are only as accurate as the data they are based on and how they are programmed.

By running computer simulations and intervening on their features, scientists can learn what would happen to a phenomenon in different circumstances or explore particular features. For example, computer simulations of the Earth's climate use dynamical equations to model the interactions of solar energy, chemicals in the atmosphere, ice, and other factors over time. These simulations can then be studied to yield insight into how climate change is progressing, its variable impact on temperature and weather in different parts of the world, and the potential of different changes to slow or reverse its progression.

Computer simulations allow researchers to have extensive control over extraneous variables. After all, in computer simulations, the researchers decide what data to base the simulations on and what factors to take into account in the program. Computer simulations often can be easily evaluated for their degree of external experimental validity, in the sense that we can evaluate how well their results generalize from the contrived conditions to real-world phenomena by comparing the results of the simulations to empirical measures of the phenomena. To the extent the simulation and the real-world system it describes have relevant similarities, we can be more confident the computer simulation produces trustworthy, generalizable results.

There was more public attention to computer simulation in science during the Covid-19 pandemic than perhaps at any previous time. Epidemiologists began to use computer simulations very early in the pandemic to predict rates of transmission, hospitalization, and mortality and to influence policy decisions on stay-home orders and school and business closures. Early in the pandemic, these simulations were by necessity based on limited data that were rapidly developing as researchers learned more about the SARS-CoV-2 virus, its transmission, and its effects. Models provided broad ranges of possibilities for pandemic progression, and different models often varied radically in their predictions. Yet sometimes these limitations were overlooked or brushed aside, and scientists, policymakers, or the public interpreted the models as making highly specific predictions and different models as providing contradictory predictions.

This instance of computer simulation in science was typical in the sense that early simulations of novel phenomena—like exploratory experiments or early scientific investigations of other kinds as well—begin with some guesswork and involve iterative improvement. But this instance was highly atypical in the extensive publicity received by early predictions of computer simulations. Because that publicity was so unusual, there is a risk that people who don't ordinarily pay attention to computer simulation in science will infer from its limitations early in the Covid-19 pandemic that it is an untrustworthy technique. As we have noted, computer simulations are limited by the existing knowledge used to develop the computer program, as well as by what factors scientists include or exclude from the simulation.

Big data and machine learning

Experiments ultimately are valuable because they generate data and because we know certain things about the conditions in which the data were generated. Observational studies, thought experiments, and computer simulations seek to approximate this approach or seek other kinds of data. A very different approach is to simply maximize data and see what you can learn.

Big data are very large data sets that cannot be easily stored, processed, analyzed, or visualized with traditional methods. Big data sets can reveal unexpected patterns, trends, and associations. Scientific researchers use big data to study weather patterns, DNA, and risk factors for disease, among many other phenomena. Cameras, medical devices, e-mails, social media, and the internet, have produced an ever-increasing stream of data in recent years. Some scientific investigations also produce a tremendous amount of data to be processed and analyzed, especially those relying on computer technologies such as telescopes in astronomy and the Large Hadron Collider (LHC) at CERN. For example, if all the sensor data in LHC were recorded, this would approach 500 exabytes per day, or 500 quintillion (5×10^{20}) bytes. And, the processing of scientific data has also gotten much faster. While it originally took ten years to sequence

FIGURE 4.6 Computer farm at the CERN Large Hardon Collider

the human genome, completed in 2003, a genome can now be sequenced in less than one day.

Generally, an *algorithm* is a procedure for obtaining some outcome that halts in a finite number of steps. *Machine learning algorithms* in particular are step-by-step procedures run on powerful computers that enable scientists to mine large data sets for patterns or to perform other tasks. Machine learning algorithms have begun to play numerous roles in science, as well as business and our everyday life. When, for example, you search the internet, access social media, or order food online, there is some learning algorithm operating in the background that has learned how to rank web pages, make personalized suggestions about content you will probably like on social media, and coordinate food orders and timely deliveries. Once the algorithm learns what to do with data, it will perform such specific tasks automatically.

In scientific research, machine learning algorithms can mine data for patterns, process images and audio files, and make predictions. For example, imagine that you want to determine general trends in food preferences, and you have at your disposal a data set containing all social media posts produced in the past year. Filtering those posts to a subset relevant to food preferences is extraordinarily valuable, as is visualizing the data about the popularity of various foods. The patterns and trends uncovered by analyzing a large amount of data about some subject can give insight into relationships among variables of interest and can be used to make predictions.

Big data and machine learning can help limit the influence of researchers' expectations on findings. For example, current classifications of mental illness, such as schizophrenia, are often criticized for being too coarse-grained and ill-defined to help clinicians and therapists make a reliable diagnosis and provide patients with adequate treatment. Because psychiatric classifications lump together several different psychiatric symptoms and variables that may have little in common, studies relying on machine learning for mining large sets of behavioral, biological, and other types of data from many patients and healthy subjects (as a control group) might discover more useful ways of classifying psychiatric conditions, even if these classifications are not intuitive to us.

The analysis of big data can even help us better understand how science works. For example, bibliometric methods, including the analysis of networks of citations in published work, can be used to investigate the level of productivity of a certain field of research, trends in the topics of scientific research, and even the social dynamics underlying scientific practice. The number of citations of a published article is an index of recognition, which is one of the primary rewards for scientists. So, citation rates and patterns can be used to quantify scientific impact and to predict what factors might affect the course of science.

But studies relying on big data and machine learning also raise distinctive challenges. One challenge is their opacity. The algorithms used in machine learning applications are sometimes unknown to outside researchers, because of security, business, or copyright reasons. This goes against the culture of open exchange that is typical of the social institution of science. Beyond this, how and why an algorithm returns a certain output—for example, a certain mental illness classification decision—can be difficult or impossible to explain even for the researchers who designed it. Finally, even the data on which the algorithm is trained may be unknown to the users of the algorithm.

This opacity is especially problematic because outputs or the underlying data may be misleading or biased, or researchers' assumptions that influenced the algorithm may be wrong. In 2008, researchers from Google claimed that they could immediately predict what regions experienced flu outbreaks based simply on people's online searches. The assumption was that when people are sick with the flu, they often search for flu-related information on Google. Unfortunately, this idea wasn't borne out. Google Flu Trends made very inaccurate predictions, significantly overestimating flu outbreaks, and was shut down.

Another challenge with big data and machine learning techniques is how they may inherit researchers' biases, an outcome they were designed to avoid. This is a negative result for scientific investigation in general, but it is especially problematic when the biases in question reflect sexism, racism, and other prevalent societal prejudices with actual real-life consequences. A prime example is a machine learning algorithm developed by the company Amazon in 2014 for employee recruitment. The algorithm was trained to review job applications automatically, indicating which applicants would be the best to hire. The algorithm had the unintended effect of systematically ranking male applications higher than female applications. Investigation revealed that the algorithm was trained on historical data about the performance of past employees at Amazon, and these employees were predominantly men. Because it was trained on this data set, the algorithm predicted that high-performing employees would likely be men. When Amazon researchers realized this in 2018, they abandoned the hiring approach. This example illustrates how the opacity of machine learning algorithms can have unintended consequences with a real impact.

Box 4.3 Algorithmic fairness and justice

In many domains like business, law, insurance, police work, and healthcare, human decision-making is increasingly supported—and sometimes replaced—by machine learning algorithms. These algorithms attempt to compute predictions based on historical data on which the algorithms were trained. Algorithmic predictions can be objective, but they can also unwittingly reflect unfairness and human prejudices without any deliberative effort by programmers to include these in the algorithm. For example, Tay was a chatbot released by Microsoft Corporation in 2016, and shutdown shortly thereafter because it produced sexist and racist content on Twitter. COMPAS (Correctional Offender Management Profiling for Alternative Sanctions) was used by US courts to predict criminals' risk of recidivism, but it produced a higher rate of false positives (non-recidivists incorrectly labelled "high risk") for Black defendants than for White defendants and a higher rate of false negatives (recidivists incorrectly labelled "low risk") for White defendants than for Black defendants. Cases like these raise difficult ethical and political questions, often grouped under the headings of *algorithmic fairness*, *algorithmic justice*, or *algorithmic bias*. One question concerns the right criterion of fairness—when it comes to recidivism, for example, should the criterion be equal rates of false positives and false negatives between two given groups, or should it be absolutely equal predictive accuracy across all individuals? A second question is whether algorithms are at least less biased than human decision-making without algorithms. Another question is whether, and why, algorithmic justice should be informed by secular, egalitarian, and liberal values rather than, say, Confucian values or politically conservative values.

EXERCISES

Watch Video 8 **4.13 Recall:** Describe and give an example of a thought experiment and a computer simulation.

4.14 Think: Describe the similarities and differences between thought experiments and computer simulations.

4.15 **Recall:** Define *big data* and *machine learning algorithm* and describe how these are used in scientific research.

4.16 **Think:** Describe how each of thought experiments, computer simulations, and machine learning algorithms are useful to gaining scientific knowledge. For each, what are the limitations in their usefulness?

4.17 **Think:** Describe three advantages and three disadvantages of scientific studies relying on big data and machine learning algorithms.

4.18 **Apply:** Algorithms and big data have an ever-increasing impact on our daily activities. Consider the following activities:
a. Online dating
b. Autonomous vehicles
c. Police profiling
d. Online trading
e. Urban planning

Choose three of these activities, then, for each of the three, write a short paragraph describing one task an algorithm might perform and what type of data the algorithm might be trained on to learn that task. Then, formulate a question or concern about the behavior of the algorithm in that domain, explaining its importance. What reasons or values are relevant to addressing the question or concern?

4.4 MULTIPLE SOURCES OF DATA

After reading this section, you should be able to:

* Describe the steps of meta-analysis and how meta-analysis improves experimental validity
* Define *methodological omnivory* and indicate the circumstances in which it's useful
* Characterize how methods of empirical investigation depend on the phenomenon under investigation, the circumstances, and the aims

Meta-analysis

As we've seen, observational studies employ various methods to control or account for extraneous variables. Nonetheless, many studies have limited utility. There may be fewer subjects than needed, confounding variables, no control group, or other foibles that limit their explanatory power. But experiments can have some of these same downsides, too; as emphasized in Chapter 3, most experiments cannot adhere to ideal experimental procedures.

The technique of *meta-analysis* offers a way to combine the results of multiple experiments or observational studies of the same hypothesis to strengthen the conclusions that can be inferred. The idea is that several studies, each with different limitations, can be combined to additionally account for extraneous variables and correct for other shortcomings. Meta-analysis is thus used to better estimate the real extent of the independent variable's influence on the dependent variable by combining the different findings of several distinct studies. Meta-analysis can also identify patterns in study results, including helping to reveal and analyze possible reasons for conflicting results.

Meta-analysis is especially common in healthcare research, but it is increasingly employed in other fields as well. To conduct a meta-analysis, researchers first identify a question that has been targeted in existing scientific research—for example, whether and to what extent patients experience the placebo effect when they know they are receiving a placebo. (See Chapter 3 on the placebo effect.) Then, researchers conduct a literature search and decide which studies should be included in the meta-analysis. For each included study, an ***effect size*** (a quantitative measure of the strength of a phenomenon) is estimated from the study's results, and then the effect sizes are combined using statistical methods that are beyond the scope of this book. This results in an estimate of the common effect size across the studies and also a measure of how much the study outcomes deviate from one another.

In the meta-analysis of the placebo effect, researchers screened 1,246 studies and selected 11 to analyze. These included randomized controlled studies of any medical condition or mental illness that compared administration of a placebo with patients' knowledge to no treatment at all. The application of relevant statistical methods resulted in the finding of a large overall effect size but also high deviation in effect sizes across studies. The studies included in the meta-analysis assessed the placebo effect on improving symptoms from back pain, fatigue from cancer, attention deficit hyperactivity disorder, nasal allergies, depression, irritable bowel syndrome, and hot flashes from menopause. From the large overall effect size, the researchers conclude that administering placebo medications, even with patients' knowledge that they are receiving placebos, may be a promising treatment for several difficult-to-treat medical conditions and mental illnesses. The researchers also note that more research is required, especially regarding efficacy of the placebo effect for different conditions, how patients are notified of the placebo, and how patient expectations influence the treatment. Comparison across studies included in the meta-analysis is used to identify these potential sources of variation to target for further study.

Meta-analyses offer a way to consider and integrate the results from many existing studies, thereby increasing the knowledge gained from them and identifying sources of discrepancies in their results. Sources of discrepancies can include differences in the specific phenomena investigated (such as which health condition patients have), the experimental design (such as how patients are notified about the placebo), techniques of data analysis, and confounding variables (variation in patients' mindset or expectations). All of this can be used to improve internal experimental validity by accounting for confounding variables as well as external experimental validity, including both ecological validity, through variation in study circumstances, and population validity, through variation in subject inclusion across studies. (See Chapter 3 for more on internal and external experimental validity.)

There are also some drawbacks and limitations to meta-analysis. First, the results of meta-analyses inherit any systematic flaws of the studies they combine. It can also be difficult to control the influence of researcher bias, as there are unavoidable judgment calls in which studies to include in the analysis and whether some study has flaws that warrant its exclusion. Finally, the whole point of a meta-analysis is to combine studies with different specifics and different findings to see what they reveal together. The measure of deviation across study outcomes can identify the extent to which findings of individual studies vary, but beyond this, meta-analysis ignores differences across studies that may provide important information. For example, it could be that the placebo effect is an especially promising treatment for some conditions but not others. Focusing on the strength of the placebo effect across a wide variety of medical conditions and mental illnesses may lead researchers to overlook or insufficiently attend to variation across these conditions.

Methodological omnivory

Recall the paleontology research into woolly mammoth and mastodon life histories from the beginning of the chapter. Experiments on living mammoths and mastodons simply aren't possible, and scientists can't conduct observational studies of these pre-historic creatures either. The targets of investigation are separated from the investigators by eons; they can't observe behavior or collect direct evidence. Instead, as we saw, paleontologists creatively employed techniques of chemical analysis on the tusks of these extinct creatures with reference to ecological data and knowledge about modern-day related animals in order to draw conclusions about long-ago happenings that we can never observe directly.

Philosopher Adrian Currie has dubbed this approach in the historical sciences *methodological omnivory*, which is the use of multiple methods and specially tailored tools to generate evidence about specific targets. This approach is identifiable from scientists combining a number of distinct methods, often from multiple scientific fields, and also from significant investment in developing special tools tailored to specific evidential roles.

For example, the mammoth and mastodon studies used highly specific isotope analyses of layers of a tusk and extensive measurements of isotope ratios in the surrounding geologic formations to piece together a specific animal's movements through its range during the different parts of its life. Comparison with migration and life history patterns of modern elephants enabled the researchers to draw other inferences from the animal's movements, such as whether it moved with a herd and that it annually travelled long distances to a mating ground. These investigations were also supported by other past research into mammoth and mastodon physiology, lifespan, and more. Across studies like these on related phenomena, paleontologists can piece together more and more evidence to develop increasingly extensive knowledge of long-ago creatures.

Another technique involved in the mastodon study was the development of a computer simulation that used the isotope data to calculate likely distances travelled and geographic locations. As this illustrates, computer simulations can be an important feature of methodological omnivory, as simulations are one way of drawing broader inferences from the data scientists are able to assemble.

When experimental data aren't readily available, a variety of techniques from observational studies to computer simulations, big data, and meta-analysis are available for potential use. As we've seen, there's no single answer to which of these methods is best. There are different recipes; and which ones are effective depend on the phenomenon under investigation, the circumstances of investigation, and the aims of the research.

Depending on the circumstances of the investigation, some phenomena lend themselves to data collection in quasi-experimental circumstances or to other forms of direct and indirect variable control. Sometimes, when ecological validity is important, these types of observational studies are better than experiments. Variability in a phenomenon and prioritizing first-hand experience increases the value of case studies and even phenomenological analysis. Thought experiments and computer simulations can provide indirect access to features of phenomena not available for experimental manipulation. And when phenomena produce a vast store of data, big data techniques and machine learning algorithms can be useful. If many empirical studies already exist, a meta-analysis of those studies can be more useful than direct empirical investigation. Finally, some phenomena are so distant in time or space that they can only be studied very indirectly, using various tools of simulation and studying their distant effects, as in paleontology and astrophysics.

EXERCISES

4.19 Recall: Describe the steps of meta-analysis. Then, describe how a meta-analysis can improve both internal and external experimental validity.

4.20 Apply: Find a published meta-analysis and look over the article (or, your instructor may provide one for you to analyze). Summarize the study, including how many studies were included, how the researchers selected those studies, if any were subsequently excluded, what conclusions they reached, and any concerns or open questions they indicated.

4.21 Recall: Define *methodological omnivory* and describe how the paleontology research into woolly mammoths and mastodons in section 4.1 illustrates its use.

4.22 Apply: Describe a new example of scientific research where methodological omnivory is used. What about this research and the circumstances in which it's conducted make methodological omnivory useful? (Hint: you might consider research in a historical science, like anthropology or cosmology.)

4.23 Recall: List all the kinds of empirical investigation without experiments we've discussed in this chapter. (By our count, there are 11.) For each, briefly describe how it generates empirical evidence and in what kinds of circumstances it's useful.

FURTHER READING

For a concise treatment of qualitative research and its methodology, see Golafshani, N. (2003). Understanding reliability and validity in qualitative research. *The Qualitative Report*, 8(4), 597–606.

For more on the role of thought experiments in science, see Horowitz, T., & Massey, G. (Eds.). (1991). *Thought experiments in science and philosophy*. Rowman & Littlefield.

For more on the use of big data in science, see Leonelli, S. (2020). Scientific research and big data. In E. N. Zalta (Ed.), *The Stanford encyclopedia of philosophy* (Summer 2020 ed.). https://plato.stanford.edu/archives/sum2020/entries/science-big-data/.

For an introduction to phenomenology in philosophy and cognitive science, see Kaufer, S., & Chemero, A. (2015). *Phenomenology: An introduction*. Polity.

For more on meta-analysis and how it varies across fields, see Kovaka, K. (2022). Meta-analysis and conservation science. *Philosophy of Science, 89*(5), 980–990.

For a discussion of methods in the historical sciences and an articulation of "methodological omnivory," see Currie, A. (2018). *Rock, bone, and ruin: An optimist's guide to the historical sciences*. MIT Press.

CHAPTER 5

Scientific modeling

5.1 THE SAN FRANCISCO BAY AND THE VALUE OF SCIENTIFIC MODELS

After reading this section, you should be able to:

- Describe how the Bay model was developed and how it was used to evaluate the Reber plan
- Define *scientific model* and *target system*, and indicate why models need to be partly similar to and partly different from their target systems
- Analyze the similarities and differences between the Bay model and the San Francisco Bay and how each is valuable or problematic

The Bay model

In an unassuming warehouse in northern California, there lies an enormous model of the San Francisco Bay and the surrounding Sacramento–San Joaquin River Delta. The San Francisco Bay area is a large body of ocean water surrounded by a large urban population living in diverse geological terrains and climates. The delta surrounding the bay is an area the size of the US state of Rhode Island, stretching from the Pacific Ocean almost halfway across the width of California. The Bay model is huge: it's over 1.5 acres in size (more than 6,000 square meters) and is made of 286 five-ton concrete slabs pieced together like a jigsaw puzzle. Still, as large as it is, the Bay model is 1,000 times smaller than the actual San Francisco Bay.

The Bay model is a hydraulic model. Pumps move hundreds of thousands of gallons of water (1 gallon is 3.785 liters) to mimic the tides and currents of the real bay. This procedure works because the model is three-dimensional and proportional; the different parts of the bay and river delta in the model are the right amount lower than sea level, and the surrounding land is the right amount above sea level. The Bay model also includes other features that affect water flow, like rivers, canals in the delta, wharfs, bridges, and breakwaters.

The Bay model is not just a toy model made for tourists. Instead, it's a scientific model. **Scientific models** are constructed to represent phenomena of interest and investigated to learn about those phenomena. This particular model is a terrific tool for

DOI: 10.4324/9781003300007-6

FIGURE 5.1 View of the San Francisco Bay model, looking toward the Golden Gate Bridge and Pacific Ocean

learning about the San Francisco Bay and how human activities can affect it. Teachers, students, and scientists use the Bay model to study geography, ecology, human and natural history, and hydrodynamics. It has been used to help answer questions about how dredging new shipping channels would affect the delta, about how mining during the California Gold Rush changed the rivers, and about what would happen if the system of dikes and levees in the delta failed.

A look at how and why the Bay model was first constructed will help us start to get a sense for the roles that models play in science. John Reber moved from Ohio to California in 1907 and set up as an amateur playwright, dramatist, and theatrical producer. Because of his work, he enjoyed social connections with numerous businessmen and politicians. In the 1940s, Reber became dismayed that the transcontinental railroad ended in Oakland rather than San Francisco, and he came to believe that the extensive bay between San Francisco and the mainland interfered with industry by isolating San Francisco from the rest of California and the US. He came to believe that this large body of water is a "geographic mistake" needing to be corrected.

Reber's career was in entertainment; he had no expertise in science or engineering. Nonetheless, he intrepidly proposed a grand plan to renovate, and then exploit, natural features of the bay that he thought would enable more efficient use of it. Reber suggested filling some parts of the bay to create additional land for things like

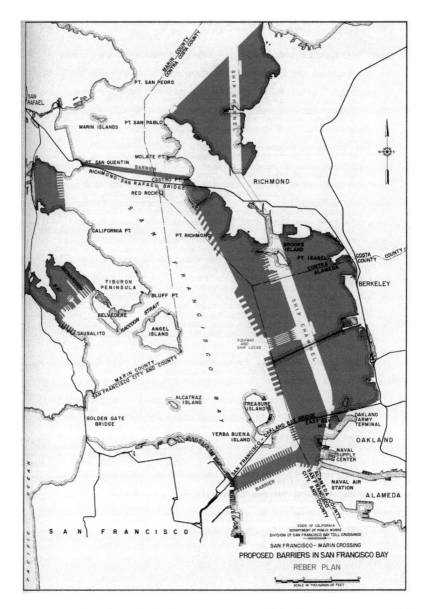

FIGURE 5.2 The Reber plan: shaded areas were parts of the bay Reber proposed filling to create land

airports and factories. This would also establish two freshwater lakes supplied by the rivers that empty into the bay. As freshwater has always been a limited resource in the San Francisco Bay area, it could be valuable to repurpose the bay for potable drinking water and irrigation.

Reber's plan was taken seriously by politicians and capitalists, and the US Army Corps of Engineers decided to test it out. An immediate problem, though, was that the Corps couldn't effectively do so in the actual bay without first implementing it, and, wisely, the Corps wasn't prepared to radically alter the bay and surrounding river delta before knowing what the results would be. They recognized that such changes might have unintended negative consequences for the local water supply, wildlife, vegetation, agriculture, and human population. What to do? How could they consider the effects of the plan without going ahead and carrying it out?

This highlights one way in which scientific models can be useful. When performing an intervention on a system of interest isn't possible, practical, or desirable, a model of the system can be used instead. The Army Corps of Engineers built a hydraulic model designed to be like the San Francisco Bay in some important respects. Once the model was sufficiently similar to the real San Francisco Bay, predictions about the bay could be made based on what was observed in the Bay model. The model could then be manipulated to determine what would happen in the real bay if the Reber plan were implemented. The scientists added scale models of Reber's proposed dams to the Bay model to create the lakes and landmasses Reber had in mind, and then they sat back to see what would happen.

As it turned out, when the Reber plan was implemented in the Bay model, its unintended consequences were disastrous. Rather than lakes, the dams created stagnant pools of poor-quality water that couldn't support ecosystems or be used for drinking or irrigation. Altering the dam configuration in the model to solve that problem just created another problem: fast currents that again destroyed ecosystems and made travel in the bay significantly more dangerous. When the Corps reported these findings, the Reber plan was abandoned.

Models and targets

The real-world system or phenomenon that scientists want to study using a model is called a **target system**, or just a target. Because scientists investigate models to draw conclusions about target systems, a model needs to be like the target system in some ways; that is, it should be similar to, or resemble, the target. And not just any similarity will do. Scientific models need to be similar to their targets in relevant ways for what is being studied. This is why the Bay model replicated tides and currents and other important features to water flow in the San Francisco Bay. If the model were being used to study traffic flow across the bridges, different features would need to be similar.

The relevant similarities between a model and target system are what makes it possible to gain knowledge about a target system from studying a model. Relevant similarity can be achieved in different ways. The Bay model uses real water and simulated tides to mimic water flow, but depth and water resistance of the model had to be adjusted before the water flow was right. Different adjustments could have been made, so long as they also mimicked the real water flow. Or, the model could have been made a different size or out of different materials.

The differences between a model and target system are just as important as the similarities. The Bay model's different spatial and temporal scales are two features that made it useful for learning about the real San Francisco Bay and delta. The model is much smaller than the real bay, with much faster tidal cycles. This allowed scientists to observe what would happen with a spatially distributed, long-lasting sequence of events in a short time and without having to leave the model's warehouse. Had the Corps tried to build a model exactly like the San Francisco Bay, it would have been too large to put anywhere, and it would have changed so slowly, they would have had to wait years to find out about the consequences of the Reber plan.

Some features of the bay either didn't matter or would have been too difficult to accurately incorporate in the model. The Bay model ignores people and buildings, since these are unimportant for water flow. And being inside a big warehouse means the model isn't exposed to changing weather like the real bay is. The model also doesn't incorporate the oceanic wind currents that affect the bay; it's tricky to see how those could be replicated and whether the outcome of doing so would be worth the effort.

The scientists who built the Bay model thus had to decide which features of the model should be similar to the real bay and which could, or should, be different. They also had to decide what to do about changing features of the San Francisco Bay, like whether the model should be like the actual bay during dry seasons, wet seasons, or some combination of these. These decisions were all important to constructing a useful model that could provide reliable information about how the bay would change if the Reber plan were implemented. As it turned out, the model they developed was sufficiently similar to the real bay not only to serve this purpose but for it to be put to other uses as well. For example, the Bay model was also used to study how a later plan of deepening water channels would affect water quality.

Some models are similar to their targets by exemplifying them. An **exemplar** is a model that is one of the target systems it is used to represent. Researchers can use exemplars to represent a broader class of targets that includes the exemplar, drawing conclusions about the whole class of targets by investigating just the exemplar. For example, the fruit fly (*Drosophila melanogaster*) is a common model organism in genetics and developmental biology. Biologists have used the fruit fly to learn how genes influence physical traits and how embryos develop from single cells to mature organisms.

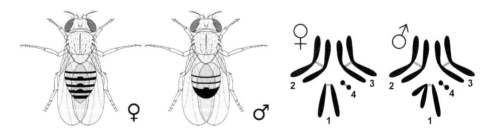

FIGURE 5.3 *Drosophila melanogaster* has a very simple genome with just four chromosomes

The fruit fly's features make it a good model for many purposes. Fruit flies are small and reproduce quickly, and large populations are easily maintained in labs. Their genome is very simple, with only four chromosomes, and scientists have it entirely mapped; so they can make precise interventions in their genes. As an exemplar, the fruit fly is used as a model organism to learn about genes in general, to show how development works in all insects, and for other applications. But, for some investigations, the differences from other organisms that make fruit flies particularly convenient models can also hinder their usefulness, like when genomic complexity is important.

EXERCISES

5.1 **Recall:** Define the terms *model* and *target system*, and indicate the model and Watch Video 9 target system(s) in one example of scientific modeling.

5.2 **Apply:** For both the Bay model and the fruit fly as a model organism, say what the model is, what the target system is, how the model is related to the target system, and what the model is useful for.

5.3 **Recall:** Describe the Bay model of the San Francisco Bay, explaining the purpose for which it was built and what one could learn about the real world from this model. Describe three similarities between the Bay model and its target and three differences, explaining the purpose of each.

5.4 **Think:** Identify two reasons why researchers often use simpler model organisms like the fruit fly to study human biology.

5.5 **Apply:** List three examples of scientific models not introduced in this section. For each, indicate what the model is used to investigate.

5.2 THE MODELING PROCESS

After reading this section, you should be able to:

- Compare and contrast the features and uses of the Lotka-Volterra model with the Bay model
- List three main steps involved in modeling and describe what happens in each
- Define *variable*, *parameter*, *assumption*, and *idealization*, giving an example of each from the Lotka-Volterra model

Specifying target systems

Scientific modeling is a way of gaining knowledge about a target system by investigating a model. The modeling process must include some initial analysis of the target system. Scientists need to decide what they are trying to learn about in order to construct a useful model. Do they want to predict the effects of proposed changes to the San Francisco Bay? Examine the genetic influences on some trait? Or, say, learn more about how number of predators influences a population of animals? All of these projects require models with different features.

An archer cannot accurately hit a target with their arrow if they don't know where the target is or what it looks like. Without some knowledge about the target system, scientists can't evaluate whether their models are similar enough, and in the right ways, to usefully study the target. So, at the beginning of the modeling process, scientists need to have a sense for what a model should be about and what they want to learn from the model. This can be preliminary and partial, just enough to get the process going. For the Bay model, the task was to evaluate the feasibility and consequences of the Reber plan for damming up the bay. This was a starting point, even though scientists didn't know exactly what would matter—currents, salinity of the water, and so on—or what they'd find—plentiful freshwater, excessive evaporation, or something else.

Once the goal of a model is settled by deciding what is to be learned about the target system, scientists also need to know something about which features of the target system are important and what those features are like. This is so they can construct a model that is similar to the target in the proper ways to be useful. When planning the Bay model, scientists figured that the tides and currents might be important to the Reber plan's effects. To calibrate the model to have the same tides and currents as the real San Francisco Bay, researchers needed extensive data about these features of the real bay. Eighty people took measurements at different locations throughout the 1,424 square kilometer (550 square mile) bay every 30 minutes through a full tidal cycle of 48 hours. They recorded tide velocity and direction, changes in the water's salinity (salt content), and the concentration of sediment. These and other data were needed in order to decide what features a model of the bay should have.

Constructing the model

After specifying the target system, researchers construct the model. This stage of modeling involves choices regarding how a model is designed to be similar to its target, what its other features are like, and to how many different circumstances or different targets a model should apply.

For the Bay model, researchers elected to construct a physical replica of the target. The San Francisco Bay is a complex system, and one advantage of a physical model is that the scientists didn't need to understand how changes occur in the bay to predict those changes. Instead, their approach was to make the replica as similar as possible to the bay in all the ways they thought might matter, and then sit back and see what happened. Still, the model required extensive calibration—comparison with the real bay followed by adjustment—before it was accurate enough to make trustworthy predictions. The engineers used their extensive measurements of the bay's tides and currents to ensure patterns of water flow in the Bay model were similar to those of the real bay. They had to tinker with the scales used for depth and width of the bay and the water resistance of the surface in order to get the proper water flow.

Other modeling approaches that don't involve constructing a physical replica offer different advantages and challenges. For example, the Lotka-Volterra model is an influential model in ecology developed independently by Alfred Lotka and Vito Volterra in the 1920s. Unlike the Bay model, the Lotka-Volterra model does not lie in any warehouse. It's a simple, abstract mathematical model. The Lotka-Volterra model uses mathematical equations to represent the interactions of predators and their prey, like foxes and hares, lions and wildebeest, polar bears and seals, and so on. Here are the equations:

$$dx/dt = \alpha x - \beta xy$$

$$dy/dt = \delta xy - \gamma y$$

One variable, x, stands for the number of prey animals (for example, seals). Another variable, y, stands for the number of predator animals (in this example, polar bears). With these equations, biologists can calculate how predator and prey populations change over time (represented in the model as the derivatives dx/dt and dy/dt) from the combination of those population numbers and a few other parameters. A *parameter* is a quantity whose value can change in different applications of a mathematical equation but that only has a single value in any one application of the equation. In this equation, α, β, δ, and γ are parameters. These help the model account for, respectively, the prey population's rate of growth without predation, the rate at which prey encounter predators, the predator population's rate of growth, and the loss of predators by either death or emigration.

The Lotka-Volterra model represents predator-prey interactions, but there's no straightforward way in which these equations are similar to animals eating other animals. Instead, the similarity is between the numbers that solve these equations for certain values of the variables and parameters and how predator and prey populations change in size over time. Recall that different choices can be made in how to achieve similarity with the target system; for models like this one, the similarity is simply in mathematical description. Mathematical models like the Lotka-Volterra model require a firmer grasp of what about the target system is important and how these features interact than physical replicas do.

Variables and parameters are explicit parts of the Lotka-Volterra model. What doesn't appear are the model's assumptions, but these are no less important. In this context, an *assumption* is a specification that a target system must satisfy for a given model to be similar to it in the expected way. In the Lotka-Volterra model, numerous assumptions must be satisfied for the numbers solving the equations to indicate the actual change in predator and prey population sizes. For instance, the model assumes that prey populations will expand if there are no predators and that predator populations will starve without prey. Both assumptions are plausible. The model also assumes that prey populations can always find food, that predators are always hungry, and that predators and prey are moving randomly through a homogenous environment. These three assumptions are idealizations. An *idealization* is an assumption made without regard for whether it is true, often with full knowledge that it is false. These idealizations enable scientists to focus on the essentials of predator-prey interactions in general, without getting lost in complicating details of specific populations' circumstances. Idealizations are discussed in more depth later in the chapter.

Because models can be similar to target systems in different ways, a single target is sometimes represented by multiple models. This can be useful when the real-world phenomenon is so complex that no single model can provide scientists with all they want to know about it. Weather patterns are a good example. Meteorological models normally represent only some of the factors needed to generate reliable predictions. Some may invoke humidity, temperature, and dew point to describe and predict certain basic weather patterns like precipitation. Other models may invoke other parameters, such as central pressure deficit and wind speed and direction, to better predict particular phenomena, like hurricanes. Sometimes meteorologists aim to make more reliable predictions by carefully cobbling together the results of different models of a given weather system.

It's also possible for a single model to have more than one target system. A model might be designed to represent a type of event that recurs often. The Lotka-Volterra model is like that; it is designed to capture something important about seal and polar bear populations, wildebeest and lion populations, and many more prey and predator populations. And the same meteorological models can be used to study different hurricanes, as well as typhoons and cyclones. In contrast, the Bay model is designed to specifically represent the San Francisco Bay and surrounding delta. A new model would need to be designed to represent any other bay.

Analyzing the model

Once a target has been specified and a model constructed with that specification in mind, the model must be analyzed to learn about the target. Models that involve equations, like the Lotka-Volterra model, can be mathematically analyzed by inputting values for parameters and variables representing specific information about a target population. Analysis may also involve manipulating a model to see the effects of changes. This kind of physical manipulation was performed on the Bay model to test the Reber plan. For model organisms like fruit flies, scientists may alter a gene and see how the offspring change to explore the gene's effects.

Such manipulations produce data that, if all goes well, can be used to learn about the target. This is a central purpose of analyzing a model: to draw conclusions about the target system(s). The Bay model revealed that freshwater lakes couldn't be maintained in the San Francisco Bay, as the Reber plan called for, and that the planned dams would have disastrous unintended consequences to the local environment. It was thereby concluded that the Reber plan shouldn't be implemented in the real San Francisco Bay. The Lotka-Volterra model, in turn, reveals that predator and prey population numbers are tightly linked. More prey leads to an increase in predators, and more predators leads in turn to a decrease in prey. This results in a cyclical relationship between predator and prey population sizes, where the population sizes go up and down together.

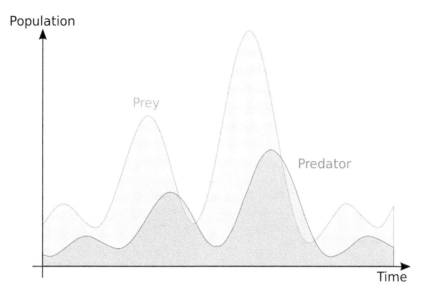

FIGURE 5.4 Visual representation of solutions to the Lotka-Volterra equations; these are values of *x* and *y* that solve the equations, which predict how prey and predator populations will covary

The main purpose of analyzing a model is to learn about the target system(s). But another important purpose is to use existing data to assess and improve the representation of target systems. Because specifying the target and constructing the model can involve some guesswork, and because models differ from their target systems, researchers may not trust that what happens in the model will happen exactly as it does in the target. Analyzing the model's behavior and comparing it with the target's behavior can be used to assess the model's success.

An example of this use of model analysis is the process of calibrating water flow in the Bay model to match water flow in the real San Francisco Bay. Engineers used their extensive measurements of tides and currents in the bay to adjust the model's depth, width, and water resistance until its water flow matched. Ultimately, they made the model bay much deeper proportionally than the real bay, which helped. But this resulted in water moving too quickly in shallow parts of the model. Researchers compensated for this by adding 250,000 copper strips to the bay floor in the model to increase water resistance. They chose how many copper strips to add to any given place by comparing the model's water flow with that of the real bay.

Different models with the same target are sometimes analyzed together to see whether and to what extent they have the same results. This technique is called ***robustness analysis***. It can help determine whether models are accurate of the target when direct comparison with the target isn't possible, like when the target is highly complex like the climate or a country's economy. If multiple meteorological models with different variables, parameters, and assumptions all predict an upcoming increase of temperature in a region, this prediction is robust and seems to have more evidence backing it. Similar predictions from different models also can help scientists identify common features of the models, which might relate to stable relationships involved in the complex phenomenon under investigation. In this way, climatologists and other scientists studying complex systems can learn whether and to what degree the predictions of a model should be taken seriously.

EXERCISES

5.6   **Recall:** List all of the parameters and variables in the Lotka-Volterra model and what each stands for. Then, describe the difference between variables and parameters. Finally, list at least three assumptions of the Lotka-Volterra model and indicate which are idealizations.

5.7   **Think:** Suppose that you are modeling interactions between alligators (predator) and ducks (prey). Make a list of five features of the target system you think your model should take into account. Then, for each feature, say how it is similar or different in other predator/prey systems. For any features that are different, can you think of a related feature that would be similar between the systems?

5.8   **Apply:** Indicate at least three important differences between the Bay model and the Lotka-Volterra model, and describe a reason for each of the differences.

5.9   **Recall:** List the three steps of modeling outlined in this section and state the goal(s) of each.

5.10   **Think:** Recall how experiments involve the three steps of generating expectations, performing an intervention, and then analyzing the resulting data. State the three main steps in modeling, and describe the similarities between those and the three main steps in experimenting. Then, describe how modeling and experimenting are different.

5.11   **Think:** Suppose that we have a model of the Earth's climate and we derive several predictions about average global temperature from the model. The assumptions of the model are somewhat unrealistic of the real-world climate, but replacing those assumptions with slightly different ones leads to similar predictions. Define *robustness analysis* and describe how this is an instance of it. What might you be able to learn from this invariance across the models?

5.3   VARIETIES OF MODELS

After reading this section, you should be able to:

- Define *data model*, say how data models are used, and list the three steps to construct one
- Give examples of models of these five types: scale, analog, mechanistic, mathematical, computer

• Describe the prisoner's dilemma and iterated prisoner's dilemma models and what scientists learned from each

Models of data

The range of things that count as scientific models is extremely broad. Scientific models can be concrete physical objects, like the Bay model, mathematical equations, like the Lotka-Volterra model of predator/prey interaction, or even implemented by computers. We'll explore this variety in this section.

To start, let's distinguish between models of phenomena and models of data. So far in this chapter we've only discussed **models of phenomena**, models used to represent target systems to investigate those systems. Models of data are also representations that are used in place of what they represent, but they represent data instead of target systems, and they play a different role in scientific reasoning by facilitating the use of data in testing hypotheses. A **data model** is a regimented representation of data, often with the aim of discerning whether the data count as evidence for a hypothesis.

Recall that data are any public records produced by observation, measurement, or experiment; because they are public, data enable observations to be recorded and compared. Thermometer readings, video recordings of capuchin monkey gestures, notes about the observed positions of planets, clicks on a website, and participants' answers on a questionnaire are all examples of data. Such records are raw data, which must be processed before they are useful to scientists. For instance, the video recordings of monkey behavior would need to be edited, organized by time and day, and rendered into a software-compatible visual format before they can be used to learn about monkeys' gestures. This process of data correction, organization, and visualization results in a model of the data.

Data models are developed by first eliminating errors, then displaying measurements in a meaningful way, and finally extrapolating from those measurements to expected data for measurements that weren't taken.

Consider measurements of the positions of a planet—say, Neptune—over a period of months. Those measurements will be influenced by more than Neptune's position. They will also reflect some combination of human mistakes, flaws and limitations of instruments used, like a telescope, and inaccuracies from changing atmospheric conditions. Scientists can try to identify such errors by calibrating the telescope and recording the atmospheric conditions along with their measurement of Neptune's position. This is called **data cleansing**: identifying and correcting errors in a data set by deciding which data are questionable and should be eliminated.

Then, the next step is to represent the data in a meaningful way. Data of Neptune's position over a period of months may be visualized as charted points. Finally, these points can be used to draw a curve of Neptune's progression. The points represent scientists' measurements, and the curve represents the scientists' best guess for Neptune's path through the sky. This curve is the data model.

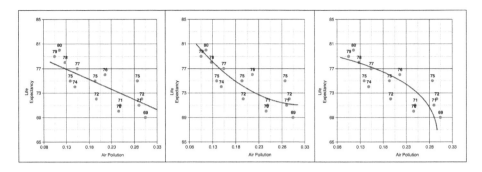

FIGURE 5.5 Illustration of the problem of curve fitting using data about air pollution and life expectancy in different cities

This is a complicated enough task that it has its own name: the problem of ***curve fitting***. Curve fitting is extrapolating from a data set to expected data by fitting a continuous line through a data plot. The problem is that data, no matter how much you collect, are always consistent with multiple different curves. Suppose that you have data for two variables—say, air pollution and life expectancy in different cities—and you want to figure out the mathematical relationship between the two. That is, you want to learn how people's life expectancy relates to the level of air pollution where they live. Your data are represented by the points in Figure 5.5. Figure 5.5 also shows different curves, representing different relationships between air pollution and life expectancy, each of which appears to fit the data pretty well. Put in terms of underdetermination, introduced in Chapter 3, the data underdetermine which curve best captures the relationship between these two variables.

There's no easy answer to how scientists should decide which curve fitting their data is best. Finding the curve that best fits all data isn't usually the best approach. Data models can fit the data too well; the problem with sticking too closely to the actual data is that those data are never perfect. There might be outliers, or values that deviate from the norm for one reason or another, and noise, influences on data that are incidental to the focus. Scientists want their data models to be better than the actual data they've collected. In the end, they must choose a data model based on their background knowledge about the phenomenon and what they want to use the data model for. New data can be used to check how well the selected data model works, just as models of phenomena can be calibrated with comparison to their targets. Big data approaches, discussed in Chapter 4, make data modeling even more challenging. A lot of data without a lot of background knowledge about the data makes it even more difficult to solve the curve-fitting problem.

Scale models

Data models are used in experiments and non-experimental studies, where the phenomena are investigated directly. In contrast, models of phenomena, our main focus in this chapter, are used to indirectly investigate phenomena: the model is studied in order to draw conclusions about its target. This can be especially useful when direct investigation, experiments and studies, aren't feasible—like how the Reber plan would affect the San Francisco Bay—or scientists want to draw broad conclusions about a class of phenomena—like how predator and prey population numbers influence each other. As we've already seen with the Bay model and the Lotka-Volterra model, there are many different types of models of phenomena.

Scale models are concrete physical objects that are downsized or enlarged representations of their target systems. Architectural models of buildings or urban landscapes are a familiar example; these are widely used in civil engineering. The Bay model is also a scale model. Its spatial scale is 1:1000 in length (one meter in the model represents

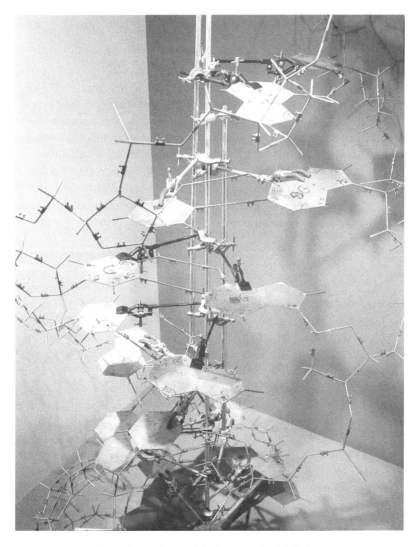

FIGURE 5.6 Watson and Crick's scale model of DNA's double helix structure

1,000 meters in the real world) and 1:100 in depth (one meter in the model represents 100 meters in the real world). The Bay model also has a shorter timescale; 14.9 minutes in the model represents a 24-hour day in the real world.

Other scale models are enlarged representations of their targets. In 1953, James Watson and Francis Crick announced the historic discovery that the structure of DNA is a double helix. Watson and Crick had spent a couple of years building scale models out of wire and tin plates in the shape of the building blocks of DNA. After several failures, the two scientists recognized, in part from the x-ray crystallography work of Rosalind Franklin and Maurice Wilkins, that a double helix structure best fit existing knowledge about DNA. Their model had a spatial scale of roughly 1 billion:1. That is, one centimeter in the double helix model represented one-billionth of a centimeter in a real DNA molecule. (See Chapter 13 for more discussion of this discovery, including Rosalind Franklin's role.)

Analogical models

Analogical models are representations with features similar to focal features of a target system. The Bay model is also an analogical model, as it shares physical properties like tides and currents with its target. But analogical models don't need to be concrete. An example of an abstract analog model is the computer model of the mind, which is based on formal similarities between computers and minds. Like computers, the human mind is often thought of as an information-processing system that can be described in functional terms—that is, without talking about its actual physical composition, or hardware, but referring to what it does. Like computers, minds can be understood in terms of the operations they carry out to solve certain tasks, or in terms of their software.

Another example of an analogical model—located somewhere between the Bay model and the computer model of the mind on the concrete/abstract spectrum—is the Monetary National Income Analogue Computer (MONIAC). Also known as a *Phillips machine*, MONIAC is a hydraulic model that uses water flow, like the Bay model, but to represent the British economy. William Phillips built MONIAC in 1949 using a collection of plastic tanks, each representing some aspect of the economy, connected by pipes and sluices and different valves. The machine used an old airplane motor to pump around dyed water, representing money, to simulate the flow of money in an economy. An overhead tank, representing a treasury, could be drained so that the water inside could flow to other economic sectors (education, healthcare, infrastructure and investment, savings, etc.). Water could be pumped back to the treasury tank to represent taxation and state revenue. Exports and imports could also be simulated by adding or draining water from the model. MONIAC is a physical model, but not a scale model. The British economy isn't operated hydraulically, of course. MONIAC used water as an analog to money; changes in water level and flow were analogous to changes in amounts of money in and transfer among various sectors of the British economy. In its day, this was an amazingly accurate tool for studying how changes in different economic sectors affect others.

One type of analogical model is the ***mechanistic model***, a model that represents the component parts and operations constituting some recurring process. Various processes

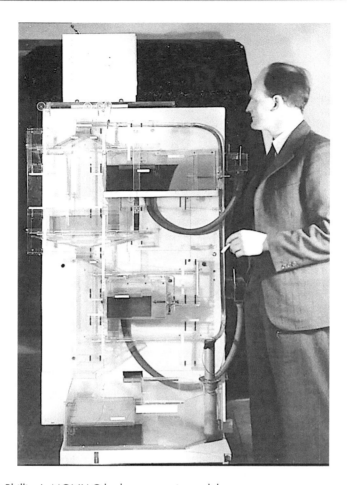

FIGURE 5.7 Phillips's MONIAC hydro-economic model

in living organisms can be construed mechanistically, such as blood circulation, protein synthesis, and cellular respiration. A mechanistic model helps illuminate how the phenomenon depends on the orchestrated functioning of component parts. Mechanistic models can be physical structures, but most mechanistic models are schematic representations of structures, functions, and the relationships among them. For example, the mechanistic model of the cellular sodium-potassium pump depicted in Figure 5.8 is a generic representation of how sodium and potassium are exchanged through cell membranes.

Relying on analogies is a particularly useful strategy in early stages of modeling, when scientists may have little or no knowledge of the target system. This enables scientists to focus on the salient features of the target and to let the discovery of analogous features guide modeling approaches. For example, spiral staircases were an inspiration to Watson and Crick, guiding their modeling efforts of DNA toward a double helix

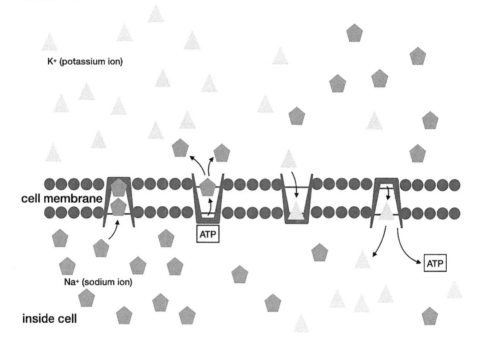

outside cell

K+ (potassium ion)

cell membrane

ATP

Na+ (sodium ion)

inside cell

ATP

FIGURE 5.8 Visual depiction of a mechanistic model of the sodium-potassium pump

structure. As knowledge about the target develops, analogical models may give way to models less obviously related to the target systems they represent.

Mathematical models

Mathematical models are mathematical equations that use variables, parameters, and constants to represent one or more target systems. The Lotka-Volterra model is an example. It uses a pair of first-order differential equations to represent changes in predator and prey populations over time. The first equation, $dx/dt = \alpha x - \beta xy$, describes the fluctuations of a prey population, dx, over time, dt, where αx represents its exponential growth and βxy represents the rate of predator/prey interaction. The number of mice at a time, for example, is determined by their population growth, minus the rate at which they're preyed upon by, say, hawks. In contrast, the number of hawks is fixed by their population growth given the supply of prey, minus their mortality rate. Hence, the second equation, $dy/dt = \delta xy - \gamma y$, describes the fluctuations of a predator population, dy, over the same time interval, where δxy represents predator population growth and γy represents loss of predators due to events like death, disease, or emigration. (When we introduced the Lotka-Volterra model earlier in the chapter, we said that x is a variable representing the prey population size and y is a variable representing the predator population size, while α, β, δ, and γ are parameters for, respectively, the prey population's rate of growth without predation, the rate at which prey encounter predators, the predator population's rate of growth, and the loss of predators by either death or emigration. Look at what we say about each part of each equation here with those definitions in mind to get a better sense for what each equation is used to represent.)

Another example of a mathematical model is a game theory model called the prisoner's dilemma. Suppose that you and your friend Dominik have been arrested for robbing a bank, and you've been placed in different jail cells. A prosecutor makes this offer to each one of you separately:

> You may choose to confess or to remain silent. If you confess and your accomplice keeps silent, all charges against you will be dropped, and your testimony will be used to convict your accomplice. Likewise, if your accomplice confesses and you remain silent, your accomplice will go free while you will be convicted. If you both confess, you will both be convicted as co-conspirators, for somewhat less time in prison than if only one of you is convicted. If you both remain silent, I shall settle for a minor charge instead.

Because you are in a different cell from your friend, you cannot communicate or make agreements before making your decision. What should you do? Assuming that neither of you want to be imprisoned, you face a dilemma. You will be better off confessing than remaining silent, regardless of what Dominik does. Look at the prior passage and think through what happens to your jail time if Dominik doesn't confess and if Dominik does confess; in either case, it's better for you if you confess. So, it looks like you should confess, right away! The problem is that Dominik is in the exact same situation. And the outcome where both of you confess is worse than the outcome where you both remain silent. If you and Dominik do the same thing, it's better to both remain silent.

The prisoner's dilemma seems to raise a puzzle for rationality. You are better off confessing, regardless of Dominik's choice, but if you both are inspired by that fact to confess, you are both worse off than if you had both remained silent. Reasoning independently, it seems you should confess. But if you could plan what to do together, you'd both choose to remain silent. What to do?

The prisoner's dilemma is usually represented using the mathematical formalism of game theory. The prisoner's dilemma we described is depicted by the payoff matrix in Table 5.1a. Although this situation may seem contrived, numerous real-life scenarios can be modeled with a generic version of the payoff matrix, as shown in Table 5.1b. The numbers represent generic payoffs or consequences of each decision. The higher the number, the more desirable the payoff. The first number in each set of parentheses represents Player 1's payoff, the second number Player 2's payoff. What's important about these payoff numbers is just that the number for defecting is always higher than the number for cooperating, but the number when

TABLE 5.1 (a) Game theory payoff matrix for the prisoner's dilemma (top); (b) Version of the prisoner's dilemma generic to any choice of cooperating or defecting (bottom)

		Dominik	
		Remains Silent	**Betrays**
You	**Remain Silent**	Each pays a minor fine	You get 3 years of prison Dominik goes free
	Betray	You go free Dominik gets 3 years of prison	Each serves 2 years of prison

		Player 2	
		Cooperate	**Defect**
Player 1	**Cooperate**	(2, 2)	(0, 3)
	Defect	(3, 0)	(1, 1)

both players cooperate is higher than the number when both players defect. In Table 5.1b, $3 > 2$ and $1 > 0$, but $2 > 1$.

These numbers capture the dilemma you do better if you defect regardless of your partner's choice, but things will go better for you if your partner cooperates. The bank robbery scenario is just a vivid illustration; the prisoner's dilemma can model lots of real-world situations involving people, businesses, nations, animals, or even bacteria. Any real-life scenario in which entities vary their strategy in ways that affect each other can be represented by the prisoner's dilemma game if the desirability of the consequences can be represented by these payoff numbers.

The prisoner's dilemma has been used to model many scenarios involving cooperation, in human societies and the biological world alike. For example, consider the symbiotic relationship of cleaner fish. Individuals of one species (the cleaner) remove parasites and dead skin from individuals of the other species (the client). Cleaner fish may choose to cooperate by cleaning the client fish or to defect by eating extra skin from the client fish. Client fish may choose to cooperate by allowing the cleaner fish to clean safely or to defect by threatening or eating the cleaner fish. Both fish types are better off if they cooperate: the client fish gets an important cleaning while the cleaner fish gets dinner. But there's a benefit to defecting for each: the cleaner fish would get a bigger dinner by eating more from the client fish, and the client fish would get to eat the cleaner fish. The prisoner's dilemma has been used to reveal the circumstances that can enable cooperative symbiosis like this to evolve.

Computer models

While many real-world situations can be modeled as cases of the prisoner's dilemma, what we've seen so far isn't enough to show why businesses, gangsters, fish, and nations so often cooperate in real life. One important reason is that, in many real-life scenarios, decisions about whether to cooperate aren't made in an isolated room, separated from your partner. Instead, businesses, gangsters, fish, and nations all tend to signal their intentions, negotiate while making decisions, or interact repeatedly over time, allowing reputations to form.

The prisoner's dilemma model can be extended to represent these kinds of interactions. One common extension is the iterated prisoner's dilemma, where we suppose that two agents play the prisoner's dilemma with each other repeatedly. This is one way in which cooperative behavior can win out over the selfish choice to defect. In the 1980s, a computer game provided insight into how this can happen. The political scientist Robert Axelrod invited social scientists to submit computer programs for a tournament of the iterated prisoner's dilemma. Fun! Each program had its own strategy governing the circumstances in which it would cooperate or defect, and these programs were pitted against one another to see which would perform best in the long run. This tournament was a computer model.

In Chapter 4, we introduced *computer simulations*: computer programs developed from data about a phenomenon to simulate its behavior. Computer simulations are a variety of model; we can equivalently call them *computer models*. Axelrod's tournament wasn't a computer simulation of a specific phenomenon of interest but a simulation of iterated exchanges in which cooperation is valuable but there's a temptation to defect.

By inviting open participation from other modelers, Axelrod made it so that the strategies deployed in his model weren't limited by his own preconceptions, and the results did turn out to be surprising. The winning strategy, accumulating the most points in the iterated prisoner's dilemma tournament, belonged to a program named Tit-for-Tat, submitted by psychologist Anatole Rapoport. The program was so simple that it had only a few lines of programming code. Tit-for-Tat cooperated in the first round of any game it played, and then it mirrored the other player's previous action in every subsequent round. So, when Tit-for-Tat played against generally cooperative programs, it behaved cooperatively and reaped the rewards of that mutual benefit. But when Tit-for-Tat played against players that frequently defected, it too played selfishly after that initial cooperative move. This protected it from exploitation by selfish programs. Axelrod's computer simulation thus demonstrated the strategic success of emulating the cooperation of others, which has been dubbed *reciprocal altruism*.

EXERCISES

5.12 Recall: Characterize models of data and models of phenomena, and give an example of each. How are these types of models similar? How are they different?

5.13 Apply: List the five types of models of phenomena described in this section, and give an example of each. For each example, indicate why it counts as a model of that type, and what target system(s) it is supposed to represent. Then, rank your examples from 1 to 5, where 1 is the most concrete relationship to the target system(s) and 5 is the most abstract.

5.14 Recall: List the three steps of data modeling. Then, describe the curve-fitting problem and indicate how it complicates those steps.

5.15 Think: Describe the prisoner's dilemma and iterated prisoner's dilemma models, and indicate what type of model each is. Considering both models, what do you take the target system(s) to be, and what do you think scientists can learn from these models?

5.16 Recall: Mathematical models are among the most abstract representations of target systems. Describe how mathematical models are nonetheless similar to target systems, using the example of the Lotka-Volterra model. (Returning to section 5.3's discussion of the Lotka-Volterra model might be helpful.)

5.4 LEARNING FROM MODELS

After reading this section, you should be able to:

- Describe how models can play an experimental role and how they can play a theoretical role
- Identify three features that all models share
- List five desirable features of models and describe tradeoffs among these features

Modeling as experimentation and theorizing

Constructing and analyzing a model shares some similarities with experimentation. Both experimenters and modelers perform interventions to test expectations based on a hypothesis; like experiments, modeling can provide evidence for or against hypotheses about target systems. Expectations about the consequences of the Reber plan for the San Francisco Bay were tested with interventions on the Bay model. Animal models like the fruit fly are used to test expectations about the genetic causes of human diseases, like diabetes and lymphoma. And studying the iterated prisoner's dilemma helps test expectations about the conditions that enable cooperative behavior to emerge among self-interested individuals.

In each example, the models were used to test scientists' hypotheses about real-world systems, sometimes with surprising results. So, models can play a role similar to experiments. One important difference is that, with experiments, scientists intervene directly on the target system, whereas with models, interventions to models are used to draw conclusions about the target system. This is why models must accurately represent their targets.

Getting models to reflect their targets more accurately is a primary task of modeling. When a model is known to accurately represent its target, it can play a role similar to a scientific theory by representing core features of that phenomenon. When a model behaves similarly to the expected target system(s) in many instances and across different circumstances, it may become accepted as part of a theory of how the target behaves. An example of such a theoretical use of modeling is the Lotka-Volterra model of predator-prey interactions. If one sets the model's parameters based on observations or estimates of specific predator and prey populations, one can then use the model to predict changes over time in the sizes of these populations. These predictions are a good account of how real predator-prey populations behave, and when they go wrong, one can usually figure out why by comparing the model's features to the target system's features.

So, models can play an experimental role by helping to investigate phenomena empirically, and they can play a theoretical role by providing an account of phenomena. Sometimes the same model can even play both experimental and theoretical roles. In Axelrod's tournament, computer simulations served as virtual environments to test which strategies would perform best in an iterated prisoner's dilemma game. While there were no expectations that Tit-for-Tat would win, this outcome accorded with an existing theory in evolutionary biology, namely reciprocal altruism. The basic idea is that it can be evolutionarily advantageous for an organism to help another at some cost to itself if there is a chance the favor will be returned in the future. The success of Tit-for-Tat was based on reciprocity, and so was consistent with this theory. Thus, the success of Tit-for-Tat in Axelrod's computer tournament confirmed the idea in evolutionary theory that natural selection can favor cooperative behavior, even when there are costs, if the behavior is reciprocal.

Common features of models

This chapter has emphasized how different models can be from one another. Still, all models share three important features. First, all scientific models are used to learn about the world. Data models represent data to enable hypothesis-testing. Constructing and investigating models of phenomena enable scientists to reason about target system(s) in hopes of gaining new scientific knowledge. In both cases, the models are used as vehicles for learning about natural phenomena investigated in science.

Second, all models are used to represent: they are about, or stand for, the phenomena they target—or, for data models, the data they characterize. For a model to represent a target, it must be like the target in the right ways. Often, this likeness is understood in terms of similarity. But we have said that models aren't exactly like the target systems they represent; they are also dissimilar from their targets in important ways—smaller or larger, mathematical or computerized. What similarities and differences are intended between a model and a target governs how the model should be interpreted and used.

Third, all scientific models involve idealization and abstraction. When constructing a model, scientists leave out some features of the target system and incorporate features the target does not have. Omitting or ignoring known features of the system is called *abstraction*. Misrepresenting features of the system with false assumptions is idealization (defined earlier). Abstraction and idealization can be used to set aside or simplify some features of target systems to focus on only those features deemed important for the purposes at hand.

The Lotka-Volterra model, for example, abstracts away from properties of prey and predators, like their speed, size, and capacity for camouflage. Those features aren't essential to how predator-prey interactions influence population size and so have been abstracted, or removed, from the model. The Lotka-Volterra model also incorporates idealizations of predator-prey interactions. That model's idealizations include the assumptions that prey can always find food, that predators are always hungry, and that both predators and prey are moving randomly through a homogenous environment. Scientists know these assumptions aren't true, but they are helpful simplifications that usually don't interfere with the model's usefulness.

These three features of scientific models—their use to learn about natural phenomena, their representation of target system(s), and their incorporation of abstractions and idealizations—all relate to one another. Abstraction and idealization are features of models that affect how they represent their targets, and the ways models represent their targets partly determines what can be learned from them. Representation, then, is at the heart of scientific modeling.

What makes a model good?

We've seen that a target system can be represented in many ways. A physical model of a hydrological system, like the Bay model, represents water flow in ways that differ from mathematical models of fluid dynamics. And both of those are different from the computer model that eventually took over the work of the Bay model. There's no one perfect model of a given phenomenon. Instead, the goodness of a model is determined by what the modelers want to learn from the model. Other factors, such as cost or ease of developing or using the model, can also be important. Sometimes one model will be enough for learning about a target system; other times, multiple models of the same target will be necessary to gain knowledge.

It is desirable for models to be accurate, general, precise, tractable, and robust. Each of these features helps make a model valuable, but attempting to maximize all of these features is futile. This is because these features trade off against one another: gaining more of one desirable feature of a model often requires losing ground on some other desirable features. For example, a model that is more general by applying to more target systems is also often less accurate of any one target system. This is because targets differ from one another in some regards. So, when constructing models, scientists must decide which desirable features to emphasize and which to compromise on.

Box 5.1 Values in Modeling

Models always are similar to their target systems—but are also different from them in various ways. This creates an opportunity, or an inevitability, really, for the influence of social values on scientific modeling. Recall from Chapter 2 that social values are the priorities and moral principles accepted in some community. One way social values influence modeling is in shaping what features of a target system scientists choose to represent accurately in a model and what features they ignore or distort. These choices depend on modelers' aims. The influence of values on these choices is salient especially (but not only) for models in social science, as how a phenomenon should be defined and measured given a certain purpose and what factors are important to understand often relate closely to social values. For example, philosopher of science Eric Winsberg has criticized predictive models of the Covid-19 pandemic that were used to justify governmental stay-at-home orders for including consideration of illness and death from Covid-19 but excluding consideration of the health impacts of school closures, social isolation, and deferred healthcare. There's not one single right way to develop a model. What's important is that similarities and differences from a target system be thoughtfully designed and open to scrutiny from others. Multiple models of one target phenomenon also can be developed; this was done for models of the Covid-19 pandemic, and it also occurs in climate change modeling and modeling of many other phenomena of widespread social importance. Multiple models of the same target can lay bare how modeling choices based on different aims and values influence what we understand about a phenomenon.

Accuracy is the extent to which a model represents the actual features of its target system; models that are more accurate better represent more features of a target system. A model that represented all and only the actual features of its target would be maximally accurate, but maximal accuracy is unhelpful; recall that models are improved by some differences from their targets. For example, the Bay model improved its representation of water flow by introducing inaccuracies of water depth. Overemphasizing a model's accuracy can come at significant costs to tractability and generality—making the model difficult to develop and inapplicable to target systems that differ in minor ways.

Generality is a model's applicability to a greater number of target systems; a model is more general when it applies to a greater number of target systems. Generality is a desirable feature of models insofar as it enables models to be reused in various circumstances. General models also highlight what various phenomena have in common with one another. Because the prisoner's dilemma can apply to nations, squirrels, and pirates, this is a general model with numerous applications. This generality reveals something that these scenarios have in common: repeated interactions can enable cooperation to spontaneously emerge. However, sacrificing some generality in a model can be worthwhile, depending on the aim of the model, if this enables the model to represent its target more accurately. A general prisoner's dilemma model might be supplemented with information about, say, how natural selection favors bacteria that can persist near one another (a form of cooperation). The resulting model will give more insight into bacteria cooperation in virtue of this additional detail. But it also will be less general—it will no longer apply to humans or corporations. Which is better depends on the modelers' aims.

Precision is the extent to which a model finely specifies features of a target system; a more precise model more finely specifies features of the target. For example, climate models that allow scientists to predict how much warmer the global average temperature will be in 30 years within a range of ±0.05° Celsius are more precise than models that allow them to predict a ±1° Celsius range of temperature increase in 30 years. Notice that precision is different from accuracy. Whereas accuracy is a matter of a value's proximity to the true value, precision is a matter of how finely specified a value is. So, a model can be very precise but inaccurate. Think of an archer loosing arrows at a target. Arrows that are scattered all around the bull's eye but near it are accurate but imprecise. Arrows that are tightly clustered together but off center, away from the bull's eye, are precise but inaccurate (see Figure 5.9). Greater precision benefits a model by enabling it to give a more specific characterization of its target and to make more specific predictions. But increasing precision usually comes at the cost of a model's generality, its tractability, and sometimes its accuracy. Highly precise models are less generally applicable and more difficult to develop. And, the more specific a prediction is, the easier it is for that prediction to be incorrect.

Tractability is the degree of ease in developing or using a model. More tractable models are easier to construct, manipulate, or analyze. This may involve different considerations, like the time it takes to run a model on a computer, whether the equations of a mathematical model have exact solutions, or even whether a modeler happens to be already familiar with one approach but not another. For example, because the iterated prisoner's dilemma involves agents having repeated encounters, this model is less tractable than the original prisoner's dilemma. One consequence of this decreased tractability is that scientists know exactly what the possible outcomes are for the original prisoner's dilemma, but they cannot easily predict the outcomes for the iterated version. This is why Axelrod ran a computer tournament to explore some of the possible outcomes. Tractability is never maximized, though: the easiest thing to accomplish is nothing at all, and more complicated models regularly result in more accurate, precise, and useful findings. For instance, the iterated prisoner's dilemma reveals how repeat encounters can overcome the dilemma, making cooperation directly beneficial.

Robustness is a measure of insensitivity to features that differ between a model and the target system, including abstractions and idealizations. Normally, scientists don't want their model's predictions to be influenced by those features, since they

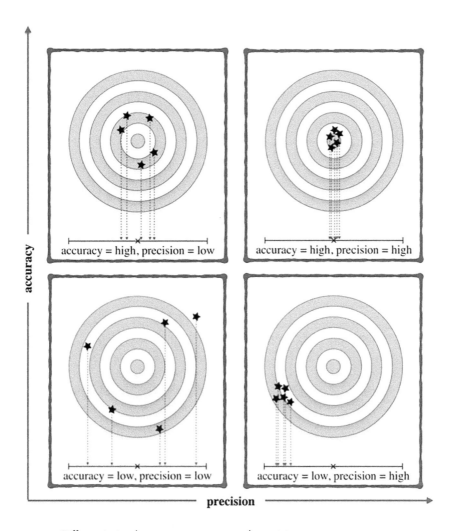

FIGURE 5.9 Differentiating between accuracy and precision

aren't shared by the target. So, a more robust model is one that changes less despite variation in its assumptions. But limited robustness is inevitable. Models incorporate assumptions, including idealizations, that are needed for them to do the tasks they are designed for. If they didn't matter, they wouldn't be needed. What scientists aim to avoid is overreliance on specific assumptions that are unlikely to be true or even known to be false. Multiple models are sometimes used to determine how robust a model's predictions are. If different models, with different assumptions, predict roughly the same result, that prediction seems more trustworthy than if it had been generated by just one model with uncertain assumptions and parameters. This is robustness analysis, introduced in section 5.2.

There is no single answer to how a model should best incorporate these desirable features, nor is there a perfect tradeoff among the features. Instead, scientists strategically develop their models to be tractable enough for their current circumstances; robust enough to be certain to some reasonable degree; accurate and precise enough to make interesting, trustworthy predictions; and general enough to be enlightening across the range of phenomena they are interested in. The balance struck thus depends in subtle ways on the phenomena under investigation, the scientists' circumstances, and the purposes to which the models are put.

EXERCISES

5.17 Recall: Describe how models can play an experimental role and how they can play a theoretical role, giving an example of each.

5.18 Recall: List the three main features that scientific models share, and articulate how these three features relate to one another. Use one example model to illustrate all three features.

5.19 Apply: Locate and investigate a scientific model not already discussed in this chapter. Classify its type of model, and describe what target system(s) it's used to represent. Describe how the elements of the model represent features of the target system(s) and what scientists have learned about the target system(s) from the model. With reference to ideas from this chapter, discuss why this model is a helpful way for scientists to investigate this phenomenon.

5.20 Recall: Define *abstraction* and *idealization*. What is the difference between them?

5.21 Think: Choose one of the models discussed in this chapter. Formulate a list of the abstractions involved in using the model to represent its target system and a separate list of the idealizations involved in using the model to represent its target.

5.22 Apply: In your own words, describe the five desirable features of models characterized at the end of this section. Then, for each feature, compare two models: the classic mathematical model of the prisoner's dilemma and the computer model of the iterated prisoner's dilemma. For each feature, write down whether you think one model is better and which one. Explain your answer.

FURTHER READING

For more on the use of models in science, see Weisberg, M. (2013). *Simulation and similarity: Using models to understand the world*. Oxford University Press.

For more on idealization, abstraction, and tradeoffs, see Potochnik, A. (2017). *Idealization and the aims of science*. University of Chicago Press.

For a discussion of computer simulation, see Frigg, R., & Reiss, J. (2009). The philosophy of simulation: Hot new issues or same old stew? *Synthese, 169*, 593–613.

For discussion of computer modeling and attention to climate change models, see Winsberg, E. (2010). *Science in the age of computer simulation*. University of Chicago Press.

For discussion of computational methods in science, see Humphreys, P. (2004). *Extending ourselves: Computational science, empiricism, and scientific method*. Oxford University Press.

CHAPTER 6

Deductive reasoning

6.1 THE AGE OF THE UNIVERSE AND SCIENTIFIC ARGUMENTS

After reading this section, you should be able to:

- Summarize Hubble's three scientific arguments regarding the size, expansion, and age of the universe
- Define *reasoning, inference, premise, conclusion,* and *argument* and describe the relationships among these concepts
- Describe the nature of a valid deductive argument

How old is the universe?

In Chapter 1, we mentioned Clair Patterson's involvement in the campaign to remove lead from gasoline because of lead's disastrous health effects. Patterson was a geochemist; using radiometric dating methods on elements like uranium and lead, he correctly calculated that our planet Earth is 4.5 billion years old. If the Earth is that old, how much older is the universe itself?

This question has been asked for a very long time, and answers to it have, of course, changed with advancing knowledge. The philosopher Aristotle argued that the universe must be eternal. He reasoned that everything which comes into existence requires some preexisting matter from which it comes. (You may have heard of the saying *ex nihilo nihil fit,* which is the similar idea that from nothing comes nothing.) This implies that if the universe came into existence, then it came into existence from some preexisting matter. But this is not possible if the universe is the totality of what there is. Further, any preexisting matter must have itself come from some other prior matter. The result is an infinite regress: no starting point is possible. Aristotle concluded that the universe must be eternal.

From the early Middle Ages to the end of the Renaissance, roughly the 7th through 16th centuries, scholars and theologians in Europe, the Mediterranean Basin, and the Middle East continued to engage with questions about the age of the universe. The structure of Aristotle's reasoning was largely kept, but the eternality of the universe was replaced by the eternality of God in order to fit with ideas about religious creation.

DOI: 10.4324/9781003300007-7

Some theologians estimated the universe to have come into existence around 4,000 BCE—that is, about 6,000 years ago. The estimate was derived from arithmetical calculations from genealogical records in various religious texts.

By the 19th century, many scientists believed that the universe—whatever its age—is in a steady state. In the 1920s, Edwin Hubble made two discoveries that were inconsistent with that belief: first, that the universe is much larger than previously thought and, second, that the universe is expanding. These discoveries also provided new evidence about the age of the universe.

FIGURE 6.1 Edwin Hubble at Mt. Wilson Observatory

Using a telescope with a 2.5-meter aperture at Mount Wilson Observatory in Southern California, Hubble observed Andromeda, an astronomical entity visible to the naked eye and identified over 1,000 years ago. He saw stars similar to those nearer to Earth, only dimmer. One of those stars is called a *Cepheid variable*, a star whose brightness changes periodically. The period of time it takes a Cepheid's brightness to change is related to the star's luminosity, the amount of energy it emits in one second. From the star's periodicity and that relationship, Hubble could calculate the star's luminosity, thereby determining how much brighter it was than the Sun.

Hubble then used his knowledge of the speed of light, which Einstein had recently shown to be constant, and the apparent brightness of the star compared to what he had calculated to be its luminosity to calculate the star's distance from Earth. Based on the distance of that star, Hubble reasoned that Andromeda is actually a different galaxy from our Milky Way galaxy. This discovery of a separate galaxy, announced in 1925, confirmed that the universe is much larger than previously thought.

Hubble also discovered that the universe has not always been this large. It's expanding. Like sound, light changes its frequency depending on the relative movement of the object emitting it and the observer. An example is the change in frequency of an ambulance siren as it moves toward, and then away, from an observer. The siren's pitch sounds higher as it approaches, and then lower pitched as it passes. This frequency change, called the *Doppler effect*, was discovered in the mid-19th century by the Austrian physicist Christian Doppler. For Hubble's purposes, the important implication is that a star moving away from Earth appears redder, while a star moving toward Earth appears bluer. The degree of redness of receding stars is called *redshift*.

Using the technique of astronomical spectroscopy, Hubble discovered that the redshift of starlight from any galaxy increased in proportion to the galaxy's distance from Earth. In 1929, he announced his discovery that galaxies are moving farther and farther away from Earth. By implication, the universe itself is expanding. According to recent estimates, the universe's expansion rate, known as *Hubble's constant*, is about 68 kilometers per second per megaparsec (km/sec/Mpc), where one megaparsec (Mpc) is approximately three million light-years—an extremely long distance!

How do Hubble's discoveries about the size and expansion of our universe help determine its age? As we've seen, knowledge of the speed of light and a star's brightness can be used to calculate the distances to faraway stars and galaxies. This distance measurement in turn indicates the amount of time the star must have been producing light in order for that light to reach Earth. Observing light from very distant stars and galaxies thus indicates their minimum age as well as the minimum age of the universe containing them. Using this reasoning, Hubble was able to show that the universe was at least 10 billion years old.

More recently, revised estimates suggest that the universe is approximately 13.8 billion years old. In 2020, astrophysicists used the 6-meter-aperture Atacama Cosmology Telescope in Chile to confirm this estimate and give a more precise estimate of the value. This finding has also been supported by convergent evidence from sciences like cosmological physics and geochemistry. The newest successor to the Hubble telescope, the James Webb Space Telescope, is set to provide additional confirmation of the universe's age.

Reasoning, inference, and argument

In past chapters, we have seen that empirical evidence is essential for developing scientific knowledge. But for observations to lead to knowledge, scientists must assess their significance and implications, and the relationships among them. In other words, scientific knowledge comes not from mere observation but from reasoning about observations. Hubble combined empirical observations with calculations of light's travel over distances and through time, established scientific knowledge about the Doppler effect, and other resources to develop arguments in favor of his conclusions regarding the universe's size, its expansion, and its age. Hubble appealed to empirical evidence in ways Aristotle did not, but like Aristotle, he also had to reason his way to conclusions.

This chapter and the next focus on inference patterns in scientific reasoning. **Reasoning** is a cognitive process of drawing inferences in support of some conclusion. An **inference** is a logical transition from one thought to another. Inferences move from **premises**, statements that provide support, to **conclusions**, statements that are supported by the premises. Inferences from premise(s) to conclusion(s) can be depicted as abstract derivations obeying rules that capture the logical transition(s) involved. So, while reasoning is a cognitive process, inferences can be thought of as simply logical relationships.

Reasoning processes can be made explicit by depicting inferences explicitly in an argument. While "argument" means different things in different contexts, here an **argument** is a set of statements in which some (premises) are intended to provide rational support or empirical evidence in favor of another (the conclusion). Recall Aristotle's reasons for concluding that the universe is eternal. These reasons can be reconstructed into the following argument:

(1) Everything that comes into existence must come from some prior material substrate.

∴ (2) If the universe came into existence, then the universe must have come from just such a prior substrate.

(3) It cannot be the case that some prior material substrate existed before the universe.

∴ (4) It cannot be that the universe came into existence.

∴ (5) The universe is eternal.

The argument is an ordered list of statements. The first four statements are the premises; the argument's conclusion is the last statement. Statements inferred from one or more premises are sometimes marked with the symbol ∴, which is notation for words like *therefore*, *so*, or *hence*. As this example shows, an argument may involve more than one inference. Here the second statement is inferred as an application of the first statement. The fourth statement is inferred from the second and third statement together. And the final conclusion is inferred from the fourth claim.

Scientific reasoning is similar to ordinary everyday reasoning but focused specifically on developing natural explanations for natural phenomena. Arguments in science also tend to incorporate empirical evidence. And then, scientific reasoning also tends to be more explicit than everyday reasoning, in part because scientists need to be able to assess each other's reasoning.

Logic is the study of the rules and patterns of inference, which is crucial for evaluating scientific reasoning. Scientific reasoning can follow three patterns of inference: deductive, inductive, and abductive. The evaluation of each of these patterns of inferences focuses on two main questions: first, do the premises rationally support the conclusion? And second, are those premises true? The premises of a good inference should provide a logically compelling reason for thinking the conclusion either must be true or is likely to be true. In a deductive argument, the truth of the premises should guarantee the truth of the conclusion, while in inductive and abductive arguments, the premises provide support for the conclusion but do not guarantee its truth. This chapter focuses on deductive inference; Chapter 7 addresses inductive and abductive inference.

Again, in a ***deductive argument***, the truth of the premises should guarantee the truth of the conclusion. If so, the argument is said to be valid. In nontechnical uses, "valid" can mean that something is believable, and, in Chapter 3, we discussed the external and internal validity of experiments. Here, in the context of deduction, *validity* has a different meaning. In a ***valid inference***, the truth of the premises logically guarantees the truth of the conclusion. For a valid deductive argument or inference, it is impossible for the conclusion to be false provided that all premises are true.

EXERCISES

6.1 Recall: How did Hubble discover that galaxies are moving farther and farther away from Earth? Describe the observation(s) he made supporting this discovery and his reasoning based on the observations.

6.2 Recall: Briefly summarize Hubble's scientific arguments regarding the size and age of the universe. For each of the two, start by identifying the conclusion, and then piece together his reasoning for that conclusion. You don't need to put these in explicit premises that comprise a valid deductive argument, but you are welcome to use that approach if it is helpful.

6.3 Think: Aristotle reasoned that the universe did not become and has always been. His argument for that conclusion as enumerated in statements (1)–(5) in this section is a valid deductive argument; so if all the premises are true, the conclusion must be true. But we know the conclusion is false, and so at least one of the premises must be false. Which of the premises might be false? Describe your reasoning.

6.4 Apply: Hubble discovered that the universe is expanding from changes in redshift. Describe what changes he observed. If, in the future, the universe began to contract, what changes in redshift should scientists expect to see? How long after the universe began to contract would scientists see this change?

6.5 Recall: Define *reasoning*, *inference*, *premise*, *conclusion*, and *argument* and describe the relationships among reasoning, inference, and argument, as well as the relationships among premise, conclusion, and argument (or inference).

6.6 **Recall:** What is required for a deductive argument to be valid? Give a simple example of a valid deductive argument—about anything!—numbering the statements and indicating conclusion(s) with ∴.

6.2 RULES OF DEDUCTIVE INFERENCE

After reading this section, you should be able to:

- Define *conditional statement, antecedent,* and *consequent* and indicate the logical relationships among them
- Identify these common patterns of valid and invalid deductive inference: affirming the antecedent, denying the consequent, affirming the consequent, and denying the antecedent
- Assess an argument's validity and soundness and, for an invalid or unsound argument, whether the argument should be revised or abandoned

Conditional statements

An important component of many deductive inferences is the ***conditional statement***, statements in which one circumstance is given as a condition for another circumstance. These are often thought of as if/then statements: if you eat your vegetable, then you can have dessert. If a star is 13.8 billion light-years away, then the universe is at least 13.8 billion years old. The first circumstance (following "if") is the antecedent, the condition for the other circumstance, while the second circumstance (following "then") is the ***consequent***, the circumstance arising from the antecedent.

You can think about the logical relationship between antecedents and consequents in terms of requirements and guarantees—or, more formally, necessary and sufficient conditions. In a true conditional statement, the antecedent is a ***sufficient condition*** for the consequent—a condition that, if met, guarantees the occurrence of the specified outcome (here, the consequent). Consider the true conditional statement: "If Lu is a dog, then Lu is an animal." Lu being a dog guarantees that Lu is also an animal; being a dog is sufficient for being an animal. But this doesn't work in reverse. In a true conditional statement, the consequent occurring doesn't guarantee the antecedent will occur. Lu being an animal doesn't guarantee that he's a dog; perhaps he's an alligator or a velociraptor. Instead, the consequent is a ***necessary condition*** for the antecedent—a condition that must be satisfied for the occurrence of the specified outcome (here, the antecedent). Lu being an animal is a necessary condition for Lu to be a dog: there's no way Lu is a dog if he's not an animal.

The meanings of *antecedent* and *consequent* relate to logical priority, not temporal succession. Sometimes an antecedent occurs before its consequent in time, as when a child has to eat their vegetable before being allowed to eat dessert. It's not good enough for the child to promise they'll eat the vegetable later. But, for the statement "if Lu is a dog, then Lu is an animal," being a dog doesn't come before being an animal; it's just that being an animal is a consequence of being a dog. The time ordering of antecedents and consequents can also be reversed. In the statement "if you are still hungry, then you must not have eaten enough dinner," the antecedent is logically prior, even though the consequent (not eating enough dinner) happened before the antecedent (still being hungry).

Box 6.1 Conditional statements

Scientific inquiry and everyday reasoning often involve the formulation and evaluation of conditional statements, like *if A then C*. But conditional statements often are expressed in nonstandard forms. Instead of *if A then C*, one might say (where A is still the antecedent and C the consequent of the conditional):

> *C if A*
> *A only if C*
> *A guarantees C*
> *Without C, A is not the case*
> *Not A unless C*

And more. Here are some tricks for navigating nonstandard forms of conditional statements:

1. Identify the candidates for being antecedent and consequent: what is being related in the claim?
2. A consequent states a necessary condition for the antecedent, or gives a requirement.
3. An antecedent states a sufficient condition for the consequent, or gives a guarantee.
4. Try to translate the statement into a standard if-then conditional and check if the same circumstances make the sentence true.

Recall that a conditional statement is only false when the antecedent can be true while the consequent is false; in all other cases, the conditional is true. For example, if parents tell their child: "if you eat your broccoli, then you'll get dessert," and she eats her broccoli, but then they withhold dessert, they were lying. (Which isn't a very nice thing to do to a little kid who wants dessert!) But if she doesn't eat her broccoli, they've made no promises about whether she'll get dessert; what the parents told her is true regardless. This kind of conditional, our focus in this chapter, is called a ***material conditional***.

To summarize, in a conditional claim, the **antecedent** is a condition that guarantees some consequence; it is logically prior to the consequent. The **consequent** is the condition that arises from, or is guaranteed by, the antecedent. In an if/then statement, the antecedent follows "if" while the consequent follows "then."

When Hubble calculated that distant stars were more than 10 billion years old, he knew that the universe itself is at least that old. This can be stated as a conditional: If some stars are more than 10 billion years old, then the universe must be more than 10 billion years old. In other words, a sufficient condition for the universe's being 10 billion years old is that some stars are 10 billion years old. If the universe were younger, it couldn't contain any objects that old. The universe being 10 billion years old is thus a requirement for any star to be that old. But again, this doesn't work in reverse: learning that the universe is a certain age would not guarantee that any star is that old. It's possible for the oldest stars to be younger than the universe containing them. The universe's having a given age is a necessary condition for there to be a star of that age, but it's not a sufficient condition.

Valid and sound arguments

Recall from section 6.1 that, in a valid deductive argument, the truth of all premises guarantees the truth of the conclusion. In other words, the truth of all premises is a sufficient condition for the truth of the conclusion. Deductive arguments for which this is not so are said to be invalid. An invalid deductive argument does not always have a false conclusion. This just means the truth of all premises does not guarantee the conclusion is true. Likewise, a valid deductive argument does not always have a true conclusion. Recall from earlier that Aristotle gave a valid deductive argument for the conclusion that the universe is eternal, but we now believe that conclusion to be false. A valid deductive argument with a false conclusion must have one or more false premises—because if the premises were all true, the conclusion would be as well (or else it's not a valid argument).

So, to assess whether an argument or inference is valid, assume or stipulate that all premises are true, and then ask whether it's possible for the conclusion to be false. If not—if the truth of the premises alone guarantees the truth of the conclusion—the inference is valid. But if this is possible, the inference is invalid. Deductive inferences cannot be made more valid, or rendered invalid, by adding more premises. Deductive inferences are thus **monotonic inferences**: inferences that cannot be invalidated by the addition of new information. Because deductive reasoning is monotonic, it is fully secure in the sense that, if your inferences are valid, you can be absolutely certain that true premises guarantee a true conclusion. (As we will see in Chapter 7, this is not so for inductive or abductive arguments.)

Some patterns of deductive inference are common enough to have names. One such pattern is **affirming the antecedent** (also known as **modus ponens**), a valid pattern of deductive inference in which a conditional statement and its antecedent are used as premises for concluding the consequent is true. For example,

(1) If the James Webb Space Telescope is at the second Lagrange point, then it will orbit the Sun at 1.5 gigameters from Earth.
(2) The James Webb Space Telescope is at the second Lagrange point.
∴ (3) It will orbit the Sun at 1.5 gigameters from Earth.

(A Lagrange point is a location where an object is equally influenced by two orbiting bodies, in this case the Sun and Earth.) The conditional statement of the relationship between the second Lagrange point and orbital distance holds regardless of where the telescope is in fact located; the observation of the telescope's location at this Lagrange point then makes the antecedent of the conditional true such that the consequent also must be true. The conclusion is, thus, just a statement asserting the consequent.

Another elementary pattern of deductive inference is ***denying the consequent*** (also known as ***modus tollens***), a valid pattern of deductive inference in which a conditional statement and the denial of its consequent are used as premises for concluding the antecedent is also false. For example,

(1) If the universe is in a steady state, then astral bodies remain the same distance from one another.
(2) Astral bodies do not remain the same distance from one another.
∴ (3) The universe is not in a steady state.

In this argument, the conditional statement relating the universe's state to relative distances of astral bodies enables an observation of increasing distance between astral bodies (as evident from changes in redshift) to be used to conclude something about the universe—namely, that it is not in a steady state but expanding.

FIGURE 6.2 Illustration of the James Webb Space Telescope, showing its mirrors and instruments

Any arguments following these two patterns are deductively valid. The premises may not be true, of course. But if they were true, they would logically guarantee that the conclusion must also be true. No matter how long and deep you think, you will not be able to find an instance of either pattern that is invalid.

Keep in mind that a valid deductive argument is more than premises and a conclusion that are all true. To have a valid inference, the truth of the premises must logically force the conclusion to be true; there must be no way around having a true conclusion if the premises are true. Consider this argument:

(1) Cats are mammals.
(2) Tigers are mammals.
∴ (3) Tigers are cats.

Both premises are true, and so is the conclusion. But the inference is invalid. Even though all three statements are true, the truth of the conclusion isn't guaranteed by the truth of the premises. To see this, substitute in *dogs* for *cats* in the argument; the two premises will still be true, but not the conclusion.

Box 6.2 Different kinds of logic

Formal logic is the study of the rules governing what can validly be inferred from a set of claims. The study of logic is deeply related to mathematics and computer science, where researchers also explore formal relations between sets of claims. Two rules of deductive logic we've encountered are affirming the antecedent (*modus ponens*) and denying the consequent (*modus tollens*). Both rules are valid, since the truth of the premises guarantees the truth of the conclusion. Formal logic employs fully precise rules like these. Yet natural languages (English and all other languages used around the world) are not fully precise. Ideas expressed in natural language can sometimes be interpreted in different ways. This means there are choices to make in how to define the precise rules of formal logic. Different choices result in different systems of formal logic. For instance, the most common logic system, called *classical logic*, has an assumption called *the law of the excluded middle*. This law stipulates that, for every claim, either the claim is true or its negation is true. For example, if it is not true that the grass is green, then it *is* true that the grass is not green. In many cases, the excluded-middle assumption accurately reflects how we reason in natural language. But sometimes it seems to go wrong. For example, it's not obvious that, if it's not true that it's raining, then it's definitely not raining. Perhaps it's sprinkling out, but we wouldn't yet call it rain. Other systems of logic make different assumptions to recognize this possibility, making it possible for a sentence to be neither true nor false or a sentence to be both true and false. These systems of nonclassical logic thus reject the law of the excluded middle.

So invalid arguments can have true premises and conclusions, while valid arguments can have false premises and conclusions. Successful deductive inferences are those that combine both validity and truth. A **sound inference** is a valid deductive inference with all true premises. If you know both that all premises are true and that the inference is valid, then you know that the conclusion must be true. No additional evidence or reasoning can change that. If it does, then either you didn't actually have a valid deductive inference, or you didn't actually know that all the premises are true. (Or something changed such that a premise that had been true became false.) If scientists know some inference is sound, they can be certain that the conclusion is true beyond a shadow of a doubt.

Uncovering bad arguments

Being persuaded is a psychological phenomenon. People can fall for bad arguments or may be unpersuaded by good ones. But whether a deductive inference is good—valid and sound—is simply a matter of logic and truth.

A valid deductive argument is unsound if one or more of its premises is false. An argument is invalid, as we've seen, if the truth of the premises does not guarantee the truth of the conclusion. You can show that an argument is invalid by giving a **counterexample**: a situation, real or imagined, in which the premises of an argument are true but the conclusion false. Even if the situation could never really occur, describing it shows the argument is invalid, as the premises being true alone can't guarantee the truth of the conclusion.

To evaluate a deductive argument, one should determine whether either of these criticisms apply: is the argument invalid? Is one or more premise false? Here, psychological reasoning intersects with logical inference. If an argument is faulty on one or both of these grounds, consider whether it can be repaired by replacing any false premises with true ones or whether additional premises could be supplied that render the inference to the conclusion valid. Sometimes promising arguments are not yet successful. Other times, an argument is confused, misleading, or just wrong and simply should be rejected.

The valid inference patterns involving conditional statements discussed earlier—affirming the antecedent and denying the consequent—have related invalid inference patterns that result from confusing the roles of necessary and sufficient conditions in conditional statements. **Denying the antecedent** is an invalid pattern of deductive inference in which a conditional statement and the denial of its antecedent are used as premises for concluding the consequent is also false. Here is an argument that commits the error of denying the antecedent:

(1) If a star is more than 15 billion years old, then so too is the universe.
(2) No star is more than 15 billion years old.
∴ (3) The universe is not more than 15 billion years old.

This is an invalid argument. Even if the first two premises are true, that's no guarantee that the conclusion must also be true. The first premise indicates that the age of the oldest star is just a minimum age for the universe. If the conditional statement is true, the antecedent guarantees the consequent but not the other way around. So, denying the antecedent, as the second premise does, provides no good reason to believe the consequent is the case, but it doesn't demonstrate that the consequent is not the case, either.

Affirming the consequent is an invalid pattern of deductive inference in which a conditional statement and its consequent are used as premises for concluding the antecedent is true. Here is an argument that commits the error of affirming the consequent:

(1) If Andromeda is 13.8 billion light-years away, then the universe is at least 13.8 billion years old.
(2) The universe is 13.8 billion years old.
∴ (3) Andromeda is 13.8 billion light-years away.

This is also an invalid argument. Both premises are true, but, again, they don't guarantee the truth of the conclusion. Some specific astral body viewed from Earth being 13.8 billion light-years away does imply that the universe is at least 13.8 billion years old, but the universe being that age doesn't guarantee the age of any particular astral body. In fact, Andromeda is around 2.5 million light-years away.

Box 6.3 The Wason selection task

Suppose you see four cards on a table. You know the cards have letters on one side and numbers on the other, and the sides you see show these symbols:

A   K   4   7

Consider this claim: "If a card has a vowel on one side, then that card has an even number on the other side." Which card(s) do you have to flip over to evaluate whether the claim is true? Briefly write down which of the cards you would flip.

   If you are like most people, you chose the card showing the A and maybe also the card showing the 4. But, according to the rules of deductive logic, to test a conditional claim, you should look for circumstances where the antecedent is true but the consequent is false—so you should have chosen the A and the 7. Rules from logic are sometimes used as standards of rationality in psychology research. While these standards are useful to generate predictions of how idealized agents *should* reason, it's a matter of debate whether deviations from these standards are a sign of irrationality. A common example comes from people's answers to this problem, called the *Wason selection task*. Deviations between ideal standards of rationality and real human reasoning raise fundamental questions about what it means to be rational and whether deductive logic provides us with the right normative standards for interpreting human reasoning. Interestingly, most people get the Wason selection task right if the question relates more to everyday experience. When you ask which of the following people's IDs should be checked to ensure no one under 21 drinks alcohol, most get the answer right—and the deductive logic is exactly the same.

16   21   Beer   Soda

The defects in reasoning seen thus far are with the form of the inference. But sometimes the problem with an inference is empirical rather than logical. Sometimes, even when an argument is valid, the world doesn't cooperate with the statements made about it. This is one place where the detective work of science comes in. For example, consider the following argument:

(1) The word *atom* means indivisible.
(2) If the word *atom* means indivisible, then atoms are indivisible.
(3) If atoms are indivisible, then atoms are the smallest type of matter.
∴ (4) Atoms are the smallest type of matter.

This argument involves multiple instances of affirming the antecedent—a valid inference pattern. Given the first two premises, it follows that atoms are indivisible; and from the conjunction of that claim with the third premise, it follows that atoms are the smallest type of matter. This argument is valid. The problem, of course, is that scientists have discovered particles even smaller than atoms. Electrons were discovered in 1897, followed by the discoveries of protons, neutrons, neutrinos, positrons, muons, bosons, and hadrons, which are all smaller than atoms. These discoveries show the conclusion to be false: atoms are not the smallest type of matter. So, the argument is unsound. But because the argument is valid, learning that the conclusion is false also tells us that at least one of the three premises is also false. Can you figure out which one?

Bad reasons to reject inferences

Being wary of invalid inference patterns and potentially false premises can help detect bad arguments. But these logical and empirical reasons to challenge some arguments should be distinguished from the negative psychological reactions that some arguments can evoke. For instance, whether someone finds the conclusion of an argument distasteful, offensive, or disagreeable is irrelevant to whether that conclusion is true. The conclusion that global warming is caused by human activity is politically and financially inconvenient for fossil fuel industries. This inconvenience has motivated them to try to subvert the scientific research leading to this conclusion, and they've been incredibly successful at sowing doubt about climate change. Taking a leaf from the Big Tobacco playbook used to instill doubt about cigarettes causing cancer in the mid-20th century, their disinformation campaigns often point to the mere occurrence of disagreement as a reason for doubting climate change, regardless of the overwhelming evidence to the contrary.

Some scientific reasoning can be counterintuitive or difficult to understand. By itself, this isn't grounds for rejection. People with limited exposure to evolutionary theory may find it difficult to imagine the course of evolution from single-celled organisms to humans. Similarly, without training in physics and cosmology, it can be difficult to comprehend the universe being billions of years old and expanding out from an initial Big Bang. But evolutionary theory and cosmological research provide solid grounds for accepting the truth of these bewildering claims—based on convergent evidence from many sources and sound arguments developed from that data. Just as a claim's intuitiveness isn't a guide to its truth, an argument's difficulty or complexity is irrelevant to the goodness of its inferential structure. Likewise, concluding that intelligent design must be true because there is no evidence proving that there is no supernatural intelligent creator is a bad argument; this is called an ***appeal to ignorance***.

Some people contend that the scientific estimate of the age of the universe is "just an opinion." Similarly, skeptics of evolutionary theory love to claim that it is "just a theory" that biological species evolved from a common ancestor. These are poor objections. Natural phenomena, and natural explanations of those phenomena, are not simply a matter of opinion. And scientific theories are developed on the basis of a tremendous amount of confirming evidence and careful inference. These criticisms are not based on disagreements about evidence or the logic of arguments but instead appeal to the trivial fact that people have different ideas about some things. Ideas that are supported by evidence and sound inference should be taken seriously.

EXERCISES

6.7 **Recall:** Give an example of a conditional statement. Label the antecedent and the consequent, and state which is a necessary condition and which is a sufficient condition.

6.8 **Think:** The following statements concern necessary and sufficient conditions. For each statement, rephrase it as a standard if/then conditional statement and state whether it's true or false.

  a.   Being a mammal is a sufficient condition for being human.
  b.   Being human is a sufficient condition for being an animal.
  c.   Being alive is a necessary condition for having a right to life.
  d.   Being alive is a sufficient condition for having a right to life.
  e.   Having a PhD is necessary to become a scientist.
  f.   It's sufficient for being awarded the Nobel Prize in immunology that one cures cancer.

6.9 **Recall:** Write out an example of each of these argument patterns in standard form (numbered statements with the conclusion last, introduced by "∴") and then label each as valid or invalid: affirming the consequent; denying the antecedent; denying the consequent; affirming the antecedent. For the valid patterns, give a short justification for the validity. For the invalid patterns, give a counterexample.

6.10 **Apply:** Review Hubble's scientific arguments about the size, expansion, and age of the universe in section 6.1. Identify an inference that fits the pattern of affirming the antecedent and an inference that fits the pattern of denying the consequent. Write each out in standard form. (One option: consider inferences leading to the conclusions that (a) the universe is expanding and (b) Andromeda is not part of our galaxy.)

6.11 Recall: List at least two good reasons to reject inferences and at least three bad reasons to reject inferences, providing justification for each being good or bad.

6.12 Apply: The following argument is invalid, but if one premise is added, it becomes a valid argument. Add a premise that makes the argument valid, and then assess the quality of the revised argument, considering the truth of each premise and the pattern of each inference.

   (1) If a star is 13.8 billion light-years away, then the universe is at least 13.8 billion years old.

   (2) If the universe came into existence 6,000 years ago, then it is not 13.8 billion years old.

∴ (3) The universe did not come into existence 6,000 years ago.

Should this argument be preserved in its revised form with the added premise? Why or why not?

6.3 DEDUCTIVE REASONING IN HYPOTHESIS-TESTING

After reading this section, you should be able to:

- Define *hypothetico-deductive method*
- Analyze an instance of hypothesis-testing using the H-D method
- Describe how the Duhem-Quine problem complicates the H-D method

The hypothetico-deductive method

Hypothesis-testing is a central component of scientific research. Chapters 3–5 focused on several strategies for gaining empirical evidence to test hypotheses, but in this chapter, we've seen that testing hypotheses also requires rational inference to see how empirical evidence bears on hypotheses. One way of thinking about the inferential relationship between evidence and hypotheses is the hypothetico-deductive method, or H-D method. This applies the logic of deductive inference to hypothesis testing.

In general, hypothesis-testing involves establishing expectations from a hypothesis, and then comparing those expectations with observations. For the **hypothetico-deductive method**, an expectation is deductively inferred from the hypothesis, and then the expectation is compared with an observation. Violation of the expectation deductively refutes the hypothesis, while a match with the expectation nondeductively boosts support for the hypothesis.

The first use of deductive inference in the H-D method is in the relationship between hypothesis and expectation. This involves determining what to expect if the hypothesis under investigation is true. That relationship can be expressed as a conditional statement with the hypothesis as the antecedent and the expectation as the consequent: "if *H*, then *E*." We don't yet know whether the hypothesis is true, but we do know that *if* the hypothesis is true, then the expectation will be true. This conditional statement can be thought of as an answer to the question: "if this hypothesis is true, what must the world be like?" Recall Hubble's reasoning about the size of the universe. The then-accepted hypothesis that the universe is in a steady state leads to the expectation that there would be no pattern in the redshift of stars: if the universe is in a steady state (*H*), then there's no pattern in stars' redshift (*E*).

After deductively inferring an expectation from a hypothesis, scientists make observations to compare with the expectations. For the H-D method, deductive inference plays a role in this step as well. If what is observed doesn't match what was expected, this enables a deductive argument for the conclusion that the hypothesis is false. That is, the observation violates the expectation, which refutes the hypothesis—proves it to be wrong. This is called **refutation** and can be represented this way:

Refutation

    (1) If *H*, then *E*
    (2) Not *E*
∴ (3) Not *H*

The applicable inference pattern is denying the consequent, which we've learned is a valid form of deductive inference. Scientists whose observations defy their expectations should reason that something is amiss. If a hypothesis guarantees some expectation will occur, and then the observation violates the expectation, one can deductively conclude the hypothesis is false. In other words, the mismatch between expected and actual observations refutes the hypothesis. When Hubble observed a distinctive pattern in stars' redshift—those further away had more redshift—this enabled him to reject the hypothesis that the universe was in a steady state.

Sometimes, of course, observations do match the expectations. This is a good sign about the hypothesis. Careful, though! If the observations and expectations match, and we conclude the hypothesis must be true, this follows the inference pattern of affirming the consequent, which is an invalid form of deductive inference. So, if observations match expectations, this does not enable a valid deductive argument for the hypothesis. Think of it this way. A match between expectations and observations is consistent with the truth of the hypothesis, sure, but it does not guarantee the truth of the hypothesis. We can think of this nondeductive inference as **confirmation**: the observation matches the expectation based on the hypothesis, providing evidence in favor of the hypothesis. Confirmation can be represented this way:

Confirmation

    (1) If *H*, then *E*
    (2) *E*
∴ (3) Probably or possibly *H*

If the evidence matches expectations, the hypothesis is confirmed; if not, it's refuted. The H-D method identifies an important difference in the logic of confirmation versus refutation. Refutation of a hypothesis is deductively valid and thus certain, while confirmation of a hypothesis is not. This is why we included "probably or possibly" in the previous argument form for confirmation.

When Hubble observed a distinctive pattern in the redshift of stars, he hypothesized that the universe is expanding. That would lead us to expect that more distant stars have a greater redshift, as they are moving away from us more quickly than stars closer by. This expectation matches Hubble's observation, and so confirms the hypothesis. But perhaps there's some other reason for this observation. Scientists needed to look for additional evidence supporting the hypothesis that the universe was expanding.

A very simple example can help illustrate this point about the different logic of refutation and confirmation. Consider the hypothesis that all swans are white. If this hypothesis is true, then the swan you observe next will be white. This is a true conditional statement: the antecedent's truth guarantees the truth of the consequent. So, you go out looking for swans, with the expectation that, if your hypothesis is true, you will see a white one. Let's say you instead encounter a black swan. This observation violates your expectation; by denying the consequent, you've shown the antecedent (the hypothesis) is false. Breaking news: it's not the case that all swans are white! However, if the next swan you see is white, then your observation matches the expectation. You haven't proven anything, but you do have a bit more evidence in favor of the hypothesis.

The H-D method thus posits a crucial difference between refutation and confirmation. **Refutation** is a valid deductive inference proving the hypothesis is false. In contrast, the logical structure of **confirmation** is a deductively invalid inference. Thus, we are not warranted in concluding the hypothesis is certainly true but only that it probably or possibly is. An observation matching what a hypothesis leads us to expect usually provides some evidence for the hypothesis, but this isn't always so, and it's surprisingly tricky to articulate how this works.

The case of puerperal fever

The philosopher Carl Hempel famously invoked the story of Ignaz Semmelweis to illustrate the H-D method. Semmelweis was a doctor working in the First Maternity Division of the Vienna General Hospital in the 1840s. At that time, many birthing women contracted a serious—even fatal—illness known as puerperal or "childbed" fever. (*Puerperium* refers to the postpartum period following labor and delivery.) A puzzling observation was that the mortality rate in the First Maternity Division was approximately three times higher than in the adjacent Second Maternity Division. These rates are shown in Table 6.1.

What accounted for the difference between the two wards? Semmelweis made several interesting observations. Women with extremely long dilation periods died of puerperal fever more often; patients in the first ward fell ill in a sequential manner, one after another; and neither patients' health nor their caretakers' skill seemed related

TABLE 6.1 Annual births, deaths, and mortality rates for all patients at two clinics of the vienna maternity hospital (1841–1846)

Year	First Clinic			Second Clinic		
	Births	Deaths	Rate	Births	Deaths	Rate
1841	3036	237	7.7	2442	86	3.5
1842	3287	518	15.8	2659	202	7.5
1843	3060	274	8.9	2739	164	5.9
1844	3157	260	8.2	2956	68	2.3
1845	3492	241	6.8	3241	66	2.0
1846	4010	459	11.4	3754	105	2.7
Total Avg.	20,042	1989	9.92	17,791	691	3.38

to the incidence of puerperal fever. Finally, not only was the illness rate in the second ward lower, but women who instead gave birth at home or elsewhere outside the clinic—even unattended on the street—did not suffer from puerperal fever.

Semmelweis used these observations to rule out several possible sources of illness. If puerperal fever were a citywide epidemic, then women laboring outside the hospital would also suffer from the illness—but they didn't. If puerperal fever were triggered by psychological concerns during childbirth, like intense modesty from being medically examined by male doctors (as had been proposed), then women delivering in the streets should also experience puerperal fever—but they didn't. Further, all of the proposed sources of illness led to the expectation of equal rates of the illness in the first and second maternity wards, but that expectation didn't match observations. As in the H-D method of refutation, Semmelweis rejected these hypotheses about the origin of puerperal fever.

Semmelweis also tried to develop hypotheses that were consistent with the observed difference in puerperal fever rates between the two wards. One difference between them was male doctors and medical students staffed the first ward, whereas female midwives staffed the second ward. Women in the first ward also gave birth on their backs, while women in the second ward gave birth on their sides. Semmelweis changed procedures in the first ward so that all women there also gave birth on their sides. If giving birth on one's back increases incidence of the puerperal fever, then that changed position will decrease incidence of puerperal fever. Alas, this expectation didn't match Semmelweis's observation: changing birth position in the first ward made no difference. Other hypotheses were similarly tested and ruled out.

In March 1847, Semmelweis learned that his colleague Jakob Kolletschka had died. Kolletschka was a professor of forensic medicine and had been performing an autopsy when a scalpel had lacerated his finger. Kolletschka had then exhibited the same symptoms as the mothers and infants who died of puerperal fever. Although distraught, Semmelweis also recognized the significance of this information for his investigation. He hypothesized that the scalpel had contaminated Kolletschka's blood with "cadaverous particles," which had caused puerperal fever leading to his death.

Semmelweis's hypothesis fit with the observed difference in illness rates between the two wards: doctors and medical students performed autopsies, whereas midwives did not. If cadaverous particles from autopsies caused puerperal fever, then only births attended by doctors and medical students who had performed autopsies would result in puerperal fever. In fact, much higher rates were observed in the ward attended by doctors and medical students, confirming this hypothesis.

Semmelweis reasoned that if the hypothesis that cadaverous particles caused puerperal fever were true, then the illness could be prevented by eliminating the particles. To test his hypothesis, he required all doctors, medical students, and midwives to wash their hands in a solution of chlorinated lime prior to examining patients. If this made no difference, then cadaverous particles weren't to blame—the hypothesis would be refuted. Instead, the mortality rate from puerperal fever began to decrease, and the incidence of illness in the first and second wards became similar.

As described here, this case involves many instances of the H-D method, and it illustrates the logical difference between refutation and confirmation. On the H-D account, refutation is decisive because it results from a valid deductive inference, while confirmation is logically weaker. And, as it turns out, Semmelweis's confirmed hypothesis was incorrect: puerperal fever does not originate from cadaverous material but from uterine bacterial infection. Luckily, chlorinated lime has antibacterial properties. Semmelweis thought the prescribed handwashing worked because it removed cadaverous material; instead, it worked because it removed bacteria. (The role of bacteria in illness was not yet widely accepted.)

FIGURE 6.3 Frieze at the Social Hygiene Museum in Budapest honoring Ignaz Semmelweis

Some other important instances of hypothesis-testing are also well captured by the H-D method. In Chapter 3, we discussed Arthur Eddington's confirmation of Einstein's theory of relativity from the 1919 solar eclipse as a crucial experiment, decisively adjudicating between two hypotheses. This is because the experiment refuted Newton's cosmological theory while fitting with Einstein's new theory. Measuring how much light bends around the Sun enabled Eddington to refute Newton's theory and confirm Einstein's.

Auxiliary assumptions

The hypothetico-deductive (H-D) method captures something important about hypothesis-testing in science, namely, the distinctive power of refutation. Data that fit our expectations are well and good, but we can really learn something from data that go against our expectations. This accords with the importance of falsifiable hypotheses, too—hypotheses must be at risk of refutation in order to have the potential to be confirmed by evidence. The power of refutation is also what makes the idea of crucial experiments compelling, as with Eddington's test of Einstein's new theory of relativity against Newton's reigning theory.

However, the H-D method also has its limitations. One limitation is that inferences from hypotheses to expectations are never genuinely deductive. Additional claims are needed to make a deductive inference from hypothesis to expectation valid. These additional claims are ***auxiliary assumptions***: assumptions about how the world works that often go unnoticed but that are needed for a hypothesis or theory to have the expected implications. In Chapter 3, we stressed the role of background knowledge in shaping expectations; this is the same point. Lurking in the background of Semmelweis's inference about handwashing, for example, was the assumption that handwashing would remove cadaverous material. Behind Eddington's refutation of Newtonian physics were numerous assumptions about the behavior of instruments, the properties of light, the location of certain astral bodies, and more.

Auxiliary assumptions often go unnoticed. Sometimes, this is because an auxiliary assumption is assumed to be true; other times, no one has even noticed the assumption is needed. But, because valid deductive inference requires the premises to guarantee the conclusion, any auxiliary assumptions are essential for the deductive inference from a hypothesis to some expectation, a key component of the H-D method. But this also impacts the conclusions one can draw. So, the schemes identified earlier for refutation and confirmation on the H-D account need to be adjusted:

Refutation (1) If H and A, then E *Confirmation* (1) If H and A, then E
　　　　　(2) Not E　　　　　　　　　　　　　　　(2) E
　∴ (3) Not (H and A)　　　　　　　　∴ (3) Probably or possibly (H and A)

In this new formulation, A stands for whatever auxiliary assumptions are required as additional premises to validly deduce E from H. These may include background knowledge about the phenomenon under investigation, information about the reliability of experimental instruments and measurement procedures, and more. But taking into account auxiliary assumptions means that all we can deductively include in refutation is that it's not the case that both *H* and *A* are true—either the hypothesis is false or an auxiliary assumption is false (or both). Given the need for auxiliary assumptions, the H-D method cannot be used to conclusively determine that the hypothesis is false.

This problem relates to *underdetermination*, introduced in Chapter 3 as the idea the evidence may not be sufficient to determine which of multiple hypotheses is true. This variety of underdetermination is known as the **Duhem-Quine problem**, the idea that scientific hypotheses can never be tested in isolation but only against the background of auxiliary assumptions. The Duhem-Quine problem is named after the 19th-century French physicist, mathematician, and philosopher Pierre Duhem and the 20th-century American philosopher Willard van Orman Quine, who both emphasized this challenge.

One consequence of the Duhem-Quine problem is that deductive logic alone is insufficient for successful hypothesis-testing. In the face of refutation, scientists need to decide whether to give up on a hypothesis or to question their auxiliary assumptions. This seems to involve an element of choice. In the face of refutation, scientists may well want to hold on to a hypothesis they like and search for other explanations for why the observations didn't turn out as expected. This makes the logic of refutation more similar to that of confirmation: neither is fully decisive.

That said, there are some guidelines for whether to reject a hypothesis in the face of refutation or instead blame an auxiliary assumption. Usually, scientists have independent evidence for many of their auxiliary assumptions. Instruments and measurement procedures have been tested and calibrated, and background knowledge about a phenomenon is based on evidence or commonsense. These considerations can help scientists decide how much to trust auxiliary assumptions and thus whether, and when, to reject the hypothesis under investigation. A refutation may also spur additional tests of auxiliary assumptions before a verdict about whether to reject the hypothesis.

Nonetheless, the need for auxiliary assumptions limits the power of the H-D method of hypothesis-testing. The Duhem-Quine problem makes clear that, just like confirmation, refutation is messier than simple deductive inference.

EXERCISES

Watch Video 10 **6.13 Recall:** The name of the hypothetico-deductive method refers to hypotheses and deductive inference. Briefly summarize the H-D method. How does the H-D method relate to hypotheses? What are the two applications of deductive inference in the H-D method?

6.14 Recall: Write out the H-D schemes for refutation and for confirmation. What is the key difference between them, and how does that difference relate to deductive inference?

6.15 Apply: Return to the description of Semmelweis's investigation of puerperal fever. Identify three inferences that can be described as uses of the H-D method (either refutation or confirmation). For each, write out the inference as an argument in standard form with numbered premises and conclusion.

6.16 Think: You might think the problem Semmelweis uncovered was that doctors weren't washing their hands. But this isn't so; the doctors in Vienna General Hospital already had been washing their hands. Consider what you know about bacteria. If doctors were already washing their hands, what do you think the problem could have been?

6.17 Recall: What is the Duhem-Quine problem? How does this relate to underdetermination? Why does the Duhem-Quine problem complicate the H-D method?

6.18 Apply: Consider Semmelweis's inference that if he required all doctors, medical students, and midwives to wash their hands in a solution of chlorinated lime prior to examining patients and this made no difference to the incidence of puerperal fever, then cadaverous particles weren't to blame. List at least three auxiliary assumptions needed for this inference to hold.

6.4 AXIOMATIC METHODS AND MODELING

After reading this section, you should be able to:

- Characterize the axiomatic method and indicate how it is used in science
- Articulate how deductive inference is used in mathematical modeling
- Describe how non-Euclidean geometry and the James Webb Space Telescope exemplify roles of deductive reasoning in science

Axiomatic methods

Deductive inference plays special roles in some fields of science. Progress in scientific reasoning is sometimes achieved through formal axiomatization, a constructive procedure by which statements are derived from foundational principles. The foundational principles, called *axioms*, are accepted as self-evident truths about some domain. A set of axioms is then used to deductively infer other truths about the domain, called *theorems*.

The most venerable example of axiomatization comes from the Greek mathematician Euclid, working around 300 BCE. Euclid's *Elements of Geometry* begins with 23 definitions and five axioms. The five axioms are the following:

(1) A straight line may be drawn between any two points.
(2) Any terminated straight line may be extended indefinitely.
(3) A circle may be drawn with any given point as center and any given radius.

(4) All right angles are equal.
(5) If two straight lines in a plane are met by another line, and if the sum of the internal angles on one side is less than two right angles, then the straight lines will meet if extended sufficiently on the side on which the sum of the angles is less than two right angles.

Together, these five axioms serve as the theoretical core of Euclidean geometry. Figure 6.4 shows a fragment—one of the oldest on record—containing a diagram that accompanies the fifth axiom of Book II of the Euclid's *Elements*. Using this set of axioms as premises, one can validly deduce theorems about the congruency of figures, parallel lines, and other results of Euclidean geometry. In turn, these theorems can be treated as premises in new arguments aimed at validly deducing new theorems.

Euclid's axiomatization of geometry was accepted as decisive for almost two millennia. This is a clear example of rigorous inference grounded in first principles, with the power to systematize all existing knowledge of geometry. Euclid's axiomatization of geometry deeply influenced Ibn al-Haytham's work in optics and Newton's physical theory of mechanics.

And yet, beginning in the 19th century, nonclassical geometries were developed that diverge from Euclid's axiomatization. These arose from questioning the fifth axiom of Euclidean geometry. The fifth axiom is notably more complicated and less intuitive than the other axioms, and over centuries, many scholars tried to prove this principle as a theorem from the other axioms. None succeeded. Mathematicians working in the early 19th century then developed different geometries by introducing deviations from Euclid's fifth axiom. Just as Euclid's geometry was central to earlier

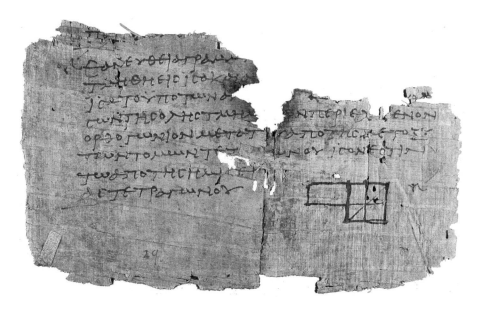

FIGURE 6.4 Fragment of papyrus with a diagram from Euclid's *Elements of Geometry*

physics and astronomy, these non-Euclidean geometries paved the way for Einstein's radical new theories of the relativity of space and time. Einstein's theories of relativity imply that the geometry of physical space itself is generally not Euclidean. A principle accepted as mathematical certainty for millennia was then rejected as not in general true. (Einstein's theories do entail that geometry here on Earth is approximately Euclidean, though.)

Another example of the axiomatic method concerns the foundations of arithmetic. Concerned with questions about the exact nature of numbers, the Italian mathematician Giuseppe Peano employed axiomatic reasoning to give a rigorous foundation for the natural numbers (that is, 0, 1, 2, 3, 4, . . .). Peano's axiomatization of natural numbers began with three primitive concepts not defined in terms of other concepts. Peano thought these primitive concepts are self-evident: the set of natural numbers, N; the number zero, a member of the set N; and the successor function S. This successor function can be applied to any natural number, and it will yield the next number after it. For example, $S(6) = 7$. Likewise, $S(0) = 1$. From here, Peano laid down several axioms:

(1) Zero is a number.
(2) If n is a number, then $S(n)$ is a number.
(3) Zero is not the successor of a number.
(4) Distinct natural numbers have distinct successors.
(5) If 0 is an element in a set of numbers and the successor of every number is in that set, then every number is in that set.

From these axioms, the basic properties of natural numbers could be described, and theorems about them, including arithmetic operations like addition and subtraction, could be deduced. To take a simple example, the supposition that there is a number preceding zero ($S(k) = 0$) would contradict axiom (3). Accordingly, the theorem that zero has no predecessor in N can be derived from axiom (3).

Deductive reasoning in mathematical modeling

This chapter began with a consideration of how Hubble reasoned from astronomical discoveries to conclusions about the size, expansion, and age of the universe. The newest chapter of the story of such cosmological discovery is the launch of the James Webb Space Telescope (JWST) in December 2021.

JWST is the successor to the Hubble and Spitzer Space Telescopes, and it is the most powerful deep space observatory ever built. Among its main goals is the delivery of data about mid-infrared light with wavelengths up to 28,000 nanometers; scientists can use that data to reason about the beginning of the universe—especially the epoch of reionization, when stars began to generate light. JWST will also investigate galactic and stellar formation and evolution, and it will measure atmospheric chemistry to search various planetary systems for the conditions that make life possible.

Constructing and launching JWST was an amazing feat of human ingenuity and effort. The endeavor is an excellent example of fruitful collaboration between vocational, scientific, and philosophical disciplines. One of the many challenges to be worked out was the path JWST should take after its launch.

Theoretical resources used to answer that question come from mathematician and astronomer Leonhard Euler and his student Joseph-Louis Lagrange, who in the 19th century explored mathematical solutions to what has come to be called "the three-body problem." In physics, this is the challenge of how the position, momentum, and gravitation of three celestial bodies affects the movement of each through space. There is no general solution to the three-body problem; there's no equation you can plug these values into and calculate the results. But a few special cases have been solved. These include when two of the bodies are motionless, and also when one of the three bodies is nearly massless relative to two massive orbiting bodies—such as a spacecraft, star, and planet.

Euler and Lagrange discovered five solutions for this special case of one nearly massless body and two massive orbiting bodies. These solutions are now known as *Lagrange points*. Lagrange points are regions of space where the gravitational pull of the two massive orbiting bodies is identical to the force needed for the nearly massless body to move with them. These points can be used to keep a spacecraft in position with less energy. In the Sun/Earth system, for the first three solutions (L1, L2, L3), the bodies are linearly aligned and the last two (L4 and L5) are the apices of equilateral triangles; collectively, these form what looks like a "peace" sign. Sending JWST to a Lagrange point offered a way to keep JWST "tethered" to the Sun and Earth, in a sense; that is, to use the gravitational force of those two bodies to speed JWST in its journey.

In January 2022, JWST journeyed to L2, which is approximately 1.5 million kilometers beyond the dark side of the Earth. In L2, JWST could harvest energy on its Sun-facing side, helping to reduce fuel needs, and this also enabled reliable communication back to Earth. But if JWST overshot L2, it would have to be abandoned or returned to Earth, exposing sensitive instrumentation to solar radiation and heat in the process. And if JWST undershot L2, it would not be able to get into a stable position. So, mid-course correction maneuvers were required to exactly reach L2. To determine the necessary course corrections, NASA scientists had to build models simulating the vectors of various thrusts in a state space. Computer simulations were used to help visualize these models, but the models themselves were essentially logical derivations from equations. These derivations enabled scientists to reason deductively about what consequences to expect from potential changes. Ultimately, scientists used these models to settle on three mid-course correction maneuvers: one immediately after launch for redirection, another just before sunshield deployment, and the final correction to insert JWST into its L2 orbit.

At L2, the force required to keep JWST in position was minimized by the balance between Earth's and Sun's gravitational pull. But our solar system is not actually a three-body system. There's also our moon's mass and gravitational pull to contend with, among other forces, and so the orbital stability of L2 is imperfect. Even minuscule deviations from equilibrium will tend to increase over time; think of a fishtailing trailer or skateboard that eventually crashes. If such deviations were left uncorrected, JWST

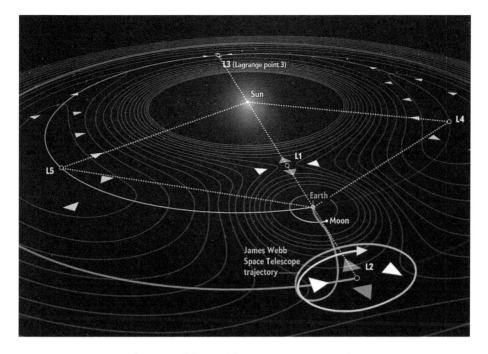

FIGURE 6.5 Depiction of system of five stable Lagrange points and JWST's trajectory

would just wander out of L2. So, the question rises anew: how can scientists maintain JWST's location over time while keeping it oriented so as to provide an unoccluded view of deep space?

NASA scientists modeled the range of possible perturbations of different orbital trajectories, examined what would result from these increasing changes without intervention, and then predicted the possible corrections to those curvatures that would allow them to maintain JWST's position. Again, these geometric and topological models involve deductive inference from various formulae to predict the consequences of various trajectories around L2. For these reasons, JWST is not actually at L2. The halo motion the scientists settled on has it moving in a very wide, tilted, and slow elliptical orbit around L2. These mathematical models also demonstrated the need for throttle adjustments or "burns" every three weeks to correct for things like rotational torque and other small perturbations. The calculations involved in determining the timing and duration of these burns again requires scientists to reason deductively from their models.

These theoretical resources and calculations required to support the position and movement of the James Webb Space Telescope in space illustrate another way in which deductive reasoning is used in science. In any scientific research, inferential reasoning is needed to connect empirical investigation to conclusions. But the role of deductive reasoning is even more central to some scientific research. This includes the mathematical modeling used to inform the management of JWST and the axiomatic methods at the heart of the development of Euclidean and non-Euclidean geometries.

EXERCISES

6.19 Recall: Describe the axiomatic method, including the roles of axioms and theorems. How does this method involve deductive reasoning? How has this method been used in science?

6.20 Apply: The triangle sum theorem in Euclidean geometry says that if you add all three interior angles—that is, the angles inside the triangle—they sum to 180 degrees. Look up a discussion of this theorem. Read the discussion, and then analyze the use(s) of deductive inference. Describe how the premises are used to infer the triangle sum theorem as conclusion. Include a citation of the source for the discussion you analyze.

6.21 Recall: Describe the three-body problem and how this relates to JWST.

6.22 Think: Why is it important that the reasoning involved in modeling the orbits of JWST be deductive?

6.23 Think: The examples of axiomatization and deductive reasoning in model building come from geometry, mathematics, and astrophysics. Why do you think examples like these come from those disciplines rather than, say, biology or sociology?

FURTHER READING

For NASA resources on the relevance of James Webb Space Telescope for science, see https://webb.nasa.gov/content/science/index.html.

For more on deductive and nondeductive reasoning, see Harman, G. (1986). *Change in view: Principles of reasoning*. MIT Press.

For more on contradictions and logic, see Priest, G. (2014). Beyond true and false. *Aeon Magazine*. https://aeon.co/essays/the-logic-of-buddhist-philosophy-goes-beyond-simple-truth.

For more on the hypothetico-deductive method, see Sprenger, J. (2011). Hypothetico-deductive confirmation. *Philosophy Compass*, 6(7), 497–508.

For more on axiomatic methods, see Cantu, P. (2022). What is axiomatics? *Annals of Mathematics and Philosophy*, 1, 1–24.

CHAPTER 7

Inductive and abductive reasoning

7.1 THE FLINT WATER CRISIS AND AMPLIATIVE INFERENCE

After reading this section, you should be able to:

- Describe inferences involved in uncovering the Flint water crisis
- Give examples of ampliative inference in science and everyday life
- List three features of inductive and abductive inferences

The water crisis in Flint, Michigan

Before 2014, Flint, an industrial city in the US state of Michigan, had received its water supply from neighboring Detroit. In April 2014, major budget deficits led city leadership to change the city's water supplier. The new supplier would provide water from the Flint River. The Flint River flowed directly through town, so it was a convenient source of water, but it had also long been polluted with industrial waste, sewage, and agricultural runoff. Immediately after the change, residents began to complain of discolored water with a foul odor and taste.

Clean drinking water is a basic human necessity, to which many people across the globe lack access. In Flint, the problem had several components. The Flint River had been badly polluted for decades, and the environmental cleanup of its contaminants and toxic waste had not been properly performed. The water previously supplied from Detroit was treated so that it wouldn't leach lead from lead supply lines, but the new water source from the Flint River wasn't treated. Further, city officials did not conduct testing as required by the US Environmental Protection Agency. And even though the problem was immediately noticeable to residents, officials ignored their concerns and repeatedly declared that the water was safe.

This led to a serious public health crisis. In the summer of 2014, there were outbreaks of *E. coli* bacteria and Legionnaires' disease, both caused by fecal contamination. Residents were instructed to boil water before drinking it, and water treatment changes were made. But the treatment wasn't conducted properly, leading to levels of chlorine byproducts that violated the Safe Drinking Water Act and resulted in high concentrations of carcinogenic trihalomethanes. But this wasn't the worst of it.

DOI: 10.4324/9781003300007-8

Officials ignored federal environmental regulations that required treatment of the water supply system with anticorrosive chemicals. The polluted water from the Flint River was highly corrosive; because it had not been treated, the corrosive water leached lead from lead supply lines into the water supply. Lead is extremely toxic and can have massively adverse health effects in everyone and especially in children. Lead concentrations in children's blood as low as five parts per billion (ppb) can result in decreased intelligence and behavioral and learning deficits.

Flint resident LeeAnne Walters' family and neighbors were suffering from rashes and extreme weakness, and her own hair was falling out. She and other frustrated community members carried jugs of brown tap water to community meetings. City officials continued to claim that the water was safe. In 2015, she had her water supply tested, and it was found to have high levels of lead. She then contacted Marc Edwards, an environmental engineer at Virginia Tech, who worked with his students to supply hundreds of testing kits to Flint residents and guidance for using them. Ultimately, 268 homes were tested, and the findings revealed that water from 85% of the homes contained lead levels over the minimum level to be reported; 17% exceeded the legal limit of 15 ppb, and 10% of the homes had water above 26.8 ppb. (Lead levels over 5 ppb is problematic, and 15 ppb is the legal limit for water.)

Another important scientific contribution was made by Mona Hanna-Attisha, a local pediatrician in Flint. Hanna-Attisha evaluated community health records and found that the percentage of young children tested to have elevated levels of lead in their blood had more than doubled in Flint since the changed water source. These different lines of research spurred by local community activists produced telling evidence that Flint's water supply was poisonous.

Once the environmental emergency was finally acknowledged, scientific research was also part of eventual remediation. One of the problems with testing was that Flint did not have reliable records of where lead service lines were located. In 2016, a group of computer scientists at the University of Michigan developed a machine learning model to predict the homes most likely to have lead pipes and thus in need of service line replacement. The model succeeded in 70% of its predictions, meaningfully improving lead pipe replacement efforts.

The Flint water crisis raises interesting and difficult issues about the relationship of science to the public good, the involvement of community members in scientific research, and the role of scientific trust and communication in designing effective public policy. We'll return to some of those topics later in this chapter and later in the book. For now, we want to draw your attention to the form of the inferences involved in determining the problem, and the extent of the problem, with Flint's water.

The aim of the water-testing research was to learn about the water quality across all houses in Flint, and the path to gaining that knowledge began with testing water quality in one family's house. The 268 homes eventually sampled is a large sample, but it's not enough for deductive reasoning. A deductive inference about the water quality in all houses in Flint would have required testing each one, but that wouldn't have been the best way to proceed during a public health emergency. Similarly, the pediatrician

FIGURE 7.1 a. Dr. Mona Hanna-Attisha visiting the University of Michigan School for Environment and Sustainability (left); b. Certified Medical Assistant Tamika Dukes tests blood samples of Flint residents for the presence of lead (right)

who raised alarm about increased lead exposure in children drew broader conclusions about more than just the children that had been sampled. Like much of scientific reasoning, the scientists investigating Flint's water instead used inductive reasoning to draw conclusions that went beyond the data they had collected.

Inductive and abductive inferences

It's sometimes said that the conclusions of deductive arguments are already contained in their premises. This saying means that the conclusions don't add any new content beyond what the premises provide, since the premises guarantee the conclusions—the requirement for any valid deductive argument. Most instances of everyday and scientific reasoning are not like this. Few scientific conclusions follow necessarily, or with any kind of logical certainty, from a set of premises.

In other words, much of human reasoning is nondeductive. Rather than beginning from general statements with universal scope or axiomatic first principles, scientists and nonscientists alike typically begin from incomplete information or data collected piecemeal. They draw tentative conclusions that may turn out to be false even when their premises are true. The community activists and scientists investigating Flint's water crisis proceeded nondeductively. Their conclusions went beyond what was guaranteed by water sampling results, the specific children who had been tested for lead exposure, and the houses known to have lead service lines.

Inductive inference is an inferential relationship from premises to conclusion that is one of probability rather than necessity. ***Abductive inference***, another type of nondeductive inference, is an inferential relationship that attributes special status to explanatory considerations. Abductive inference is also called "inference to the best explanation." Inductive and abductive inferences have three characteristics that distinguish them from deductive inferences: they are ampliative, non-monotonic, and have varying strengths.

First, inductive and abductive inferences are ampliative. An ***ampliative inference*** is an inference in which the conclusion expresses content that, in some sense, goes beyond what is present in the premises. The conclusion contains information that augments or amplifies the content of the premises. The conclusion that Flint's water supply was toxic went beyond the evidence provided by a limited number of samples from houses in Flint. Ampliative inferences are valuable insofar as they enable us to extend beyond what we already know. A downside is that the conclusions of ampliative inferences are not necessitated by the premises, and so they are not guaranteed to be true.

Second, inductive and abductive inferences are non-monotonic. Recall from Chapter 6 that a *monotonic inference* cannot be invalidated by the addition of new information. For non-monotonic inferences, then, new information may change whether or not the premises support the conclusion. Adding new premises to the existing premises of a non-monotonic inference with a true conclusion can render that conclusion false. For example, if the machine learning model predicted a group of houses in Flint had lead pipes, but then we learned that the model had recently fallen to only 20% accuracy, we would not have confidence in this prediction. One of the hallmarks of science is openness to criticism and to new evidence. Scientists can face surprising findings that lead them to abandon or update their ideas. This feature is well captured by the non-monotonic feature of inductive and abductive inferences. The addition of new information can reveal how a good inference from true premises may nevertheless be wrong.

Third, inductive and abductive inferences vary in strength, or in degree of evidential support. The **strength of an inductive inference** (or abductive inference) is the probability that the conclusion is true assuming that all the premises are true. The conclusion that all of Flint's water is toxic was strengthened when scientists sampled water from nearly 300 different homes, as compared to earlier inferences based only on problematic water samples from a single home. Deductive arguments are either valid or not, but strength comes in degrees: two arguments might both be strong, but one might be stronger than the other. Further, any inductive or abductive argument, no matter how strong, can be additionally strengthened. Sampling water from more houses or testing more children for elevated lead levels can make the respective arguments these sources of evidence relate to stronger.

EXERCISES

7.1 Recall: Summarize three uses of scientific investigation to help diagnose and resolve the Flint water crisis, and describe how each involved inductive inference.

7.2 Apply: One home in Flint had tap water tested to have lead levels of 217 ppb. How high is that amount of lead, compared to metrics discussed in this section? How serious is the problem (assuming people are drinking the tap water)?

7.3 Think: Why didn't scientists test water from even more homes in Flint? Wouldn't testing water from more homes provide more conclusive evidence?

7.4 Think: Consider decisions you've made in the last 24 hours, and describe two ampliative inferences you made in order to make one or more of those decisions.

7.5 Recall: List three features of inductive and abductive inferences described in this section, and give an example of each from the Flint water crisis.

7.6 Apply: Give an example of an inductive or abductive inference. Then, show how the inference is ampliative, non-monotonic, and can be made stronger or weaker.

7.2 INDUCTIVE INFERENCE

After reading this section, you should be able to:

- Define *inductive inference* and assess the strength of an inductive argument
- Distinguish between inductive generalization and inductive projection
- Describe the problem of induction and articulate what makes it a hard problem to solve

Strength of inductive inference

Imagine you go to the grocery store, hankering for some apples. The grocer takes one apple from the top of one box, cuts it open, and offers you a slice to taste. It tastes good! What you may not notice is that the grocer is tacitly expecting you to make the following inference:

> (1) One apple from this box is good.
> ∴ (2) All apples from this box are good.

You draw three other apples from the box at random. The three apples look good, like the one the grocer had you taste. So, you buy that box of a dozen apples. What's the inference you are tacitly making?

> (1) One apple from this box is good.
> (2) Three apples picked at random from this box look like the apple that is good.
> ∴ (3) All apples from this box are good.

Neither of these is a valid deductive inference: the truth of the premises do not guarantee the truth of the conclusions. At best, the premises make the conclusions likely or probable. Accordingly, both inferences are inductive. Recall from section 7.1 that an *inductive inference* is an inferential relationship from premises to conclusion that is one of probability rather than necessity. Even if the premises in each inference are true, the conclusion may nonetheless be false. Perhaps not all apples from the box are good, even if the apple the grocer let you taste was good and even if you checked three other randomly picked apples to ensure they looked good. For all you know, the rest of the apples in the box could be rotten.

Because the truth of the premises in inductive arguments does not guarantee the truth of the conclusion, inductive arguments are always logically invalid. Instead of logical validity, the benchmark for a good inductive inference is its strength, which is the probability that the conclusion is true assuming that all the premises are true. And, just as we assess whether the premises are true to judge deductive arguments for soundness, we should assess whether the premises of an inductive argument are true.

In our grocery store example, you saw the grocer take an apple from the box and cut it open on the spot. The premise that one apple from the box is good is certainly true. But the inductive inference the grocer wants you to make about the entire box of apples isn't very strong. After all, the grocer chose the apple he cut open, and you haven't seen any other apples in the box. Your selection of three other apples to visually examine and finding that they look just like the apple the grocer invited you to taste increases the strength of the tacit inference, and now you are satisfied that all the apples in the box are (probably) good. You're sure enough of this, at least, that you are willing to buy the box of apples.

In Chapter 6, we discussed how the hypothetico-deductive (H-D) account of hypothesis-testing contrasts the deductive validity of refutation with nondeductive confirmation.

FIGURE 7.2 Black Australian swan and white Bewick swans

The present discussion of inductive inference can shed additional light on the logic of confirming hypotheses. The H-D account makes clear that confirmation does not involve valid deductive inference. What it does involve is inductive inference. And, as we now know, inductive arguments can be judged for their strength. So, the confirmation of a hypothesis—reasoning from some observation(s) to the conclusion that a hypothesis is true—can be judged according to the inductive strength of the inference. Indeed, we will see in later chapters that scientists often employ statistical tools based on the mathematics of probability to guide their judgments of the confirmation of a hypothesis.

So, inductive inferences are better, more trustworthy, when they are stronger. Yet even very strong inductive inferences with true premises may nonetheless have false conclusions. Again, unlike deductive inference, inductive inference offers no guarantees. Until the 17th century, Europeans believed that all swans were white. Their belief was supported by strong evidence: all of the swans they'd ever seen were white, and there were swans in many parts of Europe. Further, no one they consulted had ever mentioned seeing a swan that wasn't white. Nonetheless, in 1697, the Dutch explorer Willem de Vlamingh returned to Europe with two black swans he had captured on Australia's Swan River. The strong inductive argument in favor of all swans being white was undermined by this development, and the conclusion was shown to be false.

Here's another example of a strong inductive argument that turned out to be false. The smallpox (variola) virus was completely eradicated in 1979. This does not mean that the viral particles no longer exist, however. The World Health Organization (WHO) oversees two vaults containing variola specimens: one at the Centers for Disease Control and Prevention in Atlanta, Georgia (USA) and the other at the State Research Centre of Virology and Biotechnology in Novosibirsk, Russia. Given the complete eradication of smallpox and the international oversight of WHO, it was reasonable to infer that these were the only remaining variola specimens. But, in 2014, scientists stumbled upon some 60-year-old unsecured vials of smallpox while cleaning out a storage closet at the National Institutes of Health in Bethesda, Maryland, in the US. This discovery undermined the reasonable inductive inference from the eradication of smallpox and WHO's strict control of remaining specimens to the conclusion that smallpox did not exist anywhere else. This was a strong inductive argument, at least until new evidence came to light that directly undermined its conclusion. Notice that the premises of the argument remain true; this example shows how inductive inferences are non-monotonic, as discussed in section 7.1.

Two forms of inductive inference

There are two common forms of inductive inference: generalizations and projections. An ***inductive generalization*** is an inference to a general conclusion about a class of objects based on the observation of some number of objects in that class. If the conclusion applies to all members of the class, the generalization is a universal inductive generalization. The form of inductive generalizations is something like this:

Inductive generalization

    (1) Objects O_1, O_2, O_3, . . . , and O_n each has property P.
∴   (2) All Os have property P.

The inference you made in the grocery store example was like this. You reasoned from the premise that one apple from the box is good and the premise that three other apples drawn randomly from the box also seem to be good to the conclusion that all apples in the box are good.

Inductive generalization is the kind of inference involved in making claims about an entire population based on claims about a sample. This is commonly involved in drawing conclusions from field experiments and studies. For example, in the natural experiment described in Chapter 4 investigating how women village council leaders in India affected social services, the conclusions drawn regarded not just the women who had been studied but the prospect of social service changes from women village leaders across India, and perhaps even beyond India. Drawing those conclusions involves an inductive generalization from the village leaders studied to a broader class of women in municipal leadership positions. The concept of *external experimental validity* introduced in Chapter 3—the extent to which experimental results generalize to other conditions—bears on this use of inductive generalization. (Note that this use of "validity" means something entirely different from validity in deductive inference.)

A second form of inductive inference is ***inductive projection***: an inference to a conclusion about the features of some object that has not been observed based on observations of similar objects. The form of inductive projections is like this:

Inductive projection

    (1) O_1, O_2, O_3, . . . , and O_n each have been observed to have property P.
∴   (2) The next object O_k has property P.

For example, imagine you return to the same grocery store after working your way through that box of apples. All of the apples were, in fact, delicious. The same grocer is working. This time, you don't bother tasting or visually inspecting apples, you just ask the grocer to give you a box of delicious apples. Even if you don't explicitly think it through, your inference is something like:

(1) Last time, the box of apples I bought here on the grocer's recommendation was delicious.
∴ (2) This time, the box of apples I buy will be delicious as well.

In this case, you aren't finding out about some members of a group to generalize to the whole group. Instead, you are using your knowledge of about one kind of object to draw conclusions about something similar. You don't need to try to generalize that all boxes of apples sold in this store will be delicious. All you are interested in is making a prediction—a projection—of whether the next box of apples you buy here will be delicious.

Inductive projection can also be involved in drawing inferences from experiments and studies. This is especially relevant when the experimental subjects or the conditions in which the experiment is conducted are somewhat unlike the conditions and subjects about which scientists want to draw conclusions. In Chapter 3, we saw how Newton discovered that a rainbow spectrum of light created by a prism could be restored to white light by traveling through a second prism. Newton inferred from this that all light—even light that had never been separated into a spectrum of colors—was composed of multiple colors of light. This was an inductive projection from the light studied in laboratory conditions to natural light out in the world. Drawing conclusions about a target system from a model can also involve inductive projection. In that case, the model having some attribute is used as a basis for inferring that the target system, similar in certain respects, will have a similar attribute. Efforts to calibrate the model and establish its similarities to the target system are ways to increase the strength of these inductive projections.

The problem of induction

Why do you look to the east if you want to see the morning sunrise? Why do physicians continue to prescribe acetaminophen (such as Tylenol) to reduce fevers? Why do I trust that tap water I drink is not contaminated? At first glance, the answers to these questions might seem obvious. That's where the Sun rises above the horizon every morning. Acetaminophen is usually the best approach to reducing fevers. And, throughout my life, in different homes and possibly different towns, I've had access to clean water from the tap.

Notice, though, that all three of these conclusions rest on inductive inferences. They are very strong inductive inferences, but we have seen that even strong inductive inferences from true premises can have false conclusions. It's worth asking: why think inductive reasoning in general is trustworthy?

Inductive inference is essential to our reasoning, and it seems to work very well. Inductive reasoning has led us to buy good apples and other foods based on samples, to look for the sunrise in approximately the right place, to take medication as needed, to know when water is safe to drink, and to wear a coat when we leave the house in the winter. Occasionally, we're led astray; there was that whole matter of Europeans spending centuries thinking all swans are white. But, overall, inductive reasoning has a track record of success.

That is an argument for why we can trust inductive reasoning. Do you notice anything about the form of the argument? Is it deductively valid or inductively strong?

Because our argument in favor of the trustworthiness of inductive reasoning is inductive, we can add even more premises about times when inductive inference led to true conclusions. This would make our argument that inductive reasoning is trustworthy even stronger. But there's a fundamental problem. The argument relies on inductive inference: because inductive reasoning has guided us so well up until now, we conclude that it will continue to do so. So, this just prompts the same question: what justifies the inductive inference that inductive reasoning is trustworthy?

The **_problem of induction_** is the concern that inductive reasoning cannot be logically justified, since any possible justification would need to employ inductive inference and would thus be circular. This problem was set out in the 18th century by the Scottish philosopher David Hume, who argued that the problem of induction cannot be solved. The argument goes as follows. Consider how we might justify inductive reasoning. There are two possibilities: either use deductive inference or use nondeductive inference.

First consider the possibility of using deductive inference to justify inductive reasoning. Valid deductive inference generates conclusions that cannot be false if all premises are true. Because strong inductive inference with true premises may still have a false conclusion, these inferences would be invalid if construed deductively. So, inductive reasoning cannot be justified using deductive reasoning.

Now consider the possibility of using nondeductive inference to justify inductive reasoning. To justify inductive reasoning nondeductively, we would need to show that inductive reasoning is generally reliable. But, as we just talked through, this can only be done using inductive inference. So, looking to a nondeductive justification for inductive inference leads to circular reasoning: we would need to prove inductive inference is reliable in order to justify inductive reasoning. In other words, we would have to assume the reliability of the method whose reliability we need to establish. Consequently, inductive inferences cannot be justified by reasoning nondeductively, either.

Given that deductive and nondeductive reasoning exhaust the possibilities, Hume concluded that inductive reasoning cannot be rationally justified. Notice that this is a deductive argument against the possibility of providing logical justification for inductive reasoning. The argument we tried to provide in favor of inductive reasoning was, of course, itself only inductive (hence the circularity).

Hume also noted that the circular justification of induction depends on what he called the **_uniformity of nature_** assumption. This is the idea that the natural world is sufficiently uniform, or unchanging, that we are justified in thinking our future experiences will resemble our past experiences. The uniformity of nature is an auxiliary assumption needed for the inductive inference in favor of inductive reasoning to succeed. The uniformity of nature assumption cannot itself be justified, though, since this assumption is merely based in inductive inferences from our past experience. We think that nature is uniform because, so far, it has been. But what do we know about tomorrow?

Philosophers of science have proposed several solutions to the problem of induction. One potential solution begins from the observation that inductive inferences are intended to warrant probable conclusions—not make guarantees. And there are rational grounds for inferring conclusions about the probability of something being the case based on empirical evidence. Perhaps tools of statistical reasoning, which we focus on in later chapters, can justify some varieties of inductive inference. And statistical reasoning does have a rational basis, provided by probability theory (a deductively valid axiomatic theory, as we'll see in Chapter 8).

A different approach to solving the problem of induction is simply to show that inductive inference is the best we can reasonably hope for when it comes to making reliable predictions. Either nature is uniform or it's not. If nature is uniform and we want to make reliable predictions, then a noninductive method like, say, fortune-telling may or may not work. In contrast, inductive inference will clearly work. (Remember the uniformity of nature assumption.) So, if nature is uniform, inductive inference is more reliable than noninductive methods. Now suppose nature is not uniform. In that case, inductive inference will be unreliable, but so will any alternative methods. Why? Well, suppose that fortune-telling was better than induction, such that fortune-tellers were able to reliably predict the future. This success would imply some kind of uniformity. But any uniformity in nature can be used for inductive inference. You could, for example, inductively reason to the future success of fortune-tellers from their past successes. Consequently, whether or not nature is uniform, the best approach one can take to making reliable inferences about the future or the unobserved is inductive inference.

Something we should infer from the problem of induction is that inductive inference always involves some risk. This is an inevitable aspect of ampliative and non-monotonic inference. But, on the other hand, moving beyond what we know for certain to hypothesize about what might be true is a powerful feature of science, and this requires reliance on inductive inference.

EXERCISES

7.7 **Recall:** Define *inductive inference, inductive generalization,* and *inductive projection.* Give an example of an inductive generalization and an example of an inductive projection.

7.8 **Apply:** Give a new example of a very strong inductive inference and a new example of a very weak inductive inference. For each, describe whether it is strong or weak, and label it as a generalization or projection.

7.9 **Apply:** Consider your example of a strong inductive inference in Exercise 7.8. Inductive inference is non-monotonic. What kinds of new information could you learn that is consistent with the premise(s) but would weaken support for the conclusion? Give three examples.

7.10 Think: Look back to Chapter 3, 4, or 5 and find a clear instance of inductive inference made from an experiment, study, or model. (a) Put the inference in standard argument form with numbered premises and a conclusion; you might need to make explicit some ideas that were implicit in the text. (b) Next, assess the strength of this inductive inference. (c) If you were a scientist investigating this conclusion, describe three steps you could take to find additional support for the conclusion.

7.11 Recall: Describe the problem of induction. Given how often scientists use inductive reasoning, why is it so difficult to justify?

7.12 Think: Briefly summarize the two possible responses to the problem of induction described earlier. For each, indicate whether you find it compelling, justifying your response. Can you think of any other promising responses to the problem of induction?

7.3 ABDUCTIVE INFERENCE

After reading this section, you should be able to:

- Define *abductive inference* and articulate its similarities to and differences from both deductive and inductive inference
- Indicate how scientific arguments in favor of continental drift and pan-African evolution exemplify abductive inference
- Describe five explanatory considerations that contribute to the strength of abductive inferences

Continental drift

A century ago, most scientists thought that the Earth's molten surface cooled billions of years ago and that the remnants of this cooling process are the major landmasses we see today. There were good reasons to accept the hypothesis that, once encrusted, the Earth's surface was relatively fixed and stable. Yet some surprising geological features were left unaccounted for. One such feature is the congruence of some coastlines (think of Africa and South America), which is a puzzling coincidence if the continents had always been in their current locations. Further, geologic formations that are several thousands of kilometers away from one another had been found to have several characteristics in common. And fossils of some early types of plants and animals were distributed across multiple continents.

In 1915, the German scientist Alfred Wegener advanced a systematic proposal about the geologic history of Earth. He hypothesized that there had been a single original landmass, which he called *Pangaea*, that later fragmented into the continents. This hypothesis of continental drift suggested that, even today, continents might be very slowly drifting in relation to one another. If true, that hypothesis would account for the puzzling observations that otherwise lacked an explanation.

In the 1950s, American geologists Marie Tharp and Bruce Heezen were working to generate maps of the ocean floor. They made a fascinating observation about the Mid-Atlantic Ridge, which is an extensive mountain range running the whole length (north to south) of the Atlantic Ocean, almost entirely underwater. They discovered that, at the top of that ridge, running its full length, was a valley. Further, many earthquakes originated in this valley. This observation fit with Wegener's hypothesis of continental drift. Tharp and Heezen had, it seemed, discovered that the seafloor was spreading,

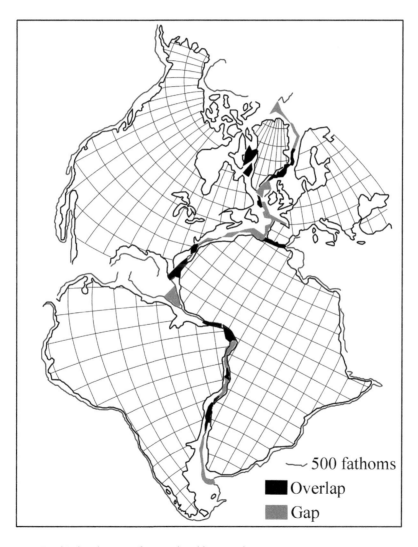

FIGURE 7.3 Earth's landmasses fit together like puzzle pieces

FIGURE 7.4 Marie Tharp and Bruce Heezen

further separating the landmasses on either side of the Atlantic Ocean. The edges of those landmasses, now quite distant from each other, were roughly congruent—a bit like jigsaw puzzle pieces fitting together.

Continental drift, if true, would account for all of this evidence. Like the shape of the continents on Earth, everything—all the observations—would then seem to fit together. Various other kinds of evidence came to light in investigations carried out in multiple fields of science, all supporting continental drift. Today, continental drift is part of the accepted theory of plate tectonics.

What kind of inference pattern was used when scientists eventually reasoned, from a variety of evidence, that the hypothesis of continental drift was true? This is clearly an ampliative inference, in that it goes beyond what's contained in the evidence. So, it's not deductive reasoning. But this doesn't correspond very well to the pattern we've seen of inductive inference either; it's not a generalization or projection from an observation of a certain kind, like the quality of an apple or level of lead in tap water, to the expectation of more observations of that kind. There's a bigger leap involved in the inference from premises about geologic features to the conclusion that landmasses have separated and moved apart over the course of Earth's history.

This is an *abductive inference*, a type of nondeductive inference that attributes special status to explanatory considerations. Abductive inference is also sometimes called *inference to the best explanation*. In reasoning abductively to some conclusion, one considers whether the conclusion, if true, would account for or explain the evidence used as premises.

The logic of abductive inference

Abductive inference is characterized by an appeal to explanatory considerations to conclude that some hypothesis is true. This is why abduction is also called *inference to the best explanation.* To reason abductively is, in a sense, to reason backwards. Its form is like this:

Abductive inference

  (1) Surprising observations O_1, O_2, O_3, . . . , and O_n have been made.
  (2) If H were true, this would explain O_1, O_2, O_3, . . . , and O_n.
∴ (3) H

Suppose, for example, that you know your roommate Theresa cut herself while preparing dinner last night. This morning, you see her walking down the hallway with stitches in her hand. The best explanation for the stitches seems to be that Theresa's cut was severe enough that it required stitches. This conclusion might not be true. The cut might not have required stitches, and stitches can treat injuries other than cuts. But if the conclusion were true, it would account for the available evidence.

Characterized in this way, abductive inference is similar to the deductively invalid inference of affirming the consequent: from the premises if H then O, and O, the conclusion H does not validly follow. This argument scheme also corresponds to the H-D method scheme for confirmation, which is not deductively valid either. (The role for deductive inference in the H-D method was in the refutation scheme.)

There is an extra element present in the abductive inference scheme though, beyond what's contained in the pattern of affirming the consequent or H-D confirmation. This extra element is the abductive inference's ability to explain observations that would otherwise be surprising. An abductive inference isn't any old instance of affirming the consequent. Instead, the thought is that, if the antecedent accounts for a consequent that would otherwise be left unexplained, then this is grounds for believing the antecedent is true. The power of a hypothesis to explain what is otherwise unexplainable is a reason to conclude that probably it is true.

Abductive inference is common in everyday and scientific reasoning. Examples include scientists reasoning from a rich variety of evidence and theory that extreme anthropogenic climate change is occurring, as detailed in Chapter 1; paleontologists using fossil evidence and the habits of contemporary creatures to infer the life histories of long-extinct species as detailed in Chapter 4; and Hubble piecing together astronomical data and physical theory to infer the approximate age of the universe as detailed in Chapter 6.

Abductive inference and explanatory considerations

Like inductive inference, abductive inferences are ampliative and non-monotonic, and their strength comes in degrees. But unlike inductive inference, abductive inference does not generalize or project from what has been observed to similar states of affairs. The strength of abductive inferences instead depends on explanatory considerations. Abductive inferences seem to rely on an inferential leap—a leap in the reasoning of one or more scientists having an "aha!" moment, of seeing how some new idea about the world might explain otherwise puzzling observations. Scientists employing abductive inference in favor of a hypothesis need to hope that their audience grasps the connection, that their audience sees how the hypothesis accounts for the observations. It's not clear whether anything definitive can be said about what it takes for a hypothesis to accomplish that task.

How should a hypothesis relate to the observations in order to explain them? One suggestion is that a hypothesis best explains a set of observations if it predicts the observations or shows why the observations were to be expected. By itself, however, the predictive power of a hypothesis isn't enough to make the hypothesis a good explanation. Explanations must also have additional qualities. Perhaps explanations should also be simple, fit with other explanations we already accept, and generate new expectations for what we will observe in the future. These qualities seem to make an abductively inferred hypothesis—an explanation—enlightening. Indeed, qualities like simplicity, coherence with other explanations, and fecundity of new ideas have been shown to play central roles in people's assessment of explanatory goodness.

Given the difficulty of giving criteria for good explanations, it's worth considering what features of abductive inferences contribute to their strength. In 1690, the Dutch mathematician and scientist Christian Huygens, in a discussion of how hypotheses about light could explain experimental findings in optics, proposed some principles that apply to strong abductive inference. He says:

> It is possible . . . to establish a probability which is little short of certainty. This is the case when the consequences of the assumed principles are in perfect accord with the observed phenomena, and especially when these verifications are very numerous; but above all when one employs the hypothesis to predict new phenomena and finds his expectations realized.

Here Huygens points out the contribution to the strength of an abductive argument of a hypothesis perfectly predicting observations, of many observations being accounted for by the hypothesis, and when a hypothesis makes predictions that later turn out to be true.

First, the degree of an observation's surprisingness and the degree to which the hypothesis dispels the surprisingness contributes to the strength of abductive inference. The finding of a rift down the center of the Mid-Atlantic Ridge with significant seismic activity would be extraordinary without different parts of the Earth's crust moving in (very slow) motion. Second, it seems the number and variety of surprising observations that a hypothesis explains contributes to its strength—the more the better. The abductive argument for continental drift became stronger over the decades, as geological observations accumulated that would be expected if continental drift had occurred and that would be surprising otherwise. Third, Huygens notes that predicting new phenomena that are then discovered to occur is valuable for the strength of an abductive inference; this is like the value of a hypothesis's fecundity noted previously. Beyond that, as we noted earlier, a hypothesis's simplicity and coherence with other theories may also contribute to the strength of an abductive inference. These might be considered two additional explanatory considerations that increase the strength of an abductive inference.

An amazing discovery in Morocco illustrates how scientists can appeal to a hypothesis's explanatory power as evidence in support of the hypothesis. In 2017, archaeologists and evolutionary anthropologists found several specimens of stone tools and human bones, including a remarkably complete jaw and skull fragments, at an archaeological site in Morocco named Jebel Irhoud. The researchers used dating techniques to determine that the remains were about 315,000 years old.

Previously, fossils from eastern sub-Saharan Africa were thought to belong to the first anatomically modern humans, the earliest representatives of our species *Homo sapiens*. From those fossils, scientists reasoned that humans evolved relatively quickly about 200,000 years ago in a region of what is now Ethiopia. This previously favored hypothesis could explain the findings at Jebel Irhoud as remains from some hominid species that lived prior to *Homo sapiens*, perhaps the Neanderthals. The Jebel Irhoud findings also prompted a new hypothesis, though: that the evolution of *Homo sapiens* was a pan-African process that occurred about 300,000 years ago—100,000 years earlier than posited by the previous hypothesis.

This new pan-Africa hypothesis was simpler than the previously favored hypothesis, as it didn't require positing an archaic hominid species in North Africa that was later replaced by *Homo sapiens*. The pan-Africa hypothesis also cohered with archaeological and anatomical observations about Neanderthals and *Homo sapiens*. For example, the teeth found in Jebel Irhoud better matched what would be expected for *Homo sapiens* than for Neanderthals. The morphology of the skull was almost indistinguishable from that of anatomically modern humans. And the pan-Africa hypothesis is consistent with geographical and ecological evidence that, 300,000 years ago, the Sahara was green, filled with rivers, and hospitable. Animals like gazelles and lions that now inhabit the East African savanna then also populated the Saharan region. Remains of plants and animals even indicate biological and environmental continuity between those regions. Finally, the new pan-Africa hypothesis explained a greater number of diverse observations about human origins than the earlier East Africa hypothesis, including the mix of anatomical features seen in the Jebel Irhoud remains and in other human-like fossils from elsewhere in Africa. It also better fits with genomic evidence collected in South Africa that seems to indicate that the lineage split between archaic hominid species and anatomically modern humans occurred more than 260,000 years ago.

Explanatory considerations, including simplicity, coherence, and fecundity, thus favored the pan-Africa hypothesis. This does not guarantee that the pan-Africa hypothesis is true, nor that the East Africa hypothesis is false. Abductive inference can't provide guarantees. Perhaps new evidence will come to light favoring the East Africa hypothesis, or perhaps some even older remains of *Homo sapiens* have yet to be discovered. But for now, the inference to a pan-African conclusion looks like our best explanation. The researchers involved in the Jebel Irhoud discovery concluded, "the Garden of Eden in Africa is probably Africa—and it's a big, big garden."

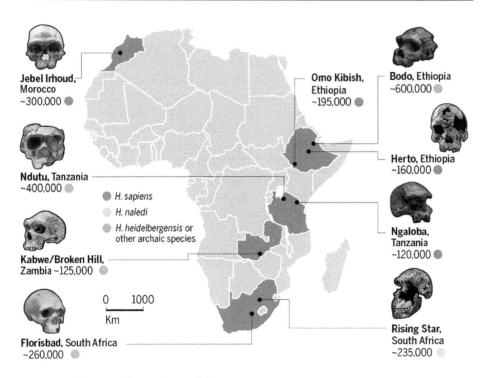

FIGURE 7.5 The pan-African dawn of *Homo sapiens*

EXERCISES

7.13 Recall: Define *abductive inference* and describe how it is similar to and different from each of deductive and inductive inference.

7.14 Apply: For each of the following inferences, indicate the conclusion, then state whether the inference is best interpreted as deductive, inductive, or abductive. Finally, indicate whether each inference is a good one, and justify your answer, appealing to validity, soundness, and strength as appropriate.

a.   The murder weapon has Pat's fingerprints on it. Therefore, Pat is the murderer.

b.   Whenever it rains, the streets get wet; and the streets are wet now. So, this guarantees it rained.

c.   These ten beans have been randomly selected from a 5-pound bag, and they are all Mayocobas instead of Cannellinis. So, all the beans from this bag are probably Mayocobas.

d.   Of the students in my class, 65% say that they prefer coffee to tea. Therefore, most people prefer coffee.

e.   Medical technologies ought to be funded if they can be used successfully to treat patients. Patients have been successfully treated with lobotomies. So, lobotomies ought to be funded.

f.   Sociologists misuse the term *equity* and cannot agree on a definition. Therefore, the term *equity* is undefinable.

g.   Without energy expenditures, disorder increases in a system. Your home is a system. So, the disorder of your home will increase unless energy is expended.

h.   The last three US Republican presidents have denied that human activities are the main cause of global warming. So, human activities don't cause global warming.

i.   Bread appears to grow mold more quickly in the bread bin than the fridge, so temperature probably determines the rate of mold growth.

j.   Leaded gasoline, paint, pipes, and other products with lead were all used throughout the 20th century but were eventually discontinued because they were toxic. So, all lead products are toxic.

7.15 Apply: Write out the logical form of abductive inference as depicted in this section. Then, choose either the continental drift or pan-Africa hypothesis, and write out the abductive inference supporting the hypothesis in this logical form. Try to include all the evidence supporting the abductive inference.

7.16 Think: Come up with an example of abductive inference in science not described in depth in this section. (a) Indicate why this should be viewed as an abductive inference. (b) Assess the explanatory strength of the inference. (c) If you were a scientist focused on this inference, what kinds of steps could you carry out to additionally support the conclusion?

7.17 Recall: Describe five explanatory considerations that contribute to the strength of abductive inferences.

7.18 Apply: Choose three of the five explanatory considerations that contribute to the strength of abductive inferences. Illustrate each with a new example of abductive inference from science or everyday life.

7.4 TESTIMONY AND TRUST

After reading this section, you should be able to:

* Describe the roles of testimony and trust in scientific reasoning
* Indicate how reliance on testimony involves abductive inference
* List three considerations influencing whether testimony should be believed

Testimony

The testimony of others plays a central role in reasoning. Many of your beliefs actually originate from what other people think is true. Consider how you know simple things, like your friend's birthdate or the customary ingredients in macaroni and cheese. Chances are, you learned things like these from things other people told you—or things people wrote down in a cookbook or published online.

This is the same in science. Scientists must rely on the testimony of others to learn the state of knowledge and accepted methods in a field, to collaborate on research, and to properly interpret others' findings. By **testimony**, we mean a spoken or written statement that something is the case. Others' testimony is regularly used as evidence for concluding that something—what they say is so—is true.

Believing others' testimony requires trust. Imagine you were a resident of Flint, Michigan, in 2014. You attended a community meeting, where the governor reported that the city water is safe to drink. To demonstrate this, he himself drinks some tap water. From this testimony, you infer that the water is safe. But your water is still discolored with a foul smell and taste, and, later, you learn water testing revealed serious lead contamination. This information undermines your earlier inference from the governor's testimony; you no longer believe Flint's water is safe to drink. After that, you will probably be less likely to take city and state officials' word for the safety of your water.

In science, belief in others' testimony is a key component involved in the balance of trust and skepticism that we saw in Chapter 1 is crucial for science. Science is collaborative: several scientists typically conduct research together. Sometimes this even includes thousands of scientists, like at CERN or NASA. When scientists collaborate, they have to rely on the specific expertise and the honesty of their collaborators in order to believe what they say (or write) is true. This trust in the testimony of collaborators is essential for many scientific projects. Reliance on the testimony of research subjects or nonscientific collaborators is also essential in some disciplines. An example is when researchers doing ethnographic fieldwork must rely on the testimony of locals to understand social or cultural practices. Trustworthy sources are essential to the legitimacy of their research findings.

To trust someone's testimony, we must believe they are knowledgeable, competent, sincere, and honest. Researchers need to be able to trust that researchers they collaborate with or learn from are knowledgeable about what they profess to know, competent at what they say they will do, sincere in what they claim, and honest enough not to make up data or results. Researchers also must be able to trust that survey participants are sincere, that scientific journals only publish vetted research, and more.

Box 7.1 When testimony justifies belief

Someone stops you in the street to tell you there is life outside our planet. You had no opinion about this before, and purely based on this person's testimony you come to believe that there is extraterrestrial life. Is your belief justified? That is, should you believe what that person says? Suppose you learn the person is a cosmologist working in a large collaborative project searching for intelligent extraterrestrial life. Would you be any more justified in believing them now? What if instead you learned they are the CEO of a social media company?

Philosophers debate the conditions, if any, in which somebody's testimony about a topic justifies belief. According to a view called *anti-reductionism* about testimony, other people's word is a basic source of justification for our beliefs, similar to perceptual evidence. On this view, unless you have evidence that a speaker is unreliable, insincere, or untrustworthy, you are justified in accepting what they tell you. The opposite view, *reductionism* about testimony, holds that testimony is not a basic source of justification. On this view, the word of others justifies your beliefs only insofar as you have independent, positive evidence for their reliability, impartiality, and expertise. Such evidence is based on your perceptual experiences, memory, and reasoning. Hence, on this view, beliefs based on testimonial evidence are justified by empirical evidence and reasoning, just like all other beliefs.

Public trust in scientific institutions and findings is equally important. Community members must be able to trust that lead testing of their tap water, for example, is carried out competently and reported honestly. People also need to be able to trust that the accumulated knowledge in textbooks and other resources is written by knowledgeable authors, based on competently conducted research, and honestly conveyed. Ultimately, public trust in science is required for people to believe the testimony of scientists and scientific institutions, which is crucial for sharing in scientific knowledge.

The problem with testimony is that sometimes people can be dishonest, insincere, or unreliable. So, under what conditions does someone's testimony give sufficient reason to believe what they say is true?

Testimonial evidence as abductive inference

When making inferences from testimony, the statement that a source makes is a premise, and perhaps even the only premise, used for concluding that it is true. Because people can lie, make up fantasies, or simply be wrong, inference from testimony is risky. This is certainly not deductive inference. But it may be plausibly described as a form of abductive inference. When we believe a statement is true based on someone's testimony, we do so because the truth of the statement is the best explanation for why the person would say it is so. We might characterize this as follows, drawing from the form of abductive inference introduced in the previous section:

(1) This individual or group asserts A.
(2) If A were true, this would explain why they asserted A.
∴ (3) A

There are a few things worth noticing about this simple characterization. First, the identity of the individual or group in question is relevant. If we have reason to trust the party in question, their testimony provides greater reason to think something is the case. One way to think about that is that a trustworthy source asserting something would be more surprising—less explainable—unless their assertion were true. And, on the other hand, if the party in question might have other reasons to provide this testimony besides its truth, this decreases the strength of the inference. This accounts for why one may be more inclined to believe a scientist's claim about Flint's water quality than the governor. The best explanation for the governor's claim of safety may not be the truth of that claim but instead his desire to reassure the public and, perhaps, protect himself from any blame.

Second, the more remarkable the assertion, the more evidence we require to rationally infer it is true, even from a trustworthy source. Testimony that violates what we think we know about the world—such as by going against established scientific theories or violating specific memories we have—may not be grounds for belief. That testimony may still be explained by its truth, but a better explanation may turn out to be that the person making the assertion is mistaken or we are misunderstanding.

Third, because abductive inference is non-monotonic, more premises can be sought to increase (or decrease) the strength of the argument. When in doubt about some testimony, more evidence can be collected. Do other trusted parties also claim the assertion is true? Does the person whose testimony you are considering have reason to be dishonest or lack credibility for any other reason? Is there independent, non-testimonial evidence for the assertion? And so on.

Thus, thinking about inference from testimony as a kind of abductive inference may help distinguish the circumstances in which testimony provides sufficient grounds for belief from when it does not. Expertise about the topic of the claim means someone is less likely to be wrong about the claim. The motivations of the person providing testimony can be taken into consideration to determine the likelihood of intentional *deception*. Remarkable testimony requires more evidence to warrant belief and, when in doubt, we can seek additional evidence, testimonial and otherwise. An assessment of a source's credibility and the credibility of their claim is essential in determining when testimony is sufficient reason for belief—and when it should instead be regarded with skepticism.

Trust in science

Not every claim a scientist makes is true, and scientific institutions have on occasion made statements that were incorrect or even intentionally misleading. Still, numerous features of science—as a set of institutions—make it worthy of public trust.

We have surveyed many of these features in this book or will do so in the chapters still to come. In Chapter 1, we saw how the social norms of science contribute to its trustworthiness. And the defining features of science also introduced in Chapter 1 contribute to its power to generate trustworthy knowledge. Techniques of collecting and analyzing data in experiments, studies, and modeling support the evidential basis for scientific knowledge (Chapters 3–5), and sound tools for inference support the development of sound and strong arguments from that evidence (Chapters 6 and 7).

For these reasons, we should take established scientific knowledge to be authoritative. Further, assertions by individual scientists with relevant expertise are generally reason to believe what they say is true.

Yet it also should be acknowledged that public trust in science faces some challenges. The health sciences have a problematic history of race and gender discrimination in research as well as in medical care, which has resulted in heightened distrust of medical advice and care. Members of some religious groups in the US are much less likely to believe scientific reports of climate change and its disastrous impacts.

In 2019, researchers finally reported that clean tap water was available in the homes of Flint, Michigan. The path to this point was a long journey of scientific research, public policy, and politics. But, $400 million of state and federal support has resulted in a clean water source, modern copper pipes to nearly every home that lacked them, and water filters for any residents who want them. Testing shows Flint's water is as good as any in Michigan. But a few weeks later, for the UN's World Water Day, the City of Flint brought in several semitrucks filled with bottled water, and Flint residents flocked to pick it up. Jim Ananich, a lifelong Flint resident and state politician, said:

> I can't tell somebody they should trust [claims that the water is safe], because I don't trust them—and I have more information than most people. Science and logic would tell me that it should be OK, but people have lied to me.

This distrust of municipal water supply isn't just in Flint, and low-income households and communities of color are most likely to purchase bottled water due to distrust of local supply. Yet bottled water is less well-regulated than tap water.

Box 7.2 Science denial

In this section, we've summarized reasons for science's trustworthiness and a few sources of public distrust of science. One contributing factor to public distrust of science is social identity—one's political affiliation, religion, ethnic community, and so on. In recent decades, there has been increasing belief polarization about scientific topics along religious and political lines in particular, such as about climate change research and evolutionary biology. Sometimes distrust in science goes even further: some people who reject some particular scientific finding use that as reason to disparage science as a credible source of knowledge in general. This is sometimes called *science denial*. Social science research about science denial has suggested that increasing knowledge about science actually doesn't eliminate science denial. Instead, increased scientific knowledge correlates with heightened belief polarization about scientific findings. Increasingly, people with different social identities are exposed only to ideas and information from others with similar social identities. This increases belief polarization. Some small groups are at the extreme—like those who advocate for the belief that the Earth is flat, despite all the evidence to the contrary. But belief polarization is something that affects all of us. This makes it even more important to seek credible sources of information and to assess the grounds for believing ideas you encounter. Evaluating the credibility of ideas' source and content is perhaps even more crucial for ideas that appeal to us because of our social identity.

Scientists are stepping up to help with distrust like this. For example, Caren Cooper at North Carolina State University, with funding from the US Environmental Protection Agency (EPA), is working with community members across the United States to test drinking water and water supply lines for safety. This initiative, "Crowd the Tap," offers resources for teachers and high school students, utility companies, faith communities, and any other members of a community to test local water supplies and to demonstrate their safety.

Watch Video 11

EXERCISES

7.19 Recall: How does scientific reasoning rely on testimony? How does relying on testimony require trust?

7.20 Recall: Describe what features reliance on testimony has in common with abductive inference.

7.21 Think: Describe a time when you relied on someone's testimony as evidence you should believe something. Then, try to characterize this as an abductive inference, specifying the premises and conclusion. Is abductive inference a good way to think about this use of testimony? Why or why not?

7.22 Recall: List the three considerations suggested in this section that influence whether testimony should be believed and give an example of how each applies.

7.23 Apply: Consider the three considerations suggested in this section that influence whether testimony should be believed. For each of these considerations, describe how scientific practices are developed to increase the strength of testimonial evidence in favor of established scientific knowledge.

FURTHER READING

For an account of the political and public transitions following the Flint water crisis, see Robertson, D. (2020). Flint has clean water now. Why won't people drink it? *Politico.* www.politico.com/news/magazine/2020/12/23/flint-water-crisis-2020-post-coronavirus-america-445459

For a philosophical treatment of inductive reasoning, see Goodman, N. (1955). *Fact, fiction, and forecast*. Harvard University Press.

For the original problem of induction, see §§4–6 of Hume, D. (1748/1999). *An enquiry concerning human understanding*. Oxford University Press.

For a helpful guide to Hume's problem of induction, see Salmon, W. (1975). Chapter: An encounter with David Hume. In J. Feinberg's (Ed.), *Reason and responsibility* (pp. 245–263) Dickenson Publishing Co.

For more on abductive reasoning, see Lipton, P. (2003). *Inference to the best explanation*. Routledge.

For an analysis of public trust in science related to health, see Goldenberg, M. J. (2021). *Vaccine hesitancy: Public trust, expertise, and the war on science*. University of Pittsburgh Press.

CHAPTER 8

Probabilistic reasoning

8.1 MEDICAL TESTING AND USES OF PROBABILITIES

After reading this section, you should be able to:

- Explain why medical tests don't provide guarantees of whether one has a medical condition
- Describe how probability relates to uncertainty and inductive reasoning and give examples of reasoning with probabilities
- Describe three ways probabilities can relate to objective features of the world and to individuals' beliefs

Rapid strep tests

You have a sore throat and so you go to the doctor. The doctor examines your throat and calls for a rapid strep test. While you wait for her to return with the results, you wonder how you should react to the news she brings.

What if she tells you the test was negative? Of course, this doesn't mean you aren't sick at all, but does it mean you don't have strep throat? Not necessarily. A negative result means there's an approximately 95% chance that you don't have strep throat. Think about it this way. If you receive 20 rapid strep tests with negative results over the years, about one of those will be a negative result when you actually do have strep throat. So, if you have all the symptoms of the illness, your doctor may want to follow up with another test, called a *strep culture*, to verify the negative result. Even with the negative result on the rapid strep test, strep bacteria might be lurking there, undetected.

What if your doctor tells you the test was positive? Then you can be pretty certain you do have strep bacteria in your throat. However, about 15%–20% of people are carriers for strep. In other words, one out of every five or six people with strep bacteria in their throat do not have any symptoms of illness. So, even if strep bacteria are present, there's a chance this isn't the cause of your sore throat. Of course, because you do have a sore throat, a positive rapid strep test is a pretty good indicator that you are suffering from strep throat. So, when a doctor sees a positive rapid strep test result, they generally prescribe antibiotics right away. Strep throat responds very well

DOI: 10.4324/9781003300007-9

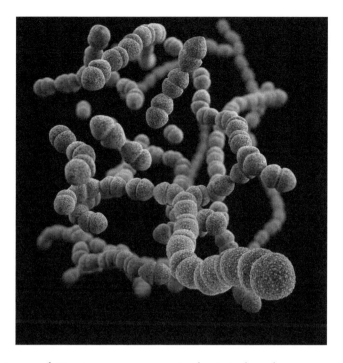

FIGURE 8.1 Image of *Streptococcus pneumoniae* bacteria based upon scanning electron microscopic imagery

to antibiotic treatment, and it's a good idea to get rid of strep bacteria even if it isn't causing your symptoms.

The rapid strep test, like most medical tests, gives you statistical data. Medical tests generally don't guarantee the presence or absence of a medical condition, but instead predict the medical condition with some ***probability***—making it somewhat or very likely, or somewhat or very unlikely. So, with any medical test result, you and your doctor need to decide how to interpret the statistical data, how strong of evidence it provides for the presence or absence of the condition, and what steps to take next.

Uses of probabilities

Uncertainty is a fundamental, unavoidable feature of life. This should be obvious from the discussion of inductive reasoning in Chapter 7. As you'll recall, induction is the main mode of reasoning both in ordinary life and in science. We have seen that, unlike deductive reasoning, induction is risky, since it does not guarantee the truth of its conclusions. For this reason, scientific knowledge based on inductive reasoning is always uncertain to some degree.

Thinking in terms of probabilities is a way to systematically and reliably reason under uncertainty and to represent the degree of uncertainty of the conclusions we reach. Tools developed to work with probabilities inform inductive and statistical techniques for quantifying uncertainty and the strength of inductive inferences, and for making inferences under uncertainty. Consider questions like: How should economists price options? How should insurance companies calculate risks? What role do random mutations play in biological processes? What's the chance that my child will inherit a particular trait of mine? What's the chance you have some illness given a positive test result? Reasoning with probabilities helps scientists answer these and many other questions. Even quantum mechanics, the fundamental laws of physics, are probabilistic: physicists can make only probabilistic predictions, and they can explain only in probabilistic terms how the world we are familiar with relates to fundamental particles.

We regularly encounter probabilities in our everyday lives, too. Probability calculations guide (or at least should guide) betting and gambling, determining insurance premiums, forecasting the weather, predicting the outcome of elections, and much more. You, just like anybody else, often make probabilistic judgments. Consider comments like: "I am 100% sure they will win the prize!"; "There is a one-in-a-billion chance I will be selected for that job"; or "What are the odds he will finish the marathon?" We typically use probabilities to ask or to describe how likely something is to occur or how confident we are that some belief or hypothesis is true.

Let's consider these uses of probabilities in a bit more depth:

(1) Figuring out the likelihood of a random mutation or of getting a particular hand of cards in poker.
(2) Predicting whether the stock market will go up or the weather will become colder.
(3) Judging whether Inter Milan will win the next Champions League, whether there is life outside Earth, or whether you'll pass the final exam.

These uses of probabilities differ from one another. In the first set of examples, (1), probability seems to be an objective feature of the world. Random mutations are just that: random. Whether one occurs is a matter of chance. We can calculate how likely a random mutation is to occur in general, but we can't predict when a particular mutation will occur. Similarly, poker and other card games are designed to start from a randomized hand of cards. This is why you shuffle a deck of cards and deal them in a standard way. In fact, these types of gambling situations first motivated mathematicians to develop a rigorous theory of probability and to systematically study situations involving uncertainty using that theory.

In the second set of examples, (2), probability seems to relate to objective features of the world, as with (1). But for (2), we have to rely on data to infer the probabilities of different outcomes. Economists use observations about the economy and markets, like the price of certain goods and past trends in the stock market (and a variety of models), to predict whether the stock market will go up and down. Meteorologists use a wide variety of data about weather patterns (and intricate computer simulations) to develop a 10-day weather forecast for locations all over the world. As new data become available, these probabilities may change to represent new states of uncertainty. This is why tomorrow's weather forecast is more accurate than the forecast for 10 days from now.

The third set of examples, (3), are judgments about outcomes that probably are not due to chance. In these instances, the probability isn't an objective feature of the world or our best estimation of probabilistic features of the world. Instead, the probability reflects your degree of confidence that some outcome will happen or that some hypothesis is true. These probabilities might be different for people with different information or even who disagree about the importance of some information. A fan of Inter Milan will have detailed knowledge of the team's recent performance and player status to inform their judgment. A coach for the team will have even deeper knowledge of these things and also know the team's game plan. In cases like these, probability seems to be subjective; it has more to do with one's beliefs than the objective state of the world. This doesn't mean we should just believe whatever we want, though. The tools of probability can still apply to our judgments about potential outcomes or ideas about the world.

These sets of examples illustrate that probabilities can be objective features of the world, they can be estimates or predictions of objective features of the world, and they can be our subjective judgments of how things are based on the data we have access to. It's a tricky business to interpret what probabilities mean, that is, how we should interpret ascribing some probability to some outcome. We'll delve deeper into how to interpret probabilities later in the chapter.

EXERCISES

8.1 **Recall:** Why don't medical tests provide guarantees of whether one has a medical condition? What do they provide?

8.2 **Think:** What are the two possibilities when you get a positive result on a medical test, and what are the two possibilities when you get a negative result on a medical test? What are possible considerations relevant for evaluating how to proceed?

8.3 **Recall:** Describe how probability relates to inductive reasoning and to uncertainty and given an example.

8.4 **Apply:** Give three examples of you reasoning with probabilities in the past week. (You might not have even realized you were doing it! Think about times you were trying to reason about something uncertain, make a prediction, or decide what to do when you weren't sure what to expect.) For each example, describe how probabilities were relevant to your reasoning.

8.5 **Recall:** Briefly characterize the three ways probabilities can relate to objective features of the world and to individuals' beliefs, and give an example of each.

8.6 **Apply:** Refresh your memory of the three ways probabilities can relate to objective features of the world and to individuals' beliefs, and then look back to the rapid strep test example at the beginning of the section. Which one of these interpretations of probability is the use of probabilities in the rapid strep test example most like? Provide a brief justification for your answer.

8.2 PROBABILITY THEORY

After reading this section, you should be able to:

- State the three axioms of probability theory
- Define the terms: *random variable, outcome space, independent outcomes, fair, total probability, mutually exclusive,* and *collectively exhaustive*
- Make simple calculations with probabilities by applying the rules of addition, multiplication, and subtraction

Random variables and outcome spaces

Thinking in terms of probabilities is a way to systematically and reliably reason under uncertainty and to represent the degree of certainty of the conclusions we reach. The tools for doing this are based in **probability theory**, a mathematical theory developed to quantify uncertainty and to reason about random events.

In the context of probabilistic reasoning, randomness does not mean haphazard or lacking aim or purpose. Instead, random events are outcomes that are individually unpredictable but that behave in predictable ways over many occurrences. These outcomes can be represented in terms of values of random variables. Recall that *variables* are anything that can vary, change, or occur in different states and that can be measured. **Random variables** are variables whose values are individually unpredictable but predictable in the aggregate.

The simplest examples of random variables are things like coin tosses and dice throws. In a normal roll of a die, you can't know whether you'll roll a one, two, three, four, five, or six before rolling the die and seeing the outcome. But even before rolling the die, you do know that if you roll the die 500 times, you almost certainly won't roll a six every time. Probability theory enables you to calculate what that probability is; it can tell you exactly how unlikely it is to roll a six 500 times in a row.

The set of all values a random variable can have is called its **outcome space** (or sample space). For a coin toss, the random variable is *figure shown on the top of the coin*. The only two possible outcomes of a coin toss are heads and tails; these are the two values of the outcome space for the variable *figure shown on the top of the coin*. We can represent all of this with the following expression:

$$X = \{h; \, t\}$$

Here X represents the random variable *figure shown on the top of the coin*, and h and t represent the values this variable can have (heads and tails, respectively). The symbols "{"and "}" are curly braces, which is the conventional notation used to indicate a **set**, that is, any grouping of elements in no particular order. So this expression shows that the outcome space for the random variable of a coin toss is simply heads and tails.

There are a few features of random variables that affect how the tools of probability apply to them. First, some variables have ***independent outcomes***, that is, the probability of each outcome does not affect the probability of the other outcomes. This is also called *statistical independence*, which we define later in this chapter and discuss further in Chapter 9. Whether a coin toss results in heads or tails on each throw is an independent outcome: how the last coin toss went won't change the probability of getting heads. But how many heads come up on a series of 10 throws is not independent from whether you get a heads or tails on each individual throw.

A useful feature of some random variables is that they are ***fair***, that is, they have independent outcomes that are all equally likely as one another. If a coin is fair, then the probability of the coin landing on heads will equal the probability of it landing on tails. This is useful, since it enables us to immediately know that the probability of getting heads or tails is 50/50. (We'll see why this is so later, when we begin calculating probabilities.)

A random variable that is not fair may be ***biased*** in favor of one or more outcomes. This means some outcomes are more likely—have a higher probability of occurrence—than other outcomes. Roulette is a casino game that developed in France. A French roulette wheel has 37 pockets, numbered 0 through 36, but an American roulette wheel has 38 pockets, two of which are 0. American roulette wheels are not fair, in a probabilistic sense, because the roulette is biased toward 0. Zero will occur more often than any other number on the wheel if we spin the roulette over and over again, because there are two ways to get a 0 but only one way to get every other number on the wheel.

A second way in which a random variable may not be fair is if its outcomes aren't independent from one another. The most common way this can happen is when previous outcomes influence future outcomes. Coin tosses, dice throws, and roulette spins are usually fair: each one is a new try, and nothing about the last one affects it. Imagine that you are spinning a roulette, and each time the ball lands in a pocket, you put something in that pocket to block it for the next spin. This way of spinning a roulette is not fair because its outcomes aren't independent. Once you've gotten a number, you know you won't get it on a subsequent spin of the wheel. This also changes the probability of getting the other numbers: they become more likely as you spin and block more numbers.

Box 8.1 The Gambler's Fallacy

Your friend plays video poker, and she reasons as follows: "Since this is a fair video poker, black and red cards come up equally often. There has been a streak of black cards. Therefore, a red card will come up very soon." This reasoning is called the **gambler's fallacy** because it is typical among aficionados of casinos. The gambler's fallacy is inferring from past variation from the expected frequency of outcomes that there will be future variation from the expected frequency in the opposite direction. Unfortunately for gamblers, this is bad reasoning. It is based on the mistaken idea that the outcomes of games of chance are not statistically independent. It can be tempting to think that unusually common occurrences will be averaged out by becoming unusually uncommon later. But, in fact, each draw of a fair video poker game is statistically independent, as are the outcomes of roulette spins, dice throws, and coin tosses. There is no way for one outcome to influence later outcomes. So, your friend is wrong that red cards become more likely after a streak of black cards. And the gambler's fallacy applies to much more than games of chance. A family with three girls still has a roughly 50% chance of a fourth child being a boy, though the parents might think they are more likely to have a boy this time around. Last week's record-breaking heat doesn't mean this week will be cooler than average.

To recap, a fair random variable must be unbiased and its outcomes must be independent. Coin tosses and dice throws are fair random variables, but lots of random variables are unfair. For example, LeBron James's free throw success is a random variable with two possible outcomes: LeBron either misses the free throw or scores. But the chance of LeBron scoring versus missing is probably not 50-50. Instead, there is a bias in favor of the outcome of scoring; for LeBron James, this is more likely than missing. The outcomes might also fail to be independent: missing a shot might make LeBron more, or less, likely to score on the next free throw. It's much more difficult to calculate probabilities for unfair variables like free throw success. So, for now, we'll stick with fair random variables, like coin tosses and dice throws.

The axioms of probability theory

Probabilistic reasoning begins with the observation of how probable it is for a random variable to take on a given value. The rules for assigning these probabilities are grounded in what are known as the *Kolmogorov axioms*, which are three basic assumptions that lie at the foundations of probability theory, introduced by the mathematician Andrey Kolmogorov. (Recall from Chapter 6 that axioms are statements accepted as self-evident truth about a domain used as a basis for deductively inferring other truths about the domain.) These three axioms are the basis for probability theory, that is, for the rules governing the calculation of probabilities.

Axiom 1 defines all probabilities as numbers greater than or equal to zero. We can express this axiom mathematically as:

Axiom 1: $0 \leq \Pr(X{=}o)$

Here X is any random variable and o is any value, and "$\Pr()$" means the probability of. Because Axiom 1 prohibits any probabilities lower than zero, this assigns the least likely outcomes—outcomes that are guaranteed not to occur—a probability of 0.

Axiom 2 says that the probability that at least one of the outcomes in the entire outcome space will occur is 1. This can be expressed as:

Axiom 2: $\Pr(X=\{o_{1-n}\}) = 1$

Here o_{1-n} represents the entire outcome space of any random variable X. No matter how many values a random variable can have, that whole set of values must have a combined probability of 1. The **total probability** of the outcome space for any random variable is always 1. Axiom 2 assigns the most likely outcomes—outcomes that are guaranteed to occur—a probability of 1.

Together, Axioms 1 and 2 ensure that all probabilities are between 0 and 1. An outcome that is guaranteed not to occur has a probability of 0. If you roll a regular six-sided die, you will never roll a 7, as this is not in the outcome space. An outcome that is guaranteed to occur has a probability of 1. If you toss a coin, then the figure on the top of the coin will always be heads or tails. If you check the weather tomorrow, it will always be raining or foggy or clear or snowing or some other precipitation status. The larger the probability of an outcome between these extremes—the closer it is to 1 and the further it is from 0—the more likely it is to occur. If you believe there's an 80% (or .80) chance of rain tomorrow, you're more certain it will rain than if you believed there's a 40% (or .40) chance of rain. If you believe there is a 110% (or 1.10) chance it will rain tomorrow, then this belief violates Axiom 2. If you assign a probability of −0.4 to the outcome of enjoying this chapter, then this also violates Axiom 1. (We hope we do better than 0, too!)

Axiom 3 says that the probability that one of several outcomes occurs is equivalent to the sum of the individual outcomes' probabilities. This can be expressed as:

Axiom 3: $\Pr(X=o_1 \text{ or } o_2 \text{ or } \ldots) = \Pr(X=o_1) + \Pr(X=o_2) + \ldots$

As earlier, X and o are any random variable and a value that variable can have. There's an important restriction on which values this axiom holds for, though. This equivalence only holds for **mutually exclusive** outcomes, or a set of values that a variable cannot take on simultaneously. On a single coin toss, for example, you will never get both heads and tails; so heads and tails are mutually exclusive values or outcomes. Axiom 3 governs the probability of one of multiple outcomes occurring; it enables the calculation of probabilities for general sets of outcomes, like getting heads or tails on a coin toss or it either raining or sleeting tomorrow.

Calculating probabilities

Recall that a fair random variable has independent outcomes that are all equally likely. For a fair six-sided die, rolling any number between 1 and 6 is equally probable. The outcome space for a die roll is $D = \{1; 2; 3; 4; 5; 6\}$, and by Axiom 2 we know $\Pr(D=\{1; 2; 3; 4; 5; 6\}) = 1$. Because these outcomes are equally probable, this means that each of the six numbers has a 1/6 probability of being rolled. Or, $\Pr(D=1) = \Pr(D=2) = \ldots = 1/6$.

We might also want to know the probability of other types of outcomes. For example, on a single die roll, how probable is rolling an even number? What's the probability of getting anything on our roll except a 4? And, if we roll two dice, what's the probability of getting two 6s? These probabilities also can be found by doing calculations based on the axioms of probability theory.

Consider the question of how probable it is to roll an even number on a single throw of a fair, six-sided die. This can be found using Axiom 3. The probability we are looking for can be expressed as: $\Pr(D=2$ or $D=4$ or $D=6)$. Each of these three outcomes has a probability of 1-6, and the three outcomes are mutually exclusive. So, by Axiom 3:

$$\Pr(D=2 \text{ or } D=4 \text{ or } D=6) = \Pr(D=2) + \Pr(D=4) + \Pr(D=6)$$

$$= 1/6 + 1/6 + 1/6$$

$$= 3/6, \text{ or } 1/2$$

Remember the importance to this calculation of the outcomes being mutually exclusive. If I ask you about the probability of rolling an even number or a 6 when you roll a single die, you cannot simply apply Axiom 3. Because 6 is one of the even numbers on the die, the outcomes of rolling a 6 and rolling an even number are *not* mutually exclusive. So, you can't simply add up the different probabilities to find the answer.

The **addition rule** is a generalization of this calculation that applies to outcomes that are not mutually exclusive to find the probability of any of those outcomes occurring. For outcomes that aren't mutually exclusive, the probabilities are still added, but then the probability of *both* events occurring is subtracted:

Addition rule: $\Pr(X=o_1 \text{ or } X=o_2) = \Pr(X=o_1) + \Pr(X=o_2) - \Pr(X=o_1 \text{ and } o_2)$

Applying this to find the probability of rolling both a 6 an even number, we get

$$\Pr(D=\text{even or } D=6) = \Pr(D=\text{even}) + \Pr(D=6) - \Pr(D=\text{even and } D=6)$$

$$= 3/6 + 1/6 - 1/6$$

$$= 3/6, \text{ or } 1/2$$

Anytime you roll a 6, you've rolled an even number, so the probability of rolling an even number and a 6 is still 1/6. Remember, the addition rule is a way to calculate the probability of any of multiple outcomes occurring.

We also posed the question of the probability of getting two 6s when we roll two dice; this isn't a question about any of multiple outcomes occurring, but a question about *all* of multiple outcomes occurring. According to the **multiplication rule**, the probability that all of a series of outcomes will occur is the result of multiplying their individual probabilities:

Multiplication rule: $\Pr(X=o_1 \ \& \ X=o_2) = \Pr(X=o_1) \times \Pr(X=o_2)$

When we ask about the probability of getting two 6s when we roll two dice, we are asking for the probability of rolling a 6 on one die roll and also rolling a 6 on a second die roll (whether the same die or a different one). The probability we are looking for is thus $Pr(D_1=6$ and $D_2=6)$. Of course, there's a 1/6 probability of a 6 for any given die roll, so $Pr(D_1=6 \& D_2=6) = Pr(D_1=6) \times Pr(D_2=6) = 1/6 \times 1/6 = 1/36$. The probability of 1/36 is a lot closer to 0 than to 1/6. That's why rolling two 6s is exciting: it seldom happens!

The multiplication rule as we've introduced it has an important limitation. It only applies to independent outcomes; recall that outcomes are independent when the probability of each outcome does not affect the probability of the other outcomes. Imagine we wanted to calculate the probability of rolling a 6 on one die roll but also a 1 on the very same die roll. We can't just multiply $1/6 \times 1/6$ because these outcomes aren't independent: if one occurs, the other is guaranteed not to occur. This means the probability in question is maximally improbable: it's 0. So, we can only use this multiplication rule to find the probabilities of a series of outcomes all occurring if the outcomes in question are independent from one another.

Let's take a moment to compare the multiplication rule with the addition rule. We saw that the addition rule is used to calculate the probability of *any* of a series of outcomes occurring. You could use it to calculate the probability of getting a 6 or a 1 on a single die roll: $1/6 + 1/6 = 2/6$, or 1/3. The multiplication rule is instead used to calculate the probability of *all* of a series of independent outcomes occurring. We used it earlier to calculate the probability of getting a 6 on two rolls: $1/6 \times 1/6 = 1/36$. (They have to be different rolls or different dice to be independent outcomes.)

What about the probability of rolling a 6 on either of two die rolls? This is about any of a series of outcomes, but the outcomes aren't mutually exclusive. So, we use the addition rule, but we need to subtract the probability of getting a 6 on both roles. And we found that probability with the multiplication rule! So, here goes: $Pr(D_1=6$ or $D_2=6) = Pr(D_1=6) + Pr(D_2=6) - Pr(D_1=6 \& D_2=6) = 1/6 + 1/6 - 1/36 = 11/36$. The probability of rolling at least one 6 goes up a lot with two rolls instead of one, but it doesn't quite double.

In our examples, the addition rule led to a larger probability (closer to 1) and the multiplication rule led to a smaller probability (closer to 0). This will always happen. Addition will always increase probability, and multiplication will always decrease probability. This is because probabilities are always positive numbers between 0 and 1, and multiplying two numbers in that range, such as two fractions, always yields a smaller number while adding two positive numbers of any kind always yields a larger number. This can provide a quick way to remember when to add and when to multiply. Do you expect the probability to get larger or smaller for the occurrence you're calculating, compared to the outcomes that generate it? It's easier (more probable) to get any of some outcomes than each outcome individually: use addition. It's harder (less probable) to get some outcomes together than each outcome individually: use multiplication. (And these outcomes need to be independent.)

Here's one more calculation tool for probabilities. Recall Axiom 2: the total outcome space always has a probability of 1. The **subtraction rule** makes use of this: you can

TABLE 8.1 Addition, multiplication, subtraction rules and their conditions

Rule	Term	Function	Condition	Result
addition rule	any	disjunction (or)		probability always increases
multiplication rule	all	conjunction (and)	independent	probability always decreases
subtraction rule	not	negation (not)	collectively exhaustive	probability can be large or small

calculate the probability of an outcome you are interested in by subtracting the probability of all other outcomes in the outcome space from 1 (the total probability). So:

$$\textit{Subtraction rule}: \Pr(X{=}o_1) = 1 - \Pr(X{=}\{o_{2-n}\})$$

For example, what is the probability of getting anything except a 2 on a single die roll? The total probability is 1, and the probability of rolling a 2 is 1/6 (as it is for any other number from 1 to 6). So, the probability of getting anything but 2 is $\Pr(D{=}\text{not-}2) = 1 - \Pr(D{=}2) = 1 - 1/6 = 5/6$.

The subtraction rule only applies to **_collectively exhaustive outcomes_**, or a set of outcomes of which at least one must occur. For a successful die roll, the die must land with one side up showing a 1, 2, 3, 4, 5, or 6—there is no other option. So, 1–6 are collectively exhaustive outcomes for the variable _die roll_. This is what makes probabilities sum to 1, which is necessary for the rule. This requirement is most easily satisfied with the use of the word "not"—rolling a 2 and not rolling a 2, rolling an even number and not rolling an even number, rolling two 6s in a row and not rolling two 6s in a row. Each of these pairs is collectively exhaustive; any possible outcome would fall in one or the other category. The word "not" can prompt you to use the subtraction rule, as it is one way of guaranteeing collectively exhaustive outcomes.

The subtraction rule can, of course, also be used in combination with the multiplication and addition rules. Previously we used the latter rules together to calculate the probability of getting a 6 on either of two dice rolls; it was 11/36. What about not getting a 6 on either of two dice rolls? This probability can be found from our earlier calculation combined with an application of the subtraction rule. $\Pr(D_1{=}\text{not-}6$ and $D_2{=}\text{not-}6) = 1 - \Pr(D_1{=}6$ or $D_2{=}6) = 1 - 11/36 = 25/36$.

EXERCISES

8.7   Recall: Write the mathematical formulation of each of the three axioms of probability theory, then say what each means.

8.8 **Recall:** Define each of the following terms: *random variable, outcome space, independent outcomes, fair, total probability, mutually exclusive,* and *collectively exhaustive.*

8.9 **Think:** Recall from this section that an American roulette wheel is not fair; its outcomes are biased because there are 38 pockets, and two of those are 0. Describe what it means for the wheel to be biased in this way. Are the spins of an American roulette wheel still independent? Why or why not?

8.10 **Apply:** There are not just cubical dice with six sides but many other polygons as well. Imagine you have a four-sided die with sides numbered 1, 2, 3, and 4. Assume the die is fair: each side is equally likely to be rolled. Write out the following in notation like we've introduced in this section for the variable *4-sided die roll (D).*
   a.   the variable's outcome space
   b.   the total probability of the outcome space
   c.   the probability of rolling an odd number

8.11 **Apply:** Consider the fair four-sided die described in Exercise 8.8. Calculate the following using the addition, multiplication, and subtraction rules as needed. Use notation like we've introduced in this section, show the calculation steps, and specify which rule you use for each step.
   a.   the probability of rolling a 3 on both of two rolls
   b.   the probability of rolling a 3 on every roll for three rolls
   c.   the probability of not rolling a 3 on a single roll
   d.   the probability of rolling a 3 on at least one of two rolls

8.12 **Think:** Imagine you are participating in a psychological experiment. The experimenter gives you the following problem: "Linda is 31 years old, single, outspoken, and bright. She majored in philosophy. As a student she was deeply concerned with issues of discrimination and social justice. Which of the following two alternatives is more probable?
A. Linda is a bank teller
B. Linda is a bank teller and active in the feminist movement."

Which one would you choose, answer A or B? Why?

If your intuitions work like those of most people, you picked B. But B cannot logically be more probable than A. Using the relevant rule for evaluating the probability of a conjunctive statement of the form "x and y," can you explain why B cannot be more likely than A? Do you think ordinary people understand the term "more probable" in this example in terms of the basic rules of probability, or do they understand it in some other way?

8.3   REASONING WITH CONDITIONAL PROBABILITIES

After reading this section, you should be able to:

*   Describe how conditional probabilities are important and define the base-rate fallacy
*   Calculate a conditional probability with the needed background information

• Apply Bayes's theorem to calculate a conditional probability with the needed background information

Conditional probability

So far, we've largely focused on independent outcomes, like individual coin tosses and dice throws, but many outcomes for random variables are not independent. For example, the probability that you get a 6 on two dice rolls goes up once you've gotten a 6 on one roll. (That probability goes to zero once you've gotten anything but a 6 on the first roll.)

Another useful role for probability is in calculating the probability of an outcome given that some other outcomes occur. For example, what is the probability of rolling a 6, given that you rolled an even number? What's the probability that global warming will be limited to 1.5° Celsius above pre-industrial levels if countries meet the goals specified in the Paris Agreement? Or—remember our initial example in this chapter—what's the probability that I have a medical condition, given that I tested positive?

The ***conditional probability*** of an event is the probability of its occurrence given that some other event has occurred. In the notation we've been developing, we can write the conditional probability of a random variable Y taking the value y, given that a variable X takes the value x as $\Pr(Y=y \mid X=x)$. The symbol "|" can be read as "given that."

For two independent outcomes, the conditional probability of one outcome given the other's occurrence will be the same as the original probability of the outcome. Indeed, the concept of conditional probability enables us to specify more exactly what independence amounts to. Two random variables X and Y are ***statistically independent*** when $\Pr(Y=y \mid X=x) = \Pr(Y=y)$ and $\Pr(X=x \mid Y=y) = \Pr(X=x)$. This means that the outcome x occurring doesn't make the outcome y any more or less likely, and the outcome y occurring doesn't make the outcome x any more or less likely.

If an outcome y is not statistically independent from an outcome x, then the probability of y occurring goes up or down if x occurs. In extreme cases, one event can result in the probability of another event becoming 1 or 0. For example, the probability of a single die roll resulting in an even number is 1/2. But the probability of an even number given that you roll a 2 is 1, since rolling a 2 is one way of rolling an even number. The probability of an even number given that you roll a 3 is 0, since 3 is odd. That is:

$$\Pr(D_1=2 \text{ or } 4 \text{ or } 6 \mid D_1=2) = 1$$

$$\Pr(D_1=2 \text{ or } 4 \text{ or } 6 \mid D_1=3) = 0$$

In other cases, statistical dependence is subtler: the probability of an event is raised or lowered by the occurrence of another event but not all the way to 0 or 1. Consider again the probability of getting two 6s when two dice are rolled. We calculated that the probability of this outcome is 1/36. But what is the probability of getting two 6s on two rolls if the first roll yields a 6? The chance has gone up, but it's still not guaranteed. Figuring out the conditional probability requires calculation. For two outcomes x and y, the probability of y occurring given that x occurs can be calculated using this formula:

Conditional probability: $Pr(Y=y \mid X=x) = Pr(Y=y \ \& \ X=x)/Pr(X=x)$

Think of this formula as a two-step procedure for finding the probability of y given x. First, you limit your attention only to cases when x occurs. This is the role of $Pr(X=x)$ as the denominator (the bottom) of the equation. Second, you look within those cases of x occurrences for occurrences of y. This is the role of $Pr(Y=y \ \& \ X=x)$ as the numerator (the top) of the equation. If the outcomes are restricted to only those cases when x occurs, this becomes the new outcome space for the variable Y. (This calculation only works when the probability of x is greater than 0, since dividing by 0 doesn't yield a real number.)

Let's try this out to find the probability of getting two 6s in two dice rolls, given that the first roll is a 6. Imagine that you decide to roll the dice one at a time, and you've rolled the first but not yet the second. Plugging this example into the formula gives us:

$$Pr(D_1{=}6 \ \& \ D_2{=}6 \mid D_1{=}6) = Pr((D_1{=}6 \ \& \ D_2{=}6) \ \& \ D_1{=}6)/Pr(D_1{=}6)$$

Before we solve this equation, take a moment to figure out why this equation is the right version of the formula for calculating conditional probabilities. Then, we can just plug in the probabilities we already know and do some simple math to solve the equation. $Pr((D_1{=}6 \ \& \ D_2{=}6) \ \& \ D_1{=}6)$ just is the probability of getting two 6s; $D_1{=}6$ is just listed twice. It shows up twice because the first roll had to be 6 for it to be possible for both rolls to be 6s. So, plugging in the probabilities:

$$Pr(D_1{=}6 \ \& \ D_2{=}6 \mid D_1{=}6) = (1/36)/(1/6) = 6/36 = 1/6$$

One nice thing about starting with this simple example is that we can check the answer. What is the probability of rolling two 6s given that you've already rolled one 6? This is the same as the probability of getting a 6 on one roll, since that's exactly what needs to happen now to get two 6s (since you already have one 6). And we know the probability of getting a 6 on a single die roll is 1/6. So, our calculation of the conditional probability gave us the right answer.

Let's try our hand at finding a slightly more difficult conditional probability for dice throws. What's the probability that you roll a number that is less than 4 on a single throw, given that you roll an odd number? This is the same as asking about the probability of rolling a one, two, or three (the outcomes less than four) given that you roll a one, three, or five (the odd outcomes). Applying our conditional probability formula, this yields:

$$Pr(D{=}1 \text{ or } 2 \text{ or } 3 \mid D{=}1 \text{ or } 3 \text{ or } 5) = Pr(D{=}1 \text{ or } 2 \text{ or } 3 \ \& \ D{=}1 \text{ or } 3 \text{ or } 5)/Pr(D{=}1 \text{ or } 3 \text{ or } 5)$$

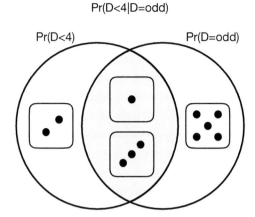

FIGURE 8.2 Visualization of the conditional probability of rolling a number less than four given that you roll an odd number

Using the addition rule, the probability of rolling a 1, 3, or 5 is 3/6: 1/6 + 1/6 + 1/6. We plug this in as the denominator. To find the probability of both rolling a 1, 2, or 3 and rolling a 1, 3, or 5, we need to notice that this can only happen when you roll a 1 or 3. Why? Because those are the only two ways of rolling both an odd number and a number less than four. Using the addition rule, the probability of rolling a 1 or 3 is 2/6. (We can't use the multiplication rule to calculate the probability, because the outcomes of rolling a number under four and an odd number aren't independent.) Plugging these into the earlier formula yields (2/6)/(3/6), which is equivalent to 2/3. The probability of rolling a number less than four given that you've rolled an odd number is 2/3; consulting Figure 8.2 should help you convince yourself that this is the right answer.

Bayes's theorem

The definition of conditional probability can be used to derive Bayes's theorem. Recall that a theorem is a statement deductively inferred from a set of axioms. This theorem is named after Reverend Thomas Bayes, an English statistician in the 18th century, and it has become central to many uses of probability and statistics in science. Bayes's theorem provides a way to calculate the conditional probability of an outcome Y given X from information about the conditional probability of X given Y (the opposite conditional probability) and general probabilities of X and Y. We'll talk through the formula, and then we'll see how this can be useful.

Recall the definition of a conditional probability we have just introduced:

$$\Pr(Y=y \mid X=x) = \Pr(Y=y \ \& \ X=x)/\Pr(X=x)$$

The same equation holds of the probability of *X* given *Y*:

$$Pr(X=x \mid Y=y) = Pr(Y=y \ \& \ X=x)/Pr(Y=y)$$

We can transform this second equation to solve for $Pr(Y=y \ \& \ X=x)$; that's just $Pr(X=x \mid Y=y) \times Pr(Y=y)$ (instead of dividing the right side of the equation by this, multiplying the left side of the equation by it). Substituting that in for $Pr(Y=y \ \& \ X=x)$ in the first equation yields **Bayes's theorem**:

$$\textit{Bayes's theorem: } Pr(Y=y \mid X=x) = (Pr(X=x \mid Y=y) \times Pr(Y=y))/Pr(X=x)$$

This equation is useful when the conditional probability of *Y* given *X* can't be calculated directly, but there's information about the conditional probability of *X* given *Y* and the general probabilities of *X* and *Y*. You can think of the *X*s in this equation, then, as information we have, and the *Y* what we are trying to determine the probability of given this information.

Medical testing is one of many useful applications of Bayes's theorem. Imagine I am screened for a medical condition that affects about one person in 1,000. I have no symptoms, and I know the test is accurate 90% of the time. This means that if I do have the medical condition, then the test result is positive with 90% probability; and if I do not have the condition, the test result is negative with 90% probability. (In other words, about 9 people out of 10 who have the medical condition get a positive result, and about 9 out of 10 who don't have the medical condition get a negative result.) After several anxious minutes, the test results come back: positive! Do you think I have good reason to worry? Why?

At the beginning of the chapter, we noted that a positive test result, such as a rapid stress test, doesn't guarantee that you have the condition tested for, and a negative test result doesn't guarantee that you don't have the condition. Bayes's theorem can be used to calculate the actual probability that you have the condition given that you received a positive test result. We want to find via Bayes's theorem the probability that you have the medical condition (*C=yes*) given that the test was positive (*T=yes*). Plugging these variables and values into the theorem shows us what we are looking for:

$$Pr(C=\text{yes} \mid T=\text{yes}) = (Pr(T=\text{yes} \mid C=\text{yes}) \times Pr(C=\text{yes}))/Pr(T=\text{yes})$$

Now we need to find the probabilities on the right-hand side from the information provided. The general probability that I have the medical condition, $Pr(C=\text{yes})$, is .001, since we know that about one person in a thousand is affected by the condition. If you pick 1,000 people randomly from the human population, you'd expect to find one person with this medical condition. The chance that I have the medical condition thus is .001, which just is 1/1,000. This is often called the **prior probability**, since it's the probability of the outcome we are interested in, not taking into account the new information (in this case, the result of the medical test).

We also know the probability of getting a positive test result given that one has the medical condition, Pr(*T=yes* | *C=yes*); this is .90. Remember that this medical test is 90% accurate. If one has the condition, the probability of a positive test result is .90. This number is often called the **likelihood**—how likely the outcome we've observed (positive test result) would be if the outcome we're curious about (having medical condition) has in fact occurred.

The last probability we need to use Bayes's theorem is Pr(*T=yes*), or the probability of receiving a positive test result regardless of whether one has the medical condition. We don't immediately have this probability, but we can figure it out. Either one has the medical condition or not, and in each of these circumstances, there's some likelihood of getting a positive test result. What are those likelihoods? We've already said that the likelihood of a positive test result if one has the medical condition, Pr(*T=yes* | *C=yes*), is .90. The likelihood of getting a positive test result if one does not have the condition, Pr(*T=yes* | *C=no*), is .10, since if one does not have the condition, there's a 90% chance of a negative result. This leaves a 10% chance of still getting a positive result. But it's not equally likely that one does and doesn't have the condition; recall that the medical condition affects just 1/1000 people, or .001; 999/1000 people, or .999, don't have it. We can put all of this together to find the probability of a positive test result as follows:

$$\Pr(T{=}yes) = (\Pr(T{=}yes \mid C{=}yes) \times \Pr(C{=}yes)) + (\Pr(T{=}yes \mid C{=}no) \times \Pr(C{=}no)$$

$$= (.90 \times .001) + (.10 \times .999)$$

$$= .0009 + .0999 = .1008$$

Now we have all the ingredients we need to apply Bayes's theorem. Plugging in the probabilities we've found, we get Pr(*C=yes* | *T=yes*) = .001 × .90 / .1008 = .0089, or .89%. This result is called the **posterior probability** since it is conditional on an observed outcome, namely the positive test result. It turns out that the probability that I have the medical condition given the positive test result is small: it's still less than 1%! So perhaps I should not yet get too worried. The probability that I have the medical condition has gone up a lot, from .001 to .0089, but it's still very likely that I don't have the condition. Bayes's theorem just gave me an incredible sense of relief.

We called the general probability of having the medical condition the *prior probability*. Prior probabilities are also known as *base rates*. These terms both refer to the probability of some outcome in the absence of other information. The proportion of the population in a country that is employed by the government in civil service, the proportion of all humans struck by lightning, and the proportion of all people with a certain medical condition are all examples of base rates. For example, in the US, 1.9% of the population works for the federal government, and 98.1% do not. The base rate of civil servants in the US is 1.9%, or 19/1,000. The base rate of people struck by lighting is estimated to be 1/15,300.

In scientific and everyday reasoning, base rates provide us with important information for making correct probability judgments, though it's easy to forget about base rates. The ***base-rate fallacy*** is neglecting the base rate and focusing only on the individual information. For example, this happens when people, sometimes including healthcare professionals, assume someone has an unusual medical condition because they have received a positive test result. Consider that, for the medical test example we just worked through, many more of the people testing positive for the condition do not have the condition than do have the condition, simply because the condition only affects 1/1,000. In situations where some event is particularly striking, like a new disease outbreak, a terrorist attack, or a lightning strike, ignoring base rates can result in mistaken judgments and in wasting money and time trying to intervene on very unlikely events. Base rates and new evidence should both be taken into account in probabilistic reasoning, and Bayes's theorem is an important tool for doing so.

Box 8.2 Probabilistic biases

Several cognitive biases influence probabilistic reasoning and judgment. We have already encountered the base-rate fallacy and the gambler's fallacy. Three other examples are the conjunction fallacy, the neglect of probability bias, and the zero-risk bias.

The ***conjunction fallacy*** is the error in judgment when we judge a conjunction of two events to be more likely than either one of the events on their own. For example, if you meet a random person in their twenties and know nothing about them, do you think they are more likely to be a student or a student struggling for money? Sure, students are often poor, and so that might strike you as likely. But it's always more probable to be a student than to be a student and broke.

The ***neglect of probability*** is a bias in judgment or decision under uncertainty that comes from disregarding probabilistic information. People's purchase of lottery tickets, anxiety about terrorist attacks, and opinions about the safety of vaccines all tend to neglect relevant probabilities—all of these are extremely unlikely outcomes.

The ***zero-risk bias*** is a preference for policies—especially about health and the environment—that entirely eliminate a risk instead of alternatives that more effectively reduce risk (by being cheaper, easier to implement, etc.) Many people asked if they prefer (a) winning $100 for sure, or (b) 10% chance of winning $500, 89% chance of winning $100, and 1% chance of winning nothing, opted for the zero-risk alternative (a)—even though (b) is 99% likely to yield at least the same earnings and has a good change of yielding lots more.

Watch Video 12

EXERCISES

8.13 Recall: Write out the formula defining a conditional probability. In your own words, describe how the formula can be used to find the probability of getting two 6s in two dice rolls, given that the first roll is a 6.

8.14 Apply: Use the formula for conditional probability to find the probability of getting an odd number on a single die roll given that you did not get a 6. Show each step of your reasoning and indicate any rules you use (addition, multiplication, subtraction).

8.15 Recall: Write out Bayes's theorem. Label the parts of the formula that are the prior probability, the likelihood, and posterior probability. In your own words, describe how we used the theorem to find the probability that you had a medical condition given a positive test result (with the numbers involved in our example).

8.16 Apply: Imagine you arrived in a new town and ask a random stranger about the best bar in town.

You know in this town 10 out of every 100 people will lie. You also know of the 10 people who lie, 8 have a red nose. Of the remaining 90 people who don't lie, 9 also have a red nose. Suppose you meet a group of 100 people in the town with a red nose. Use Bayes's theorem to calculate how many people out of this group of 100 will lie.

8.17 Think: Describe how conditional probabilities are important. Then, define and give an example of the base-rate fallacy.

8.18 Think: Consider a fair 10,000 ticket lottery that has exactly one winning ticket. You are justified to believe that any particular ticket of the lottery will not win, because the probability of any one ticket winning is 1/1000. But if you are justified to believe that ticket #1 (#2 & #3 & . . . #10000) will not win, then you should also believe the paradoxical conclusion that no ticket will win in that lottery!

There are three individually plausible ideas that together seem to generate this "lottery paradox":
(a) One is justified to believe a proposition that is very likely true.
(b) If one believes that A is true and they also know that A deductively entails B, then they should also believe B.
(c) One is never justified to believe a contradiction.

First explain why you should believe the paradoxical conclusion no ticket will win in that lottery, and then reflect on how to solve the lottery paradox. Do you think that the trouble may be with one of the three prior ideas? Which one and why?

8.4 INTERPRETING PROBABILITIES

After reading this section, you should be able to:

- Describe the problem of interpreting probability and why it is challenging
- Describe the subjectivist interpretation of probability and the two main challenges for it

- Describe the frequency and propensity interpretations of probability and indicate the main challenge for each

The problem of interpreting probability

The first section of this chapter gave three types of examples of probabilities:

(1) Figuring out the likelihood of a random mutation or of getting a particular hand of cards in poker.
(2) Predicting whether the stock market will go up or the weather will become colder.
(3) Judging whether Inter Milan will win the next Champions League, whether there is life outside Earth, or whether you'll pass the final exam.

You have surely faced situations like these. And you have made and heard claims like, "A coin is just as likely to land heads as tails," "There's an 80% chance of snow tomorrow," and "I will probably pass the exam." But what do these kinds of claims about probabilities mean? And what makes them true or false?

In the beginning of the chapter, we said of these sets of examples that (1) seems to be objective features of the world, (2) seems to be our estimates of objective features of the world, and (3) are our guesses of what outcomes will arise for processes that probably aren't actually based on probabilities. Now we'll delve deeper into the question of how probabilities should be interpreted, including whether situations like (1), (2), and (3) should be interpreted in the same way. Different interpretations of probability provide different accounts of what kind of things probabilities are and of what makes probability claims true or false. Given the ubiquity of probability claims in science and everyday life, it is important to understand how probabilities should be interpreted and why.

There are two broad families of interpretations. The first family is called *epistemic* (this word means relating to knowledge, knowing, or rational belief) and grounds interpretations of probability in the notion of ***rational degree of belief***, or the extent to which a rational agent should believe some claim. Within the family of epistemic interpretations, we will focus on the subjective interpretation. The second family is called *objective* and understands probability as an objective feature of the world, just like rocks, sunshine, and water—on this interpretation, probabilities, just like these things, exist regardless of any opinion you or anybody else may have about them. We will focus on two kinds of objectivist interpretations: the frequency interpretation and the propensity interpretation.

Subjective interpretation

According to the subjective interpretation of probability, the probability of an outcome is the subjective, rational degree of belief of a suitable agent that the outcome will obtain. If I say I will probably pass the exam or that the weather forecast predicts an 80% chance of snow tomorrow, for example, these probabilistic judgments express some suitable agents' rational degrees of confidence or certainty about certain outcomes—not just about what my confidence happens to be but about the confidence someone should have. On this interpretation, probability claims aren't made true or false by how the world is but instead by the rational degrees of belief of suitable agents.

To understand this view, we need to be clear on what a degree of belief is and when a degree of belief is rational. A ***degree of belief*** is the amount of confidence in the truth of a given hypothesis. But how can you find out about somebody's degree of confidence that some hypothesis is true? Probability theorists have tried to quantify this by thinking of it in terms of possible bets agents would accept or reject. The idea behind this is suggested by ways of talking like "You can bet she will be late to the meeting" or "I am willing to risk all my salary on the result of the political election," which express the conviction something will happen or will not happen.

We can figure out somebody's degree of belief in a hypothesis or degree of confidence in an outcome by identifying the odds on a bet about the hypothesis such that the person would be equally willing to take either side of the bet. To bet on the truth of a hypothesis at odds of 10:1, say, means to be willing to risk losing $10 if the hypothesis is false while gaining only $1 if the hypothesis is true. This would be a bad bet on a coin toss because each time you happen to guess correctly you only gain $1 while you lose $10 for each (equally likely) incorrect guess. But if you are really certain about something, this bet gets you easy money, since you know you aren't actually risking $10. So, the higher odds you are willing to accept on a bet corresponds to increased confidence that the hypothesis you are betting on is true.

At some odds, you aren't going to be sure whether you should bet for or against a hypothesis. Imagine you gain $10 if you correctly guess whether the coin will land heads and lose $10 if not. Should you guess that it will or won't land heads? If you think the coin is fair, you probably can't decide. Guessing it will land heads is not obviously a bad bet (risking more than you stand to gain) or a good bet (gaining more than you stand to lose). When I am equally willing to take a bet that a coin will land heads and a bet that the coin will not land heads at certain odds, this means those are subjectively fair odds for me. The odds are subjective, and so they might be different for someone else.

These subjectively fair odds for me indicate my degree of belief in the hypothesis the bet is about. (I don't know about you, but for me, that's 50/50 for a coin toss.) In general, if my subjectively fair odds for a bet on some hypothesis is $X:Y$, then my degree of belief that the hypothesis is true can be calculated as $X/(X + Y)$. This is the ratio between my stake X (how much I could lose) and the total amount staked $X + Y$. For example, if you believe that a 3:1 odd is fair for a bet that you will pass this course, then I can conclude that your degree of belief you will pass this course is 75% (or .75). You are willing to lose $3 if you fail in return for $1 if you pass, because you're pretty sure (but not incredibly sure) you will pass. Your degree of belief that you will pass the course is $3/(3 + 1) = 3/4 = 75\%$. You might not have been able to specify this, but your betting behavior revealed it.

Now we have a way to find degrees of belief through (imagined) betting behavior. But we've emphasized that these are subjective degrees of belief—what each of us happens to choose, with no right or wrong choices. The next step is to explore what makes a degree of belief rational. This is often understood as probabilistic coherence: a rational degree of belief respects the axioms of probability.

The basic idea for why it's rational for your degrees of belief, and betting behavior, to conform to the axioms of probability is that not doing so leads you to accept bets that guarantee you lose money. This is called *the Dutch book argument*; a Dutch book is a set of bets that guarantees the person establishing the betting terms, the book-maker, will profit. The only way not to lose money (or, not to be wrong more often than you are right) is to have subjective degrees of belief that align with the rules of probability. This alignment with the rules of probability is taken to be what makes a degree of belief rational.

Imagine my degree of belief that a coin toss will land heads is 0.6, and my degree of belief that the toss will land tails is also 0.6. These beliefs violate the rules of probability theory because the full outcome space, heads or tails, must have a probability of 1, but I've assigned it a probability of 1.2. Someone offers to bet me $10 to my $15 that the outcome will be heads and also $10 to my $15 that the outcome will be tails. Given my degrees of belief about the outcomes of heads and tails, I should accept both bets as fair. If the coin lands heads 60% of the time, as I believe it will, then I'll win six out of every 10 coin tosses, so the $15 losses will be made up for by getting more $10 wins. And the same for the coin landing tails, given what I believe. (To check this, remember that for odds $X:Y$, my degree of belief can be calculated as $X/(X + Y)$. So to accept 15:10 odds, my degree of belief should 15/25, or 0.6, which it is.) It's easy to see I'm going to lose money on this bet! And this is because my degrees of belief fail to be rational because they violate the rules of probability theory.

So, on the subjective interpretation, probabilities just are the rational degrees of belief of suitable agents. This interpretation is intuitive for many uses of probability, especially situations like those in (3) earlier, where it seems probabilities are describing our judgments of what is likely to occur. But there are at least two challenges for the subjective interpretation.

One challenge is that, if the only constraint on rational degrees of belief is coher-ence with the rules of probability theory, this does not prohibit rational degrees of belief that are very odd. For instance, you may judge that the probability that the Netherlands will disappear under water tomorrow is 0.9. Given what you know about climate change and the Netherlands, this belief of yours would be very odd and not informed by relevant evidence, but it doesn't violate probability theory, so it counts as rational. In other words, mere coherence with probability theory is a very weak criterion for rational belief. Another challenge with the subjective interpretation concerns the link between belief and betting behavior. In particular, betting behavior doesn't seem to be a good guide about what one should believe. One might dislike gambling, be more or less attached to money, or value money differently in differ-ent circumstances. This would change one's betting behavior, but it wouldn't change what one believes to be true.

Objective interpretations: frequency and propensity

Suppose an honest person tells me, "Draw a card from a standard 52-card deck. If you draw an ace of hearts, you pay me $1. If you draw any other card, I pay you $1000." Should I accept?

The answer seems obvious. I should accept because it is less probable to draw an ace of hearts than to draw some other card in a standard 52-card deck. The probability of drawing an ace of hearts is only 1/52 (about 0.019). This seems like an objective feature of the world that does not depend on anybody's opinion or beliefs. Probabilities like this one, like the situations described in (1), seems to refer to probabilities that are objective features of how the world is. How can we make sense of probabilities as objective features of the world?

According to the *frequency interpretation*, the probability of an outcome is the limit of its relative *frequency*, or how often the outcome occurs, in a very long series of trials. Consider the random variable *Card drawn from a standard 52-card deck*. This variable can take the values *ace of spades, ten of clubs, king of diamonds, queen of hearts,* and so forth. Each draw is a trial: each draw returns a value for that random variable. If you put the card you drew back in the deck, shuffle the deck, and draw another card, and you repeat this operation over and over, that will be series of trials. Over a very (very) long series, you will observe the outcome *ace of hearts* once for every 52 trials. According to the frequency interpretation, this proportion is the probability of the value *ace of hearts* for the random variable *Card drawn from a standard 52-card deck*.

One challenge for this interpretation of probability is that it does not assign probabilities to one-off events, that is, for outcomes where there couldn't possibly be a long series of trials. For example, there will never be a series of trials for whether humans will land on Venus by 2025 or for whether the LA Lakers will win the NBA Championships next year. These events only happen once. So, on the frequency interpretation, it would be meaningless to say that the probability my party will win the next elections is 10% or that the probability you'll pass the exam is 75%. These are like the scenarios in (3) earlier.

A different objective interpretation of probability is the propensity interpretation. On this view, the probability of an outcome is the propensity of the physical conditions to produce the outcome. We can understand propensities as causal dispositions to produce certain outcomes, or the tendency to behave in certain ways under certain circumstances. Consider properties like being fragile or being soluble. Fragility is the causal disposition of glass to shatter when struck; solubility is the causal disposition of salt to dissolve when put into water. So, according to the propensity interpretation, a standard 52-card deck has a causal disposition to produce the outcome *ace of hearts* when a card is drawn from it. This causal disposition is the propensity to produce the outcome *ace of hearts* about once every 52 draws. On the propensity interpretation this is the meaning of the claim that there is a 1/52 probability to draw an ace of hearts from the deck.

The propensity interpretation can assign probabilities to one-off events since it doesn't rely upon the idea of series of trials. But it faces its own challenges. One

Subjective interpretation	Objective interpretation: Frequency	Objective interpretation: Propensity

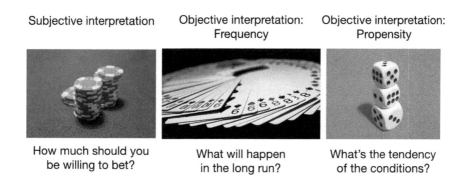

How much should you be willing to bet?	What will happen in the long run?	What's the tendency of the conditions?

FIGURE 8.3 The subjective, frequency, and propensity interpretations of probability differ in how they interpret what a probability claim means

challenge is to make sense of the relationship between causality and probability. If propensities are causal tendencies, then the propensity interpretation does not seem to allow us to make sense of some conditional probabilities. We can specify the probability of having a headache given that one drank too much wine, and we can also specify the probability that one drank too much wine given that one has a headache. But, while drinking wine has some propensity to produce a headache, headaches don't cause one to have consumed wine in the past. Some probabilities, like the probability that you drank too much wine given that you have a headache, don't seem to be causal dispositions.

Each of these interpretations of probability can account for some uses of probability and has difficulty making sense of other uses. It might be that the mathematical tools of probability theory can be employed in multiple ways. Perhaps these tools capture how truly random outcomes play out, like the scenarios in (1), and as suggested by the frequency and propensity interpretations, and the tools also apply to how we reason, or should reason, about the likelihood of outcomes, like the scenarios in (3), and as suggested by the subjective interpretation. Consider again the scenarios in (2): predicting whether the stock market will go up or the weather will become colder. These seem to be in a gray area between (1) and (3), where the probabilities are about occurrences in the world but are shaped by our knowledge and reasoning processes. It's unclear whether one interpretation of probability can be developed to accommodate all these uses of probabilities, and others still, or if probability means different things in different uses.

EXERCISES

8.19 Recall: Describe the problem of interpreting probability. Why is this a challenge?

8.20 Recall: Describe the subjectivist interpretation of probability and the two main challenges facing it.

8.21 **Recall:** Describe each of the frequency and propensity interpretations of probability, and then indicate the main challenge for each.

8.22 **Apply:** Consider the scenarios described in (1)–(3) at the beginning of this section (also introduced in section 8.1). Choose one scenario from each numbered group and describe how subjective, frequency, and propensity interpretations would each interpret the probability claim. Put an asterisk next to the interpretation you think is most successful for each of the three scenarios you've focused on, and briefly say why you think it's the most successful.

8.23 **Think:** Based on the interpretations introduced in this section and the discussion of their pros and cons, what do you think is the right way(s) to interpret probability? Do you believe there must be only one correct interpretation of probability? Why, or why not? Use simple examples to support your answers to these questions.

FURTHER READINGS

For a comprehensive treatment of various formal approaches to uncertain reasoning, see Halpern, J. Y. (2017). *Reasoning about uncertainty*. MIT Press.

For more on how people (mis)use probability for reasoning and decision-making under uncertainty, and what this means for human rationality, see Tversky, A., & Kahneman, D. (1993). Probabilistic reasoning. In A. Goldman (Ed.), *Readings in philosophy and cognitive science* (pp. 43–68). MIT Press. Also see Samuels, R., Stich, S., & Bishop, M. (2002). Ending the rationality wars: How to make disputes about human rationality disappear. In R. Elio (Ed.), *Common sense, reasoning and rationality* (pp. 236–268). Oxford University Press.

For a historical perspective on how the modern concept of probability emerged, see Hacking, I. (2006). *The emergence of probability: A philosophical study of early ideas about probability, induction and statistical inference* (2nd ed.). Cambridge University Press.

For more on the leading philosophical interpretations of probability, see Hájek, A. (2019). Interpretations of probability. In E. N. Zalta (Ed.), *The Stanford encyclopedia of philosophy* (Fall 2019 ed.). https://plato.stanford.edu/archives/fall2019/entries/probability-interpret/

CHAPTER 9

Statistical reasoning

9.1 WORLD RELIGIONS AND TWO USES OF STATISTICS

After reading this section, you should be able to:

- Characterize what statistical information reveals about religious belief worldwide
- Describe what statistics is and what purposes it serves
- Indicate two differences between descriptive and inferential statistics

Is religion a thing of the past?

Do you believe religion is going to disappear in the near future? Do you believe Christians are less oppressed than other religious communities? Depending on where you live and what communities you participate in, you might believe one or both of these ideas. But the data indicate otherwise.

According to a series of studies conducted by the Pew Research Center, a nonpartisan policy institute based in Washington DC, all major religions around the world are currently growing and will continue growing in the future. As of 2022, compared to nonbelievers, religious people are generally younger, have more children, and live outside western Europe and North America. Atheists, agnostics, and others without any religious affiliation are indeed increasing in numbers in countries like France and the US, but the populations of those countries are a declining share of the world's total population. Islam is growing faster than any other religion, though Christianity will remain the largest religion for the next few decades. There are now about 2.3 billion Christians, of which more than half are Catholics. And, though it may sound surprising, Christians are harassed and persecuted in more countries than any other religious community.

If you wonder how we can possibly know all of this, the answer is that we know based on statistical evidence.

Up until the 18th century, *statistics* referred to any data relevant to running a state, like data about birth and death rates, the size of different religious groups, individual and national wealth, the level of employment, and the share of the population eligible for military service. Today, **statistics** is a set of tools broadly used in science to

DOI: 10.4324/9781003300007-10

TABLE 9.1 Size and projected growth of major religious groups

	2010 Population	% of World Population in 2010	Projected 2050 Population	% of World Population in 2050	Population Growth 2010–2050
Christians	2,168,330,000	31.4	2,918,070,000	31.4	749,740,000
Muslims	1,599,700,000	23.2	2,761,480,000	29.7	1,161,780,000
Unaffiliated	1,131,150,000	16.4	1,230,340,000	13.2	99,190,000
Hindus	1,032,210,000	15.0	1,384,360,000	14.9	352,140,000
Buddhists	487,760,000	7.1	486,270,000	5.2	−1,490,000
Folk Religions	404,690,000	5.9	449,140,000	4.8	44,450,000
Other Religions	58,150,000	0.8	61,450,000	0.7	3,300,000
Jews	13,860,000	0.2	16,090,000	0.2	2,230,000
World total	**6,895,850,000**	**100.0**	**9,307,190,000**	**100.0**	**2,411,340,000**

Source: Pew Research Center

systematically collect, curate, analyze, present, and interpret data. The data don't have to be about human populations—they can be about literally anything targeted in scientific investigation, from subatomic particles to long-extinct species.

Sound statistical reasoning and the analysis of others' statistical reasoning are incredibly important for knowing what to believe and why. This is even more crucial because scientists, journalists, politicians, pundits, and most people are sometimes sloppy, or even deliberately misleading, in how they use statistical methods and evidence. The ability to correctly assess statistical information (and misinformation too!) is crucial.

Descriptive and inferential statistics

There are two main kinds of statistical reasoning. The first kind, the main focus of this chapter, is ***descriptive statistics***: summarizing, describing, and displaying data in a meaningful way. Finding a class's average score on an exam is a common use of descriptive statistics. Finding patterns in a data set—such as averages, trends, and correlations—and graphically representing them are forms of descriptive statistics with many applications, from scientific research to business and politics.

The second kind of statistical reasoning, the topic of Chapter 10, is ***inferential statistics***, or using statistical reasoning to draw broader conclusions on the basis of limited data. This kind of statistical reasoning is used to make inferences from a data

% of global population, 2010-2050

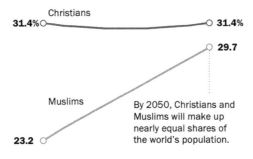

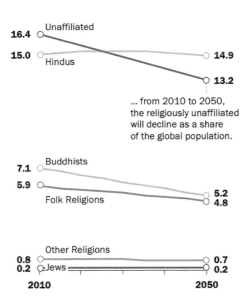

FIGURE 9.1 Projected percent of global populations for different religions, 2010–2050

set that "go beyond" that data set to capture features of the broader phenomenon the data represent. You may use tools of inferential statistics to estimate the distribution of grades you would expect in your whole course based on the exam grades from only some students or to predict the grades of students in other sections of the course from the grades in one section. We saw in Chapter 7 that drawing conclusions that "go beyond" what is known is a feature of inductive inference, where the relationship between premises and conclusion is one of probability rather than certainty or necessity.

This is true for inferential statistics as well. In fact, inferential statistics is also known as *inductive statistics*.

To illustrate the distinction between descriptive and inferential statistics, consider our example of statistics about religious belief. *Religiosity* is a variable that can take different values: Hindu, Jewish, Christian, Muslim, Rastafarian, atheist, and so on. If you survey a group of people to measure that variable, you will probably find some variation in their answers. Some will tell you they are Hindus, others that they are atheists, others that they are Muslim, and yet others may prefer not to disclose that information.

Simple tools of descriptive statistics enable you to summarize with percentages the share of each value of the variable *Religiosity* for the group you surveyed. Suppose, for example, that you poll 30 people in your town or at your university, and you find that four are Jews, eleven Rastafarian, five atheists, two Buddhists, one does not say, and seven are Catholics. This means that five people among the 30 you polled are atheists, or 1/6 (16.6%) of your total sample. It also means the most common religion in your data set is Rastafarian. There are, of course, other conclusions you could draw. You may decide to summarize the data you collected graphically, perhaps in a pie chart, to make understandable at a glance how religiosity varies across the people you polled. You may also want to investigate the relationship between religiosity and other variables like age in the same group of 30 people; you may discover that almost all the seven Catholics among the people you polled are over 40 years old.

But many times when we conduct a survey, we are not really interested in the particular people we survey. Instead, we want to draw broader conclusions, about religiosity, for instance, on the basis of surveying a sample of people. You might want to draw conclusions about religious belief in your town or even across the world's population. Scientists, policymakers, and business analysts also regularly aim to generalize from the groups they study to a larger group, which they call a *population*.

In statistics, a population need not be the world population nor even a human population. A **population** is a large collection of entities that shares some characteristic. Some researchers are interested in populations of humans, but others study populations of bacteria, stars, subatomic particles, or more abstract objects like companies, households, and nations. In most cases, it is impossible to collect data about each individual in a population—think about surveying all the people in India on some question or collecting data about all stars of the Milky Way galaxy. For this reason, scientists regularly use inferential statistics to draw conclusions about a population on the basis of data about a subset of the population, or **sample**.

Researchers on world religions rely on census data, as well as polling, phone canvassing, media content analyses, and other sample data to answer general questions about the world population like: How many Hindus are there now? Is the number of nonreligious people growing? Will Catholics outnumber Evangelicals in 50 years in Brazil? These and other questions can be answered with the help of tools from inferential statistics. The answers can then be displayed with percentages, or graphically with charts and plots, using tools from descriptive statistics. The information on world religions we began the chapter with is based on inferential statistics. Instead of simply summarizing data, inferential statistics uses the data as evidence relevant to evaluating hypotheses about the larger population. This evidence is defeasible and is open to criticism and revision, just like any other type of evidence in science and everyday life.

A second difference between descriptive and inferential statistics concerns the kinds of error each involves, where *error* refers to the difference between the values recorded in a data set and the true value of the variable one is measuring. Both descriptive and inferential statistics involve measurement error, since all measurements are limited in precision and accuracy. Any time one tries to measure something—the weight of a person, the number of Buddhists in the world, or the economic performance of a company—the value obtained will vary across measurements, and most of the measurements will be slightly off the true value. For example, in polling 30 people about their religion, you might accidentally misreport one or two answers, or some of the people may lie or not to respond.

In contrast, only inferential statistics involves sampling error. Unlike descriptive statistics, inferential statistics is used to generalize from the observed sample to the mostly unobserved population. **Sampling error** is differences between the features of a sample and a population due to the unrepresentativeness of the sample. If the sample is not representative of the population you want to learn about due to sheer chance variation or nonrandom sampling, the conclusions you draw about a population from a sample may be incorrect. If you want to know about global trends in religiosity, for example, it's probably a bad idea to rely solely on a sample within a particular country. Even a randomized worldwide survey can lead to results that vary from the population to some extent by sheer chance.

EXERCISES

9.1 **Recall:** List a few features of religion worldwide that we know through the use of inferential statistics.

9.2 **Apply:** Define *sample* and *population*. Then, state whether each of the following statements refers to a sample or to a population, briefly justifying your answer:

a. Researchers found that 2% of the Americans they interviewed believed they had seen a UFO.

b. Based on their survey data, the researchers concluded that 1 in 3 of all car crashes in the country are linked to alcohol impairment.

c. Two thirds of the butterflies we observed were pink.

d. After reading four essays, the teacher expects that 85% of the class will pass the exam.

e. 25% of the planets in the solar system have no moons

f. More than one billion people in the world live on less than one dollar a day.

9.3 **Recall:** What are the two main differences between descriptive statistics and inferential statistics?

9.4 **Apply:** Indicate whether each of the following statements is based on descriptive or inferential statistics, and explain why.
   a. As of 2023, the director Quentin Tarantino has received a total of two Academy Awards.
   b. Students with an undergraduate GPA of 3.00 are expected to have a starting salary of $50,000.
   c. In 2020, the population of São Paulo, Brazil, was 12.33 million.
   d. The mean score of the class was B+.
   e. A study stated that British adults are nearly 12 kilograms heavier now than they were in 1960.
   f. Economists say that mortgage rates may soon drop.
   g. The gross national income per capita in South Sudan in 2015 was $1,040.
   h. According to the latest WHO data published in 2020, life expectancy in Bangladesh is 72 years.

9.5 **Apply:** Find a news article or opinion column published in the past week that uses statistical reasoning of some kind. After citing the source, write a paragraph describing:
   a. The main point of the article or column
   b. What statistics are provided
   c. How the author makes use of the statistics in their reasoning
   d. How good this use of statistical reasoning seems to be, and why

9.6 **Think:** Statistical reasoning pervades our lives, often in ways we don't realize. After reflecting on your daily routine, write out a list of four ways in which statistical reasoning is part of that routine, either explicitly or implicitly. In light of your examples, briefly explain how *incorrect* statistical reasoning could make a concrete difference to your life.

9.2 STATISTICAL DISTRIBUTION AND CORRELATION

After reading this section, you should be able to:

- Identify and evaluate pie charts, bar charts, histograms, and scatterplots
- Characterize different types of statistical distributions
- Define *statistical independence* and *correlated variables* and evaluate direction and strength of correlations from a scatterplot, regression analysis, or correlation coefficient

Visualizing statistical distributions

As mentioned in section 9.1, descriptive statistics involves summarizing, describing, and displaying data in a meaningful way. As a reminder, *variables* are anything that can vary, change, or occur in different states and that can be measured—that is, that take on different values. *Random variables* are variables whose values are individually unpredictable but predictable in the aggregate. Descriptive statistics can display such aggregate information. For instance, while you can't guess the religious belief of someone you don't know, you can use statistical data to describe what religions are most common and how religions vary in their distribution across the world.

Charts, tables, and graphs provide visual representations of the statistical features of random variables. This is a common form of descriptive statistics in scientific research, as well as in newspapers and magazines. With the graphical representation of statistics, the key to success is a simple, clear, and appropriate presentation of the data. Different kinds of visual representations are helpful in different circumstances, depending in part on the kind of data and the needs of the audience.

Consider *pie charts*, in which a circle is divided into different-sized slices to depict how much of the outcome space for a variable falls into different categories of values. The area of each slice represents the percentage of outcomes associated with that value of the variable: the bigger the percentage, the bigger the area. To avoid confusion, there shouldn't be too many categories shown in a pie chart, and the values of the variable those categories depict should be mutually exclusive and collectively exhaustive. This is needed so that the pie slices don't overlap with one another (mutually exclusive) and so they add up to a whole pie (collectively exhaustive). A pie chart is a useful way to represent *qualitative variables*—variables with values that are not numerical but descriptive, like the variable *sport*, with values *basketball*, *hockey*, and so on. For example, a pie chart can be used to represent the percentage of a coffee shop's sales for different beverages—say, americanos, cortados, cappuccinos, chai tea lattes, espresso macchiatos, and smoothies. Notice that if this list doesn't include every item on the menu, then there needs to be an "other" category as well so the values are collectively exhaustive.

Pie charts are most effective when representing a variable that can take a small number of distinct values. The values must be distinct, so you know where to draw the lines between them, and too many narrow slices makes a pie chart difficult to read and understand. Besides coffee shop beverages, pie charts also can be used to display the distribution of votes received by candidates in some election or the distribution of quiz grades earned by students. Determining the proper variable can sometimes be tricky; in the latter example, the variable is not student but grade (outcome space = {A, B, C, D, F}). This also shows how it can be necessary to group values into broad groupings: a pie chart of percentage grades would have too many categories for an effective pie chart, while letter grades are broad enough groupings to be depicted in this way.

Bar charts use bars of different heights to show the amount for different values of some variable. The values are typically placed horizontally and equally spaced; then vertical bars are used to represent the size of each value. This size can correspond to an absolute number, like the number of students who got an A, B, C, D, or F on the most recent test, or a percentage, like the percent of the class who got each grade. Like pie charts, bar charts are great for categorical variables that can take discrete values. The values should still be mutually exclusive, but they don't need to be collectively

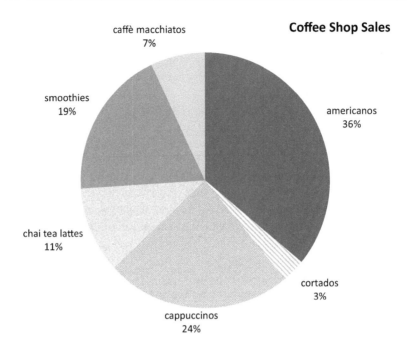

FIGURE 9.2 Example pie chart showing an imagined coffee shop's sales

exhaustive. (For a pie chart, the categories have to add up to a whole pie, but bar charts are not limited in this way.) Bar charts are better than pie charts for comparison of proportion between values or when there is a larger number of values. Figure 9.3 shows an example bar chart of per capita (per person) beer consumption across several countries. This wouldn't work as a pie chart, since the values aren't collectively exhaustive: more beer than this is sold in the world, and also each value is a per-person average rather than a national total.

Like bar charts, **histograms** are graphical displays of data that use bars of different heights. Unlike bar charts, the bars of a histogram are not distinct categories. Instead, the values are grouped into numeric intervals. Histograms are effective ways to visually depict values of **quantitative variables**—variables with numerical values, such as height or percent correct on an exam— that don't obviously fall into discrete categories. One example of such a variable is the height of students in your class. People's height varies continuously. If you develop a histogram of students' height, it is up to you to decide what numeric intervals to use to group the continuous variable of height—for example, you might group together all heights within 1-foot (30.48-centimeter) intervals or only within 10-centimeter intervals. Your decision will partly depend on the range of values, that is, the difference between the largest and smallest values in the data set. If everyone in your class is between 5 and 6 feet tall, grouping that full interval together

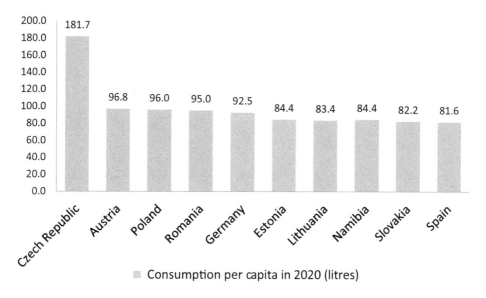

FIGURE 9.3 Example bar chart showing beer consumption per capita for several nations

will result in an uninformative histogram. Like bar charts, bar height in histograms may reflect the total number in each interval or their percentage.

Unlike bar charts, bars aren't necessary for a histogram. Histograms may use other approaches to visualizing the same information, such as color or color intensity. For example, consider the variable *ocean temperature*, with values presented as different shades of blue. Darker blues might represent colder temperatures near the poles, while lighter shades might represent warmer temperatures near the tropics and equator. And rather than a two-dimensional graph with an *x*-axis and *y*-axis, imagine projecting the values of those variables onto a map of the globe. Such a histogram is very different from a bar chart.

Bar charts and many histograms depict statistical information about the relative frequency of different values for a variable as relative height of bars or points. This can reveal features of the overall statistical distribution of values of the variable. If a histogram has a single peak, this shows that one value is the most common. The most common value is called the **mode**. So, a data set with just one peak, in which one value in a range is the most common, is called a **unimodal distribution**. If a histogram has two different peaks of similar height, then it reveals a **bimodal distribution**, where two values in a range are the most common; the two peaks correspond to the two most common values of the variable.

A histogram for class grade percentages nicely illustrates the difference between these distributions. A common distribution of grades is unimodal: it has one peak where the most common grades occur—often somewhere in the range of B to C. Math and logic courses often have bimodal distributions instead: they have two

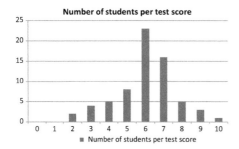
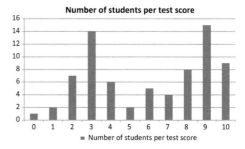

FIGURE 9.4 (a) Example histogram of a unimodal grade distribution (left); (b) Example histogram of a bimodal grade distribution (right)

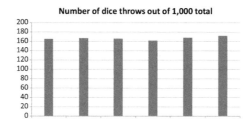

FIGURE 9.5 Example histogram of a uniform distribution of dice-throw outcomes

peaks, one at the top of the grading scale and the other in the middle or lower part of the scale. See Figure 9.4 for example histograms of unimodal and bimodal distributions.

If the height of the different bars in the histogram is the same for all values, then it shows a ***uniform distribution*** where all values are equally likely. Grading distributions are rarely uniform. Fair random variables should have uniform distributions across their values over lots of trials. For example, 1,000 dice throws should have an approximately uniform distribution across the individual outcomes of one, two, three, four, five, and six. See Figure 9.5 for a histogram depicting this uniform distribution.

Another important feature of statistical distributions visible in histograms and bar charts is symmetry—that is, whether the portions to the right and left of the mode(s) are the same. Symmetric graphs can have a uniform distribution, a U-shape, or a ∩-shape. A flat line—uniform distribution—is symmetric without a mode. A U-shape is a symmetric bimodal distribution where large and small values are the most common, while a ∩-shape is a unimodal distribution with the most common values clustered around the middle, with decreasingly common outcomes as the values get higher and lower. (See Figure 9.6.) As we will see in the next chapter, this ∩-shaped distribution, called a ***bell curve*** or ***normal distribution***, is especially important in inferential statistics. Note that a symmetric histogram might not be perfectly symmetric, just approximately so.

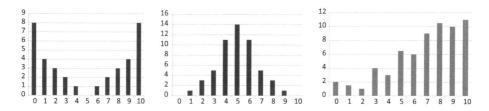

FIGURE 9.6 (a) Example symmetric bimodal histogram (left); (b) Example symmetric unimodal histogram (center); (c) Example asymmetric histogram (right)

Correlations

Most scientific research is concerned not just with variables but also with the relationships among them. For instance, some years ago French researchers studied whether people drink more alcohol when they hang out in loud bars. They found a positive correlation between the variable *decibel level in bar* and the variable *alcohol consumption*. If you ask whether level of marijuana consumption is different in different states in the US, you are interested in the relationship between the variable *marijuana consumption* and the variable *state of the USA*. Or, if you wonder whether being able to read at a younger age predicts salary level in adulthood, you are again asking about the correlation between the values of two variables.

Recall the definition of *statistically independent* in Chapter 8. When two variables are statistically independent, the value of one variable does not raise or lower the probability of the other variable taking on any given value. Variables that are not statistically independent are **correlated variables**: the value of one raises or lowers the probability of the other having some value. For example, the correlation found by those French researchers between loud bars and alcohol consumption means that a person going into a loud bar is more likely to have more alcoholic drinks than is a person going into a quiet bar.

When greater values for one variable are related with greater values for a second variable, these variables are said to be **positively correlated**. Decibel level of bar and alcohol consumption were found to be positively correlated. When greater values for one variable are related with smaller values for a second variable, these are said to be **negatively correlated**. Perhaps level of alcohol consumption on a given evening is negatively correlated with waking up early the following morning: the more alcohol someone drinks, the less likely that person is to wake up early.

For quantitative variables, scatterplots can provide a visual representation of whether and how the variables are correlated. A **scatterplot** is a graph in which the values of one variable are plotted against the values of the other variable. For example, the horizontal axis of the plot, the *x*-axis, may report the decibel level in different bars, and the vertical axis, the *y*-axis, the average number of drinks consumed in those different bars, as shown in Figure 9.7.

A scatterplot that shows a positive correlation between variables will have dots that tend to form an upward-sloping line from left to right. As the values of one

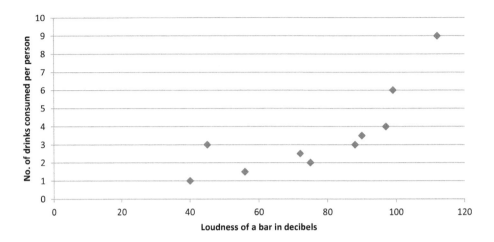

FIGURE 9.7 An example scatterplot of the relationship between alcohol consumption and decibel level in bars

variable get larger, the values of the other variable also tend to get larger. However, there can be exceptions—dots that vary from that general pattern. Some very quiet bars may serve a lot of drinks, and some very loud bars may serve few drinks. But this needn't eliminate the general correlation between decibel level and alcohol consumed.

A scatterplot that shows a negative correlation between variables will have dots that tend to form a downward-sloping line from left to right. As the values of one variable get larger, the values of the other variable tend to get smaller. Of course, there can be dots that vary from this pattern as well without interfering with the negative correlation.

What would you expect for a scatterplot of two variables that aren't correlated? Well, there won't be an upward sloping line, and there won't be a downward sloping line. What you usually see are dots all over the place, with no pattern between the values of one variable and the values of the other variable.

One compact way to summarize the relationship between variables is called *regression analysis*. The basic idea is to find the best-fitting line through the points on a scatterplot. Modern regression analysis was invented in the late 19th century by Sir Francis Galton. Galton was a geographer, meteorologist, tropical explorer, inventor of fingerprint identification, eugenicist, best-selling author, and half-cousin of Charles Darwin. He was obsessed with measurement. In 1875, Galton began to investigate heredity: why do successive generations remain alike in so many features? And how do offspring vary from their parents? One of his projects was to measure the diameter and weight of thousands of mother and daughter sweet pea seeds (see Table 9.2). He plotted his results and hand-fitted a line to his data as best as he could.

TABLE 9.2 Width of parent/offspring sweet pea seeds

| Seed | Diameter of Sweet Pea Seed (1/100ths of inch) | |
	Mother	Daughter
#1	15	15.3
#2	16	16.0
#3	17	15.6
#4	18	16.3
#5	19	16.0
#6	20	17.3
#7	21	17.5

Source: Galton 1889: 226

Galton wanted to find the line that best fit his data. Intuitively, this is the line that runs closest to the points scattered on a plot. Galton aimed to draw a line that minimized the sum of the distances of the points on the plot from that line, while still maintaining a straight line. In this sense, it can be considered the best fit. Figure 9.8 shows the best-fitting straight line for Galton's data.

As you might be able to guess from what we've already said about scatterplots, when there is a positive correlation, the line that best fits the dots on a scatterplot has an upward-sloping trajectory as it moves right, and when there's a negative correlation, the line has a downward-sloping trajectory as it moves right. IQ and SAT scores show a positive correlation: the slope of the regression line that describes the relationship between IQ and SAT scores is from the bottom left to the top right of a scatterplot. (See section 9.4 for discussion of what we can, and can't, conclude from that correlation.) In contrast, speed and accuracy in carrying out a task are negatively correlated: as speed increases, accuracy decreases. In this case, the slope of the regression line goes from upper left to lower right on the scatterplot.

A regression analysis also gives information about the **_strength of correlation_**: how predictable the values of one variable are based on the values of the other variable. The closer the dots are to the best-fitting line, the stronger the correlation, that is, the more linked the values of the two variables. (Notice that the slope of the line is not related to correlation strength; the slope only gives information about how the values of the variables tend to relate to each other.) A maximum strength correlation, often called a _perfect correlation_, will have all the dots directly on the regression analysis line. A very weak correlation will have dots that almost look uncorrelated; they fall all over the place, far from the line, but there's just a hint of a relationship between the values of the two variables. In Figure 9.9, you can

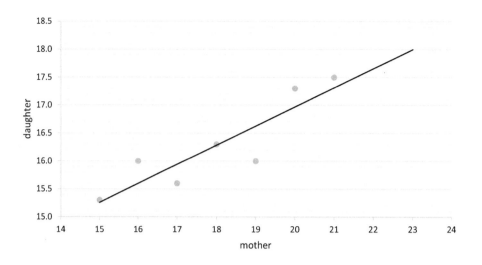

FIGURE 9.8 Regression analysis of Galton's data on the diameter of sweet pea seeds

see examples of a very strong correlation and a very weak correlation that have the same relationship among variables, so identical regression analysis lines.

From his regression analysis, Galton saw that, as the size of a mother sweet pea seed increased, so did the size of its daughter sweet pea seed. However, the daughter seeds tended to be less extreme in size compared to their mother peas: they "regressed" back towards average pea-size. Extremely large mother seeds grew into plants whose daughter seeds tended not to be as extremely large, and extremely small mother seeds grew into plants whose daughter seeds tended not to be as extremely small. Galton called this loss of extremity the **regression to the mean**. It can be explained as just an effect of variability: if a variable has an extreme value, then most other values that variable can have will be less extreme. So, even though mother and daughter pea sizes are positively correlated, extreme-sized peas tend to have less extreme-sized daughter peas (but even this is something that can vary). The same also holds true in reverse: extreme-sized peas usually have less extreme-sized mother peas.

Galton also determined a correlation coefficient for mother and daughter pea-size. A **correlation coefficient** provides information about the direction and strength of correlation. It has two parts: a positive (+) or a negative (−) sign to indicate positive or negative correlation respectively, and a number between 0 and 1 to indicate the strength of the correlation. This is a measure of the dispersion of the points on the scatterplot. The stronger the relationship between the two variables, the closer the correlation coefficient is to 1, when the value of one variable is a perfect predictor of the value of the other variable. A value of 0 means that the points on the plot are randomly scattered and the two variables are statistically independent: the value of one gives no information about the value of the other.

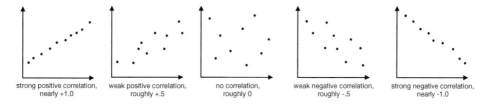

FIGURE 9.9 Example scatterplots with different correlational strengths and directions

---

Box 9.1 The history of statistics and eugenics

Statistics emerged as a distinctive discipline concerned with the collection, analysis, summary, and representation of information drawn from data in the 19th century, in response to the needs of industrializing and colonizing states like the UK. Researchers who made substantial contributions to the emergence of statistics included Francis Galton, Karl Pearson, Charles Spearman, and Ronald Fischer. Born and raised at a time when and in places where Anglo-Saxon White supremacism was widespread, these men held virulently racist views. Those views influenced their work in statistics on the measurement and interpretation of group differences. One of their explicit social aims was to create a society with the most desirable traits by suppressing reproduction in people deemed to be inferior. Galton coined the term **eugenics** (from the Greek: "well-born") for the idea that a human population can be improved by controlling breeding. He advocated against mixed-race marriages and in favor of incentives to encourage able, upper-class White couples to have children. In the 20th century, eugenics movements in several countries implemented policies restricting human liberties and threatening human dignity, including forced birth control, marriage restrictions, racial segregation, compulsory sterilization, and even genocide. The historical development and political abuse of tools in statistics is just one example of how scientific research has been misused to justify unethical and abhorrent practices.

EXERCISES

9.7 **Recall:** Indicate the defining features of each of the following ways to depict statistical information graphically: pie chart, bar chart, histogram, and scatterplot.

9.8 **Think:** Divide the following list of variables into qualitative and quantitative variables. For the quantitative variables, say which are discrete and which are continuous.
   a. the height of a mountain
   b. the color of starfish
   c. the breed of a dog
   d. the mass of a planet

    e.   the winner of Wimbledon

    f.   the population of a city

    g.   the outcome of a throw of a die

    h.   the GDP (gross domestic product) of a country

    i.   type of pizza

    j.   the number of pizzas one person eats per week

    k.   the amount of salt in the Atlantic Ocean

9.9   **Apply:** Label each of Figures 9.2, 9.3, 9.4, and 9.7 as a pie chart, bar chart, histogram, or scatterplot. Then, for each, write a few sentences interpreting what each chart or graph depicts about the data.

9.10 **Apply:** Draw simple histograms depicting each of the following:

    a.   a normal distribution

    b.   a unimodal distribution that is not normal

    c.   a bimodal distribution

    d.   a uniform distribution

    Then, draw a small line on the x-axis to indicate the mode(s) of any distribution that has one or more mode. Finally, describe (in words) what feature(s) distinguish the distributions in (b), (c), and (d) from the normal distribution in (a).

9.11 **Think:** A great many variables seem to be distributed *normally*. People's heights, examination grades, IQ scores, milk production of cows, errors in measurement, and so forth all appear to be captured by the normal (or Gaussian) distribution. And yet this doesn't mean all other distributions of frequency are *abnormal*. Do some research to find out why the normal distribution is called "normal." Report your findings in a brief paragraph, and explain how the label "normal" might be misleading.

9.12 **Apply:** Label each of the following as a scatterplot, regression analysis, or correlation coefficient. Then, evaluate the direction and strength of the correlation depicted in each. That is, say whether each correlation is positive or negative, and rank them from strongest to weakest.

    a.   −0.99

    b.   +0.28

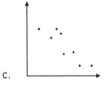

    c.

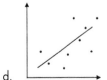

    d.

9.3 CENTRAL TENDENCY AND VARIABILITY

After reading this section, you should be able to:

* Explain the difference between the central tendency and the variability of a data set
* Define three measures of central tendency and find each for a data set: mode, median and mean
* Define three measures of variability and find each for a data set: range, variance, and standard deviation

Measures of central tendency: mean, median, and mode

Consider a ∩-shaped histogram, like pictured in Figure 9.6b. There is one peak at the center, corresponding to the most common group of values of the variable. There are also two "tails," the values that are less and less common the further away they are from the most common group in the middle. As we've discussed, the peak is the mode; this is one measure of central tendency, the most common or typical value(s) for a random variable. The distribution of occurrences across values, even the uncommon ones, is the *variability* (or dispersion) of the data set. Measures of variability capture the extent to which the distribution of values in a data set is stretched or squeezed.

Measures of central tendency and variability are summary measures that can compactly describe a whole data set. Let's consider measures of central tendency first; in the next subsection, we'll consider how to measure variability.

Imagine this situation. Your instructor has just returned the class's first quiz. You see that your grade is 6/10; your percentage of correct answers is 60%. How should you react? Perhaps you judge that you performed poorly—a 60% is quite low, isn't it? Yet another reaction might be to withhold judgment until you have additional information. You might want to compare scores with your classmates or inquire how the class performed as a whole or "on average." This additional information about the distribution of scores would help you know whether you did poorly on the quiz, and if so, how poorly.

Imagine the students' grades are as outlined in Table 9.3. Your instructor can provide you with three different answers to the question of how the class on average did on the quiz. These correspond to three different measures of the central tendency of a distribution (of grades, or anything else). These measures are the mode, median, and mean of the distribution.

The *mode* is the most frequent or most numerous value in the data set. As you can see from Table 9.3, the mode of the class's scores is 7/10. Four students scored a 7, which was the score that was more common than any other score.

The mode can be informative, and for qualitative variables, it may be the only measure of central tendency that can be employed. However, even for a unimodal normal distribution, or bell curve, the mode may not reflect the central tendency of a distribution well. Notice from the list of ordered scores that a 7, although the most frequent score, is lower than half of the students' grades. Some distributions have

TABLE 9.3 An imagined data set and central tendencies for 17 student scores on the first 10-point quiz

Student	Score	Student	Score	Central Tendency	
#1	0.0	#10	8.0	mode:	7.0
#2	5.0	#11	8.0	median:	7.5
#3	5.0	#12	8.5	mean:	7.1
#4	6.0	#13	8.5	variance:	5.0
#5	7.0	#14	9.0	standard deviation:	2.2
#6	7.0	#15	9.0		
#7	7.0	#16	9.0		
#8	7.0	#17	10.0		
#9	7.5				

more than one mode, which also limits the ability of the mode to capture the central tendency. Finally, if all values were different, then none would be more common than any other; such a distribution would have no mode at all.

The ***median*** is the very middle value in a distribution when the values are arranged from the lowest to the highest (or from highest to lowest). By "very middle," we mean that the median value splits the distribution exactly in half: half of the values are on one side, the other half on the other side. In our example, the median will be whatever score was earned by the student who did the ninth best/worst; the reason is that there were 17 students total, and so eight scored above that ninth student while eight scored below. Student #9 earned a 7.5 on the quiz. When the distribution has an even number of values, the median is the average of the two middle values.

The median is the preferred measure of central tendency when the distribution is not symmetrical. This is because the median is not strongly affected by ***outliers***, that is, by data values that are remarkably different from the rest like student #1, who scored a 0 on the quiz. But that strength is also a weakness, depending on the nature of the information you want to capture. You might want the central tendency measure to be different when some students bombed the quiz instead of all having scores grouped around the middle value. Further, unlike the mode, the median cannot be identified for qualitative variables, since the values these types of variables cannot be ordered from lowest to highest. (Is *cappuccino* lower or higher than *cortado*?)

The word *average* is also used to refer to the ***mean***, which is the sum of all values in the data set divided by the number of outcomes. The class's scores summed to 121.5, and so dividing that sum by 17 gives us a mean grade of 7.1 on this imagined quiz. Like the median, the mean cannot be calculated for categorical or qualitative data, as such values cannot be used in addition. Unlike the median, the mean is affected by outliers; the mean is pulled in the direction of the distribution's longer tail. The student who scored a 0 on the quiz pulled down the mean score by nearly half a point compared to the median.

When a distribution is unimodal and perfectly symmetrical, the mode, median, and mean coincide, and they are all exactly in the middle of the distribution. Asymmetric and multimodal distributions can lead these measures of central tendency to be radically different from one another.

To calculate a population's mean (μ) you sum all the data, x_i, in your data set and divide by the total number of data, N:

$$\mu = \frac{x_1 + x_2 + \ldots x_n}{N}$$

To find either the mode or the median, you should begin by ordering all the values in the data set from least to greatest. To find the mode, count how many times each value occurs; the value that occurs most often is the mode. There may be no mode, if no value appears more often than any other, or there may be two or more modes if two or more values occur the most often. For example, if your data set is {1, 3, 3, 4, 5, 5}, then 3 and 5 are the modes.

To find the median, search for the value in the very middle of the ordered data set; the middle value is the median. If there is an even number of data, and thus no single middle value, then the average of the two values in the middle is the median. For example, if your data set is {12, 14, 15, 17, 17, 19}, then 16 is the median.

Measures of variability: range, variance, and standard deviation

Variation is at the heart of what it is to be a random variable. And, like central tendency, variation can be measured. Measures of variation provide us with a summary of the spread of the values in a data set—that is, the degree to which they vary.

Information about variability can differentiate data sets that have the same central tendency, that is, the same mean, median, and mode. Let's return to our simple imaginary example of quiz grades. Suppose the next quiz has the exact same mode, median, and mean as the scores shown in Table 9.3. This suggests the class did equally well on this next quiz. And, on average, they did. But this isn't the whole story; compare the quiz grades in Table 9.4 to those in Table 9.3. What differences do you notice?

This second data set may have the same mean, median, and mode as the quiz scores from Table 9.3, but there is much less variation in scores. Visualizing the two data sets with a histogram makes it easier to spot the differences (see Figure 9.10). There is more variation in the grades on the first quiz (Table 9.3) than on the second (Table 9.4). As this illustrates, measures of central tendency don't capture all the information about a statistical distribution; you also need measures of variability.

There are three primary measures of variability: the range, variance, and standard deviation of a distribution. The **range** is the difference between the smallest and largest

TABLE 9.4 An imagined data set and central tendencies for 17 student scores on the second 10-point quiz

Student	Score	Student	Score	Central Tendency	
#1	4.0	#10	7.5	mode:	7.0
#2	5.0	#11	8.0	median:	7.5
#3	5.0	#12	8.0	mean:	7.1
#4	6.0	#13	8.0	variance:	1.9
#5	7.0	#14	8.5	standard deviation:	1.4
#6	7.0	#15	8.5		
#7	7.0	#16	8.5		
#8	7.0	#17	9.0		
#9	7.5				

values in a data set. To find the range, you can just subtract the smallest value from the largest value. For the first quiz, the range was 10, since the lowest score was 0 and the highest score was 10; for the second quiz, the range was 5, since the lowest score was 4 and the highest a 9.

Range does not take outliers into account very well, since it doesn't specify anything about the distribution of scores within the range. In other words, range won't tell you whether the distribution's tails are skinny or thick—the "spread" of the data. This can be done with a measure of the distance of values from the mean, such as variance or standard deviation. Both of these measures summarize the spread, or how close the various values are to the mean.

Population variance (σ^2) is the average of the squared differences of values from the mean:

$$\sigma^2 = \Sigma(\text{value} - \text{mean})^2 \, / \, N$$

The sigma (Σ) indicates that you should sum all instances, and N is the number of values in the data set. Notice that calculating variance requires knowing the mean of a data set. After calculating the mean, the first step to finding the variance is to find the difference of each value from that mean; this is the distance between the mean and each value in the data set. Finding this difference will show whether the values tend to vary a lot or only a little from the mean. Next, each difference is squared. (Otherwise the differences on either side of the mean would cancel each other out, since the difference for values greater than the mean is positive and the difference for values less than the mean is negative.) Finally, find the average of those squared differences by adding them together (Σ) and then dividing by the number of values (n).

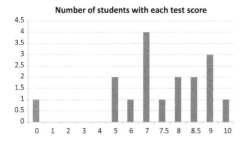 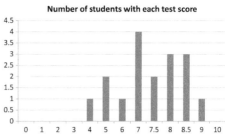

FIGURE 9.10 Histograms for student scores on the first exam (data in Table 9.3; left) and the second exam (data in Table 9.4; right)

Let's find the variance for our population of scores in the first and second quizzes. For both, the mean is 7.1 (rounded to one decimal point).

For the first quiz, the population variance would look like this:

$$(0.0 - 7.1)^2 + (5.0 - 7.1)^2 + (5.0 - 7.1)^2 + (6.0 - 7.1)^2 + (7.0 - 7.1)^2 +$$
$$(7.0 - 7.1)^2 + (7.0 - 7.1)^2 + (7.0 - 7.1)^2 + (7.5 - 7.1)^2 + (8.0 - 7.1)^2 +$$
$$(8.0 - 7.1)^2 + (8.5 - 7.1)^2 + (8.5 - 7.1)^2 + (9.0 - 7.1)^2 + (9.0 - 7.1)^2 +$$
$$(9.0 - 7.1)^2 + (10.0 - 7.1)^2] / 17 = 5.0$$

For the second quiz:

$$(4.0 - 7.1)^2 + (5.0 - 7.1)^2 + (5.0 - 7.1)^2 + (6.0 - 7.1)^2 + (7.0 - 7.1)^2 +$$
$$(7.0 - 7.1)^2 + (7.0 - 7.1)^2 + (7.0 - 7.1)^2 + (7.5 - 7.1)^2 + (7.5 - 7.1)^2 +$$
$$(8.0 - 7.1)^2 + (8.0 - 7.1)^2 + (8.0 - 7.1)^2 + (8.5 - 7.1)^2 + (8.5 - 7.1)^2 +$$
$$(8.5 - 7.1)^2 + (9.0 - 7.1)^2] / 17 = 1.9$$

Comparing the variances for the two quizzes makes clear that scores on the first quiz had more variation than on the second quiz.

The final measure of variation to discuss is the **standard deviation** (σ), which is calculated directly from the variance. For a population, the standard deviation is just the square root of the population variance:

$$\sigma = \sqrt{[\Sigma(\text{value} - \text{mean})^2 / N]}$$

Returning to our quiz example, the standard deviation for the first quiz is 2.2 (the square root of 5.0). For the second quiz, it is 1.4 (the square root of 1.9). The standard deviation provides us with a sort of "yardstick" for measuring variation. It is a number against which you can assess individual values or groups of values to see how far they are from the mean, relative to total variation in the data set.

If the histogram describing our data set is bell-shaped (unimodal and roughly symmetric), then around 68% of the values fall within one standard deviation of the mean,

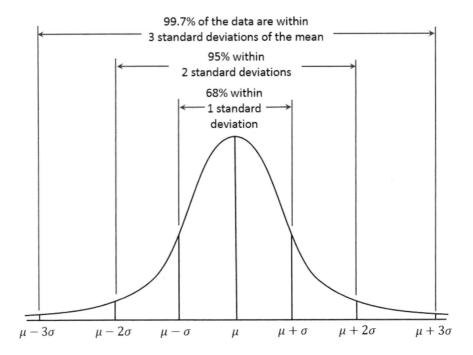

FIGURE 9.11 Standard deviation in a normal distribution; the values within one standard deviation of the mean account for 68.27% of the values in the data set, while those within two standard deviations account for 95.45%, and those within three standard deviations account for 99.73%

and around 95% of the values fall within two standard deviations of the mean—that is, fall within the distance that's twice as long as the standard deviation value. And, virtually all of the values lie within three standard deviations of the mean. Look at Figure 9.11: this plot shows the locations of one, two, and three standard deviations for a probability distribution associated with a bell-shaped histogram. The standard deviation distances will, of course, change depending on the "spread" of the data. The standard deviation value reflects this by being a relatively large number (lots of spread) or a relatively small number (little spread).

Mean and standard deviation are the most commonly reported summary statistics for a data set. Together, the mean and standard deviation capture central tendency and variability around that central tendency in a way that is informative and—as we will see in the next chapter—central to statistical inference.

You can calculate mean and standard deviation in four steps. First, find the mean. Second, for each value in the data set, subtract it from the mean and then square each result. Third, calculate the average of the resulting values (that is, sum the results and divide by the number of values) to find the population variance. And fourth, find the square root of the population variance, which is the standard deviation.

EXERCISES

9.13 Recall: List three measures of central tendency and three measures of variability. What roles do the concepts of central tendency and variability play in statistics?

9.14 Apply: Consider the following three data sets:
A = {4, 10, 11, 7, 15}
B = {10, 10, 10, 10, 10}
C = {1, 1, 10, 19, 22}
a. Without doing any calculations, guess which data set has the largest standard deviation.
b. Calculate the mean, mode, and median of each data set.
c. Calculate the range, variance, and standard deviation of each data set.

9.15 Think: In Israel in 2022, the average person earns about 12,000 Israeli shekels (more than 3,000 US dollars), their height is about 1.75 meters, and weight 85 kilograms. Do you think this "average person" really exists? Why or why not?

Now, consider that medical drugs work a certain way on average. For example, for adults, drug dosage is set for all adults by considering average weight. How should we understand this feature of medical drugs? What are the advantages and disadvantages of medical drugs working certain ways on average?

9.16 Think: Suppose the waiting time to be seated at a restaurant is an average of 20 minutes and has a standard deviation of 2 minutes. What can you conclude about when people are seated?

9.17 Apply: On one x- and y-axis, draw three histograms approximately depicting normal distributions for each of the following pairs of means and standard deviations. Label each with (a), (b), or (c), depending on which pair it depicts.
a. mean = 4, standard deviation = 2
b. mean = 8, standard deviation = 0.5
c. mean = 8, standard deviation = 4

9.18 Apply:
(a) Reconsider Table 9.2, which provides a simple data set used by Galton in his studies of heredity, and represent the data set with either a pie chart, bar chart, or histogram. (Choose carefully!) You might want to review what characteristics of a variable's values makes it well suited for each of these types of charts.
(b) Find the mean, median, and mode of this data set. Next, calculate its range, variance, and standard deviation.

9.4 SLOPPINESS, HYPE, AND MISUSES OF STATISTICS

- List three ways in which visual statistical representations, correlations, and numerical statistical representation can go wrong
- Describe the difference between absolute and relative risk

- Analyze visual and mathematical representations of statistical information for accuracy and potential problems

Misleading presentation of statistical information

So far in this chapter, we've seen how statistical data can be meaningfully represented visually in charts, tables, and graphs and mathematically with measures of central tendency and variability. These visual and mathematical techniques of representing trends in the values of variables and correlations among them are powerful. With these techniques, complex data can be depicted in ways that are simpler and easier to understand. Unfortunately, these same features of visual and mathematical representation of statistical data also make these techniques susceptible to misuse—unintentionally and even intentionally, when they are used to exaggerate or mislead.

Imagine you just read on social media that a vegan diet increases the risk of a rare skin disease by 100%. Learning this scary news is based on a study by a group of nutritionists; so you and many other vegans in the country decide to change your diet. Puzzled by your decision, your friend goes and actually reads the original study. The study says that, out of 20,000 omnivores, the researchers found that one person developed the rare skin disease, and, out of 20,000 vegans, two developed it. What the study found is that the absolute increase in the risk of developing the disease is only 1 in 20,000. Nothing to worry about—you may stick to your vegan diet.

The attention-grabbing, and scarier, 100% figure is the relative increase in risk found in that study. You may have thought you understand what 100% means, but you were misled. Although the absolute and relative risk in that study quantifies the same effect, you may not be aware of the difference between the two. An ***absolute risk*** is the number of individuals experiencing some condition—say, a rare disease—in relation to the relevant population—say, the general population of vegans. A ***relative risk*** is a comparison between the incidence of a condition in two groups—say, the groups of vegans and omnivores—and is generally expressed as a ratio or percentage. If the media had reported the findings of that study in terms of absolute risk, nobody would have paid attention.

Like the media, scientists also compete for attention. After all, their research is often dependent on grants, which can prize impact and showy findings rather than modest studies that only contribute a tiny bit to the existing knowledge. This type of pressure can lead the popular media and some scientists to misleadingly hype or misrepresent scientific findings. Statistical information can also be misleading simply due to sloppiness. Sometimes media outlets or reporters misunderstand the implications of scientific research they are reporting on.

All of this can make statistical information opaque and hard to understand for the general public. For example, bar charts (introduced in section 9.2) often feature in scientific work, as well as newspapers, magazines, and social media. They afford an automatic feeling of understanding. But, like numbers, graphs and diagrams can easily be exploited for communicating statistically information misleadingly. In an article on

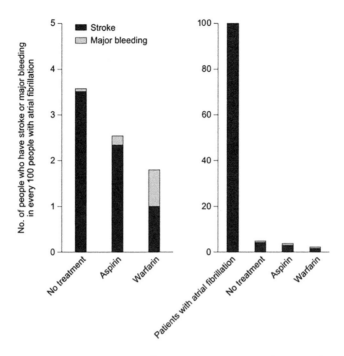

FIGURE 9.12 Two bar graphs with very different representations of the benefits of aspirin and warfarin; reproduced from Kurz-Milcke et al. (2008)

how statistical information can be made more transparent, Kurz-Milcke and collaborators illustrate this point with two bar graphs presenting the same type of statistical information from a study on heart conditions.

The bar chart on the left compares the incidence of two conditions, stroke and major bleeding, in patients with a heart condition who took either aspirin or warfarin and patients who did not have any treatment. You can see clear differences between the heights of the three bars, showing that aspirin is better than no treatment at all, but warfarin is better than aspirin at preventing these conditions. But this representation invites the same confusion between absolute and relative risk reduction we described earlier. The problem is apparent from the bar chart on the right, where the same statistical information is presented in terms of a reference group of 100 people with the heart condition. With the addition of this reference class, we get a sense for absolute risk: it's true that warfarin decreases risk of stroke compared to aspirin or no treatment at all, but that reduction is risk is merely from 3%–4% risk (no treatment) or 2% risk (aspirin) to 1% risk. This may still be valuable risk reduction, but it is a much more modest improvement than it seemed from the chart on the left. This is just one example of how communicating statistical information transparently can be challenging and can offer perverse opportunities to mislead the scientific and public audiences.

Thinking critically about correlations

In section 9.2, we noted there is a correlation between IQ and SAT scores. But there is also a strong correlation between a child's socioeconomic status and their IQ scores, as well as between their socioeconomic status and their SAT scores. So, we can ask whether IQ is responsible for SAT scores at all, or if socioeconomic status is just responsible for both.

There's a familiar adage: correlation is not causation. If you find a correlation between higher IQ scores and higher SAT scores, for example, you cannot automatically conclude that higher IQ causes higher SAT scores. That's a sloppy and misleadingly hyped way of presenting your finding. There are other possible reasons for this correlation.

It's true that we can learn a lot from correlations. We'll focus on causal reasoning in Chapter 11, but it's already important to clarify that, although correlation differs from causation and does not guarantee causation, the presence of a genuine correlation among a set of variables is a defeasible clue to causal relationships. Having an accurate picture of correlations among variables helps scientists understand interrelationships in complex phenomena—like education and scholastic performance. An accurate pattern of correlations can also guide the design of experimental studies aimed at finding out about the causal structure of those phenomena. Yet, despite these and other uses, "correlation is not causation" is a familiar saying because it's all too common for scientific research and science reports in the popular media to assume causation from mere correlation. Given this tendency, you should think critically about what correlation demonstrates, especially correlations uncovered with non-experimental studies involving a great many variables.

Here's an example. Researchers randomly picked 50 ingredients from recipes in a cookbook and, for each, looked for published nutrition science research concluding that the ingredient affects one's risk of cancer. They found that 40 of the 50 ingredients—including tea, tomatoes, carrot, parsley, nuts, and bread—were claimed to have an impact on cancer risk. But most of the claims were based on single, non-experimental studies reporting implausibly large correlations. Some of these correlations might have been spurious too, since the larger the number of variables involved in a study, the higher the chance of finding a correlation between some ingredient and some health outcome. The sheer volume of research on foods and cancer risk—and popular media reports of this research—can lead to outsized attention to specific nutrients compared to other well-established and significant influences on cancer risk, such as alcohol and tobacco use.

Box 9.2 Panic headlines

Behavioral economist Emily Oster quickly gained a following for her no-nonsense analysis of scientific research bearing on everyday decisions, especially decisions related to pregnancy and parenthood. She has an electronic newsletter, and some of her newsletters discuss what she calls "panic headlines." Oster is referring to the tendency of popular media to report scientific findings—especially those bearing on health—in dramatic or even hyperbolic ways. One such headline from early 2023 was, "Wine before pregnancy changes baby's face." Panic headlines make us more likely to read the article, but they can also provoke misplaced fear and overstate scientific findings. Fortunately, some of what we've learned in this book can be used to help put panic headlines in context. Here are some questions to ask about any popular reporting of scientific findings:

1. Do the scientific findings reported fit with the headline? Many headlines are misleading, overstate conclusions, or focus on one small detail instead of the big picture.
2. Is the report based on one research study? If so, what is the consensus in the field? Are there any meta-analyses of studies you can consult?
3. Can you tell whether the research attempted to control variables? If not, is there reason to suspect the correlation is spurious or influenced by confounding variables?
4. Is there any indication of the strength of the correlation or effect size? Identifying a correlation or an effect doesn't guarantee the relationship is strong.

Let's now consider the correlation between IQ, SAT, and socioeconomic status. IQ scores and SAT scores are highly correlated, and there's a straightforward reason for why: both are standardized tests designed to measure similar features of intellectual ability and aptitude. Focusing only on that relationship might lead to an assumption that IQ measures innate ability, and in turn causes SAT performance. And indeed, this view dominated, at least in the US, in the late 20th century. But this view of IQ is too simple. IQ scores and SAT scores are both influenced by a variety of life circumstances. Schooling—from preschool through adulthood—influences intelligence as measured by IQ scores. Experiences associated with poverty, beyond access to quality schooling, also influence intelligence as measured by IQ scores, including healthcare, social services, family resources, and levels of stress.

Can the numbers lie?

We've seen that visual presentation of statistical information can be misleading, and correlations misused or misinterpreted. You might think that numerical statistics themselves, including measures of central tendency and variability of a data set, cannot lie, since these are straightforward, simple-to-calculate summaries of a data set. Alas, this is not always true.

One issue with statistical information to keep an eye out for is statistical data that look too neat. As we've seen, the values of random variables are inherently noisy. Statistical patterns can emerge in aggregate data, but variability and exceptions still arise. Statistical information appearing too neat can mean that the data are presented misleadingly, or even that the data are partly or entirely fake. We introduced the idea of data cleansing—identifying and correcting errors in a data set by deciding which data are questionable and should be eliminated—in Chapter 5's discussion of data models. But that process can go too far, resulting in choices about data inclusion that artificially lead the data to fit with expectations. Data from different samples or collected at different times should vary somewhat, in terms of their central tendencies like the mean and also in terms of their variability like range. The absence of such variability is a reason to be suspicious of the data.

In 2010, economists Carmen Reinhart and Kenneth Rogoff published research providing telling evidence in favor of austerity. Using data from many countries, the two economists focused on the relationship between the rate of growth of a country's economy, the public debt of the country and its gross domestic product, or GDP. (GDP is a measure of the added economic value created with new goods and services a country can produce during a certain period.) Their main conclusion was that it's bad for economic growth when the debt-to-GDP ratio of a country is above 90%. The paper received wide coverage in the media and played a role in many countries' policies of austerity measures, that is, spending cuts and tax increases to reduce deficits in the face of recession. But it later emerged that the results of that paper were based on a mistake in countries' reported debt level in the Microsoft Excel spreadsheet the two economists had used for their statistical analyses. Though the paper had concluded that average growth above the 90% ratio was negative, the average was actually positive! This entirely undermined the researchers' conclusion that had been so influential over global economic policies.

Summaries of a data set are only as good as the data set and calculations performed on them. We have seen that data are affected by measurement error and other sources of inaccuracies, and both data and analysis are subject to human error. Sometimes, as in the case of Reinhart and Rogoff's research, they are simple, honest mistakes; other times they are due to bias or incompetence. The psychologist Michèle Nuijten worked with a group of statisticians and methodologists to analyze how common numerical errors are in scientific research, focusing on more than 30,000 papers published in major psychology journals over three decades. These researchers used a computer algorithm to flag potential mistakes, and they found that about half of the papers in their sample had at least one inconsistency. Further, the inconsistencies tended to make the results of a study more likely to support the study's hypothesis; so at least some of the errors seemed to have been consciously or unconsciously incorporated to increase support for the research findings.

The upshot of this section is that, even as statistical information is highly valuable in presenting information and assessing claims, its power is also a liability. Statistical information, presented visually or numerically, appears decisive but can, through sloppiness or intentional misuse, invite misinterpretation and mistaken conclusions. The appearance of statistical information isn't enough to guarantee trustworthiness: in scientific research and popular reporting alike, the use and presentation of statistical information needs to be carefully interpreted and assessed for its accuracy.

EXERCISES

9.19 Recall: List three ways in which visual statistical representations, correlations, and numerical statistical representation can go wrong, giving an example of each.

9.20 Apply: Characterize the difference between absolute and relative risk. Then, describe a new example of relative and absolute risk in health research. (An internet search is one way to find an example, but be discerning about the source and content you choose!)

9.21 Recall: Describe how information about correlations can be misleading, using the example of IQ, SAT scores, and social factors.

9.22 Think: Briefly summarize the discussion of correlation and causation in this section. Then, describe how this relates to confounding variables (see Chapter 3).

9.23 Apply: Imagine a company advertises the toothpaste they produce by declaring, "More than 80% of dentists recommend our toothpaste." (a) Describe the most reasonable interpretation of this claim, explaining the type of survey evidence that would support it. (b) Suppose the actual survey given to dentists by the company asked them to indicate several toothpastes and brands they might recommend instead of only one. In light of this additional piece of information, explain how the prior advertisement may be misleading.

9.24 Apply: You can find many examples of misleading mathematical and visual representations of statistical information in the news and social media. Consider this example from September 2021, where a news program in the US used the chart pictured here to show how the number of Americans claiming to be Christians is decreasing. (a) What does the graph visually suggest? (b) Why is this graph misleading? (c) How should it be redrawn not to mislead the public?

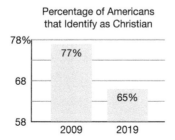

FURTHER READINGS

For a comprehensive history of statistics and statisticians, see Stigler, S. M. (2002). *Statistics on the table: The history of statistical concepts and methods*. Harvard University Press.

For more on the nature of averages and the normal distribution, see Stigler, S. M. (1997). Regression towards the mean, historically considered. *Statistical Methods in Medical Research*, 6(2), 103–114; Lyon, A. (2014). Why are normal distributions normal? *The British Journal for the Philosophy of Science*, 65(3), 621–649; and Røislien, J., & Frøslie, K. F. (2019). Jane and John do not exist. *Tidsskrift for Den norske legeforening*. https://doi.org/10.4045/tidsskr.18.0824

For a focused discussion of how to understand correlations and what can be learned from them, see Anjum, R. L., & Mumford, S. (2018). What's in a correlation? In R. L. Anjum & S. Mumford (Eds.), *Causation in science and the methods of scientific discovery* (ch. 4, pp. 29–36). Oxford University Press. https://doi.org/10.1093/oso/9780198733669.003.0004

For a comprehensive survey of sloppiness, bias, hype, fraud in statistical—and more generally, scientific—practices, see Ritchie, S. (2020). *Science fictions: Exposing fraud, bias, negligence and hype in science*. Random House.

CHAPTER 10

Statistical inference

10.1 THE HIGGS BOSON AND DRAWING INFERENCES WITH STATISTICS

After reading this section, you should be able to:

*   Identify three roles of statistical inference and illustrate them with examples
*   Describe how probability theory extends the reach of statistics
*   Distinguish frequency distribution from probability distribution

From a "bump" in data to a new fundamental particle

In 2012, scientists at CERN (European Center for Nuclear Research) in Geneva, Switzerland, announced the discovery of a new boson, a very tiny particle whose existence is crucial to our understanding of the fundamental structure of matter. Using CERN's Large Hadron Collider (LHC), a 27-kilometer (17-mile) tunnel straddling the Swiss-French border, scientists had been repeatedly observing the outcome of collisions among protons (one kind of particle in the nucleus of atoms). These collisions produce a shower of new particles, most of which are unstable and decay into other particles in a tiny fraction of a second.

New particles formed in proton-proton collisions in the LHC can be identified by tracking their trajectory, energy, and momentum. Detecting their mass is particularly relevant to distinguishing between different types of particles. However, because subatomic particles are so small, it's a challenge to distinguish the signature properties of new particles from background events.

In the summer of 2012, scientists recorded a "bump" in their data, corresponding to a particle with a mass between 125 and 127 GeV/c^2 (one gigaelectronvolt, or 1.783×10^{-27} kg). This is about 133 times heavier than protons. It was thought that this recorded bump could provide evidence of a new particle—perhaps of the long-sought Higgs boson. Using methods for statistical inference, the scientists calculated that this bump emerging from background events in the collider would occur by chance, without the presence of a boson, only once in three million trials. Because this probability was so low, the scientists rejected the idea that the bump occurred purely by chance and concluded it had been caused by a Higgs boson.

DOI: 10.4324/9781003300007-11

FIGURE 10.1 Portrait of Fabiola Gianotti, project leader and spokesperson for the ATLAS experiment at CERN involved in the discovery of the Higgs boson in July 2012

Scientists all over the world were thrilled with this news. The discovery of the Higgs boson could lend additional support to the standard model of particle physics, which is scientists' current best theory of the most basic forces and building blocks of the universe. The hypothesized Higgs boson was supposed to be like the glue of the universe: it's what joins everything together and gives it mass. It seemed this hypothesis now had been tested and confirmed.

The groundbreaking discovery of the Higgs boson illustrates how fundamental statistical inference is to scientific findings. Very often, techniques of inferential statistics are required to know what conclusions are supported by the empirical evidence.

From description to inference

Descriptive uses of statistics enable scientists to summarize and represent data sets in meaningful ways. In Chapter 9, we saw how to do this visually with charts and graphs and also numerically with means, standard deviations, and correlation coefficients. However, merely describing data sets regularly falls short of what scientists and everyone else are interested in. We are also interested in generalizing, making predictions, and testing hypotheses based on data sets.

Data collected in pre-election polls are used not merely to describe one group of voters but to make predictions about the eventual outcome of the elections. Basketball fans want to predict, from his past record, whether LeBron James's free-throw success will improve over time. Health researchers want to test, from observed treatment effects in a medical trial, whether a medical drug is efficacious against a certain ailment in the general population. For these kinds of interests—predicting the future, generalizing from a sample, and testing hypotheses—we need inferential statistics.

Inferential statistics is an important form of inductive reasoning that extends the reach of descriptive statistics with the use of probability theory. Coin tosses, dice throws, LeBron's free throws, people's voting intentions, effects of medical drugs, and more can all be treated as random variables. Inferential statistics can be used to analyze data sets to predict yet-to-be-measured values of those variables. For example, one might infer from a sequence of heads and tails whether the coin is unfair, predict from his past record whether LeBron's free-throw success will improve over time, predict from an opinion poll which candidate will win the election, and infer from observed treatment effects to the efficacy of a medical drug.

Probability theory, introduced in Chapter 8, is used to expand statistical tools from description to inference. The basic idea is that observed frequencies are used to estimate probabilities.

A *frequency distribution* is how often a variable takes on each value in some data set. This might be depicted in a table or histogram, for instance. A *relative frequency distribution* records the proportion of occurrences for each value instead of the absolute number. For example, suppose you have a bag containing 35 M&Ms of different colors: *M&M Color* = {brown, red, yellow, green, blue, orange}. The second column of Table 10.1 depicts the frequency distribution of each color, and the third and fourth columns of Table 10.1 depict the relative frequency distribution—in terms of proportion in column 3 and in terms of percentage in column 4. Each of columns second, third, and fourth is a way to display the proportions of the colors in your particular bag of 35 M&Ms.

TABLE 10.1 Frequency Distribution of Bag of 35 M&Ms (second column); Relative Frequency Distribution (third and fourth columns)

Color	Frequency	Proportion	Percentage
Blue	1	1/35	2.86%
Orange	3	3/35	8.57%
Yellow	4	4/35	11.43%
Red	5	5/35	14.29%
Green	5	5/35	14.29%
Brown	17	17/35	48.57%

Relative frequency distributions can be used to estimate the ***probability distribution*** for a variable—that is, how often the variable is expected to take on each of a range of values. In this example, this isn't a description of your bag of M&Ms, but a prediction about other bags. Some bags will have a distribution of colors similar to yours; others will vary more. Based on your sample bag of M&Ms, you may estimate that if you take a random bag of M&Ms, open it, and choose an M&M at random from that bag, the probability of getting a blue M&M is about 3%.

Let's consider a very simple probability distribution to illustrate how probability distributions support statistical inference: the probability distribution of the number of times heads are expected to come up over 100 coin tosses. The range of possible outcomes is 0 to 100: heads might come up as few as zero times and up to a maximum of 100 times. In other words, these are the possible values of the variable *Heads per 100 coin tosses*.

In theory, we could calculate the probability of each outcome in that range using probability theory developed in Chapter 8. Notice that $\Pr(heads = 0)$ is equivalent to $\Pr(tails_1$ and $tails_2$ and . . . $tails_{100})$. If every coin toss comes up tails, then there are zero instances of heads. What is the probability of that ever happening, assuming the coin is fair? The probability of getting tails on a throw is always .5, and we multiply probabilities to calculate the probability of multiple independent events all occurring. The probability of getting 100 tails would be a really, really tiny number: $1/2 \times 1/2 \times . . . \times 1/2$, or $1/2^{100}$. Notice that it's also the same as the probability that heads comes up 100 times.

In between 0 and 100, the calculation for the probability of every value of number of times heads comes up is much more complicated, though in principle we know enough probability theory to do these calculations. We won't carry out the calculations, but considering how they would go gives us a sense for how the probability changes for intermediate numbers of heads.

Notice that there is only one way to get zero heads and only one way to get 100 heads: for one, the coin never lands heads, and, for the other, the coin lands heads all 100 times. Now consider getting exactly one head out of 100 tosses. There are 100 different ways that could happen! It might be the first toss, or the second toss, or the third toss, or the 37th, or any other single toss. So, $\Pr(heads=1)$ is equivalent to $\Pr[(heads_1$ and $tails_2$ and $tails_3$ and . . . $tails_{100})$ or $(tails_1$ and $heads_2$ and $tails_3$ and . . . $tails_{100})$ or . . .] and so on, up until the circumstance of getting heads only on the 100th toss. Using our previous calculation, and because we add when calculating the probability of one of several mutually exclusive events occurring, this is $1/2^{100} + 1/2^{100} + . . . + 1/2^{100}$, or $100 \times 1/2^{100}$. This is still a tiny number, but it's 100 times bigger than the probability of heads coming up no times or every time. The same calculation gives us the probability that heads comes up 99 times; so we're building our probability distribution from both ends at the same time.

There are even more ways for heads to come up twice (or 98 times), and even more ways than that for heads to come up three times (or 97 times). Each time we add another outcome of heads the calculation becomes more complicated, and the probability of getting that number of heads increases. The increasing probability of each of these outcomes isn't linear; the increase gets bigger each time.

The probability distribution will be symmetric, since the calculation is the same whether the number of heads = 0 or 100, whether the number of heads = 1 or 99, and so forth. The middle of the distribution, the most probable outcome, is thus 50: that you get heads on 50/100, or .5, of the coin tosses. Figure 10.2 shows a histogram of the whole probability distribution. As the number of tosses increases, the shape of the histogram becomes a bell curve. Recall from Chapter 9 that a bell curve is a perfectly symmetric, unimodal distribution for continuous variables—what is also called a *normal distribution*.

The normal distribution—unimodal and symmetric—is especially important for statistical reasoning. Like coin tosses, the behavior of random variables over many repeated, independent trials tend to have a probability distribution that is normal. This results from what is known in statistics as the **central limit theorem**: the statistical claim that samples with a large enough size will have a mean approximating the mean of the population. What varies for different random variables is the central tendency and variability of the normal distribution, which—as we saw in Chapter 9—can be described with mean and standard deviation. Whereas the mean value of heads on 100 coin tosses is 50, the mean value of 6s on 100 dice rolls is 16.67 (1/6 × 100). The

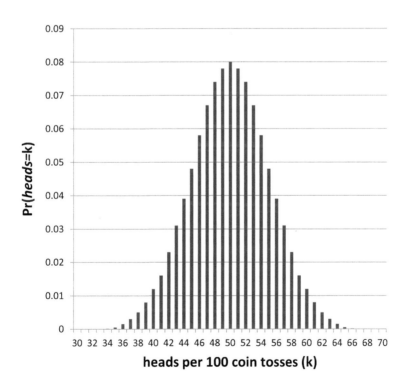

FIGURE 10.2 Probability distribution of heads for 100 coin tosses; the mean of this distribution is 50 and its standard deviation is 5

standard deviation, but not the mean, is also influenced by the number of trials. You are more likely to get none or all heads in five coin tosses than in 100 coin tosses; the standard deviation is larger for the former.

Probability distributions for fair coins and dice can be calculated directly from the probabilities of the individual outcomes, as sketched earlier for 100 coin tosses. This is not so for variables like success rate at free throws and proportion of blue M&Ms. For variables like those, the frequency distribution of a data set is used to predict the probability distribution. The predicted probability distribution can then be the basis for our expectations for other, relevantly similar data sets, like future basketball games and unopened bags of M&Ms.

EXERCISES

10.1 Recall: Describe how inferential statistics can be used to make predictions, generalize from an example, and test a hypothesis, and give an example of each.

10.2 Recall: Define *frequency distribution*, and describe in your own words how mean and standard deviation in descriptive statistics relate to frequency distributions. Then, define *probability distribution*, and describe in your own words how inferential statistics makes use of probability distributions.

10.3 Think: Define *descriptive statistics* and *inferential statistics*. In your own words, describe how probability theory is used to make inferential statistics possible.

10.4 Apply: In class, or with a group of several classmates, find a coin for each person and carry out the following steps, each person recording the answers individually.

a. Agree in your group about your expectation for the mean percentage of coin tosses that will land heads up; record this percentage.

b. Each person should toss their coin four times, recording each result as either heads or tails. Summarize your individual result as the ratio of heads to tails. This will be either 0:4, 1:3, 2:2, 3:1, or 4:0.

c. Record how many people in your group got each of the possible ratios: 0:4, 1:3, 2:2, 3:1, and 4:0. Draw a histogram showing these results.

d. Some people didn't get a ratio of 2:2. Why not, when you predicted 50% heads?

e. Should we expect each person to get the same ratio of heads to tails on the next four coin tosses as they did on the first four? Why or why not?

f. Each person should toss his or her coin four more times, recording each result and then summarizing as a new ratio of heads to tails.

g. Add these ratios to the data set of the first round of coin tosses. There should now be twice as many series of four-coin tosses. Draw a new histogram describing the complete data set.

h. Compare the histogram from (d) to the histogram from (g). What has changed? Why?

10.5 Apply: Convert the following data about the outcomes of 11 rolls of a die to a relative frequency distribution:

1st roll: 4; 2nd roll: 3; 3rd roll: 4; 4th roll: 6; 5th roll: 2; 6th roll: 1; 7th roll: 6; 8th roll: 4; 9th roll: 5; 10th roll: 5; 11th roll: 6

Then, draw a histogram showing the relative frequency distribution for this data set.

10.6 Apply: Construct a histogram of the probability distribution for an experiment in which a die is thrown five times and the number of occurrences of 4 is recorded.

10.2 STATISTICS IN ESTIMATION

After reading this section, you should be able to:

* Define *sample mean* and *sample standard deviation* and indicate how these differ from mean and standard deviation for descriptive statistics
* Describe how inferential statistics is used to estimate population values from a sample
* Indicate how sample size and representativeness matter for estimation

Estimating from a sample

One common application of statistical inference methods is estimating the features of a population based on data about a sample. Generalizing from a sample to a population is a very common way to learn about the features of a population, especially for populations that are very large. The population in question might be a human population, but it might be different organisms, other entities, or even events instead, such as tree frogs, nations, corporations, and election outcomes. So, statistical inference can be (and has been) used to estimate prevalence of different religions worldwide, variation in tree frog calls, and voter turnout in local versus national elections—among many other things.

The basic idea of statistical **estimation** is to use the observed frequency distribution for a sample as the basis for estimating the probability distribution for that variable in the general population. The measures introduced in Chapter 9 of mean and standard deviation can be adapted to represent the probability distribution for the variable in the population.

Imagine scientists have a sample of 100 university students, and they want to use that sample to estimate the range of political views among all students at that university. They might administer a questionnaire to the individuals in the sample, with each individual's responses scored between 1 and 10, where 1 is most politically conservative and 10 is most politically liberal. Imagine the questionnaire scores are as shown in Table 10.2. This data set gives the scientists all the information they need to estimate

TABLE 10.2 Imagined questionnaire scores of 100 university students

Questionnaire Score	Number of Individuals
1	0
2	2
3	5
4	7
5	10
6	15
7	22
8	18
9	13
10	8

the mean degree of liberality (or, equally, conservativeness) in the full population of university students.

The **sample mean** is used to estimate the average value for the variable in the population; it's an estimate of the population mean. It is called *sample mean* because it is based on the mean found in the sample. The sample mean might not turn out be the mean value in the population, but it's the most likely value and thus our best guess.

Now, recall from Chapter 9 how to calculate the mean: you sum the scores, and then divide by the total number of students. For this data set, the mean score is 6.82. (College students do tend to be a rather liberal bunch, on average.) This score is the sample mean. If the sample is representative—this idea was introduced in Chapter 3 and is discussed further later—this score is most likely to be the mean value in the population of university students.

Just like there's a sample mean, there's also a **sample standard deviation**; this is an estimate of the spread of the probability distribution for the random variable. Standard deviation, as introduced in Chapter 9, is a measure of the frequency distribution for the data set, while the sample standard deviation is instead an estimate of the probability distribution. The word *sample* is included, as with sample mean, to signal that this is a prediction about the population on the basis of data about the sample.

Sample standard deviation is also calculated in a slightly different way from the standard deviation formula introduced in Chapter 9. The sample standard deviation (s) is calculated as follows:

$$s = \sqrt{[\Sigma(\text{value} - \text{mean})^2 / (N - 1)]}$$

For comparison, standard deviation (σ) is calculated from $\sqrt{[\Sigma(\text{value} - \text{mean})^2 / N]}$. The change is the denominator: $N - 1$ (for the sample standard deviation, s) instead of N. This is a way of correcting for systematic underestimation about the population mean. We won't ask you to perform this calculation here, but for our example of estimating the political views of university students, the sample standard deviation of the data set works out to be 1.98.

The probability distribution estimated with the sample mean and sample standard deviation is the "middleman" so to speak: a statistical model enabling a prediction of the characteristics of interest in the population. A helpful rule of thumb for getting a rough probability estimate of a characteristic of interest, given its standard deviation, is **68–95–99.7 rule**. This rule can be used to remember the percentages of probabilities within a certain range around the mean in a normal distribution. It says that about 68%, 95%, and 99.7% of the values lie, respectively, within one, two, and three standard deviations of the mean. (The other 32%, 5%, and 0.3% are equally scattered on either side of these ranges.) Look back at Figure 9.11 in Chapter 9 to see a visual depiction of this.

Applying the 68–95–99.7 rule to our example of political views of university students (taking into account sample mean and sample standard deviation) yields the following results. Any given student has a 68% probability of having a score between 4.84 and 8.80 on our conservative/liberal scale, which is calculated by subtracting/adding the sample standard deviation (1.98) from/to the sample mean (6.82). Given that 5.00 is the dividing line between liberal and conservative, a student thus has about a 68% chance of being more liberal than conservative. Any given student has a 95% probability of having a score of 2.86 to 10.00, which is calculated by subtracting/adding two standard deviations (3.96) from/to the mean (6.82). We can be much more confident that some student will fall within this range than within one standard deviation, but it is also a wider, and thus less informative, range. The only thing this tells us is that most (95%) of college students are predicted to be outside the most conservative part of the scale.

Managing errors in estimation

A complication in using a sample mean to make predictions about the mean of a population is that the sample mean can vary from sample to sample. Were you to sample a different 100 students at the university, you could get a sample mean of 5.00 or 6.93 (for example) instead of 6.82. The distribution of the sample means you would get from repeated sampling is called the *sampling distribution of the sample mean*. How much this varies from the true population mean can be estimated with the **standard error**. Standard error measures sampling error. This can be calculated as:

$$\text{SE} = s/\sqrt{(\text{sample size})}$$

As earlier, s is the sample standard deviation. The standard error measures the precision of the sample mean, that is, how much uncertainty there is about the estimated mean of the population. The formula shows that the standard error, and hence uncertainty about the population mean, decreases as the sample size increases. This is because a large sample size helps control for chance variation. For our imagined survey of political views, the standard error is $1.98/\sqrt{100}$, which is .198.

The standard error can be used to calculate a confidence interval. A sample mean score is a point estimate: it gives you a single value to serve as your best guess of the true population mean. If the sample is large, your point estimate will be good, but you can't know how good. Confidence intervals provide this type of information. A *confidence interval* is an interval within which the value of the variable should lie for a given percentage of possible samples. For our political survey with a standard error of .198, 68% of sample means will fall within .198 of the true population mean, and 95% will fall within .396 of the true population mean. (This is for the same reason as the 68–95–99.7 rule introduced earlier.) So, for the sample mean of 6.82, the interval 6.42–7.22 will contain the true population mean 95% of the time.

Standard error and confidence intervals are ways to measure sampling error due to chance variation. (Recall from Chapter 9 that sampling error is the difference between the features of a sample and a population due to the unrepresentativeness of the sample.) Larger samples can be expected to have less sampling error than smaller samples, that is, to be more representative.

In Chapter 3, we introduced the idea of random sampling, using a chance method for selecting a sample to investigate from the population so every member of the population has an equal chance of being selected for participating in an experiment. Besides chance variation, a sample might be unrepresentative due to nonrandom sampling. Samples chosen in ways that make some individuals in a population less (or more) likely to be included than others will introduce bias in the inferences made about the population based on the sample. A poll that only solicited the political views of the students in a particular club, for example, may not be representative of the views of the full student population.

Here's a historically significant case of nonrandom sampling resulting in serious sampling error. In 1936, a magazine, *Literary Digest*, sent out 10 million postcards asking Americans how they would vote in the year's presidential election. They received almost 2.3 million back, which is a very large sample. In that sample, Alfred Landon had a decisive lead over Franklin Roosevelt: 57% to 43%; Figure 10.3 shows the headline of the story they ran about the poll's result. The *Digest* did not gather information that would allow it to judge the representativeness of its sample. A young pollster, George Gallup, utilized a much smaller sample of 50,000 (which is still larger than most modern political polls). His sample was representative, and it showed Roosevelt on track to win by a landslide. That was, of course, the eventual outcome of the election. The *Literary Digest* closed down soon after, and Gallup's name lives on in the well-known Gallup poll approach to measuring public opinion based on surveying a sample.

In many research contexts, random sampling and large sample sizes are very difficult to accomplish. For example, in a telephone poll of voter preference prior to an election, the phone numbers dialed can be randomly selected. But who picks up the phone, whether a person hangs up immediately or answers the questions, and even who has a phone and who doesn't are all nonrandom influences on the people sampled, and those influences might correlate with voter preferences. And, even when sampling is truly random, smaller samples have greater sampling error.

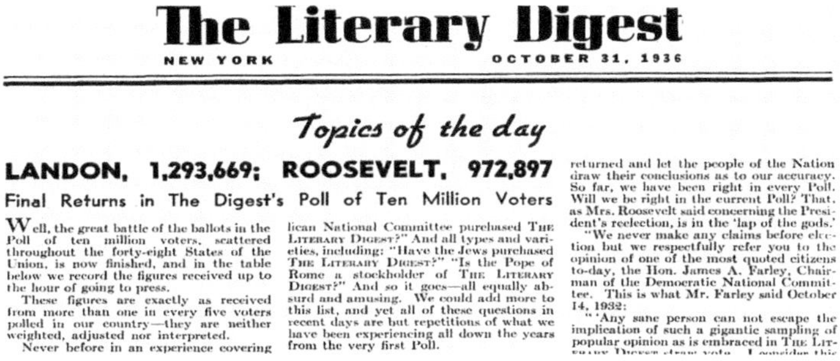

The Literary Digest

NEW YORK   OCTOBER 31, 1936

Topics of the day

LANDON, 1,293,669; ROOSEVELT, 972,897

Final Returns in The Digest's Poll of Ten Million Voters

Well, the great battle of the ballots in the Poll of ten million voters, scattered throughout the forty-eight States of the Union, is now finished, and in the table below we record the figures received up to the hour of going to press.

These figures are exactly as received from more than one in every five voters polled in our country—they are neither weighted, adjusted nor interpreted. Never before in an experience covering

lican National Committee purchased THE LITERARY DIGEST?" And all types and varieties, including: "Have the Jews purchased THE LITERARY DIGEST?" "Is the Pope of Rome a stockholder of THE LITERARY DIGEST?" And so it goes—all equally absurd and amusing. We could add more to this list, and yet all of these questions in recent days are but repetitions of what we have been experiencing all down the years from the very first Poll.

returned and let the people of the Nation draw their conclusions as to our accuracy. So far, we have been right in every Poll. Will we be right in the current Poll? That, as Mrs. Roosevelt said concerning the President's reelection, is in the 'lap of the gods.'

"We never make any claims before election but we respectfully refer you to the opinion of one of the most quoted citizens to-day, the Hon. James A. Farley, Chairman of the Democratic National Committee. This is what Mr. Farley said October 14, 1932:

"'Any sane person can not escape the implication of such a gigantic sampling of popular opinion as is embraced in THE LITERARY DIGEST

FIGURE 10.3 Announcement of results of 1936 *Literary Digest* poll

Still, as with so many decisions in science, real-world limitations need to be factored into sampling procedures. Sometimes researchers can't eliminate confounding variables but only limit them as much as feasible and note that confounding variables are possible. As important as representative samples are, simply pointing out that a sample isn't truly random or is somewhat limited in size isn't usually reason to entirely dismiss a prediction or estimation based on a sample. Careful analysis is needed of sampling procedure, potential confounding variables, and statistical properties like sample standard deviation and sampling error.

EXERCISES

10.7 **Recall:** Define *sample mean* and *sample standard deviation* and indicate how these differ from mean and standard deviation for descriptive statistics, considering what each measure means and how it is calculated.

10.8 **Think:** Summarize the steps of estimating population values on the basis of a sample.

10.9 **Apply:** There are 3,000 people at a party. (It's a very large party!) 100 are interviewed at random, and it is discovered that 80 are philosophers, 10 are geologists, and 10 are artists.

   a. What's the percentage of philosophers in this sample of 100 party guests?

   b. What's the point estimate for the percentage of philosophers at the party?

   c. Suppose the standard error for this data is .08, or 8%. What's the probability that the percentage of philosophers at the party is in the range of 72-88%? (Hint: consult the discussions above about confidence intervals.)

   d. You're 95% sure that the percentage of philosophers at the party falls within a certain range. What range is that?

   e. Can you conclude most of the partygoers are philosophers? Why or why not?

10.10 **Think:** Consider the following statements according to the information provided in Exercise 10.9. For each, say whether the data support the conclusion. Motivate your answers with reference to the information provided and what you know about statistical estimation.

a. It's highly probable that the majority of party guests are philosophers.
b. 80% of the people at the party are philosophers.
c. It's more likely than not that at least 8% of the guests are non-philosophers.
d. It's highly likely that the geologists are outnumbered at this party.
e. It's highly probable that most people in the world are philosophers.

10.11 Recall: Describe how sample size and representativeness matter for estimation.

10.12 Apply: Find an article in a newspaper, magazine, or reputable online source that draws conclusions from a poll. Alternatively, your instructor may provide one article for the whole class to use for this exercise. Answer the following questions; if you can't find the answer, say so, providing your best guess if possible. If you selected your own article, please submit a copy or printout of it with your responses.

a. What variable was under investigation? What were the researchers interested to know?
b. What was the sample size? How was the sample selected?
c. Is the sample likely to be representative? Why or why not?
d. What data did the researchers collect about the sample?
e. What conclusions about the population did the researchers draw from the sample?
f. Assess the poll, the results, and the researchers' conclusions. Are there any problems with any of these? How could the poll or the conclusions be improved?

10.3 STATISTICS IN HYPOTHESIS-TESTING

After reading this section, you should be able to:

- List the steps of statistical hypothesis-testing and describe the role of each
- Assess whether to reject the null hypothesis from data and a probability distribution, supporting your reasoning with statistical considerations
- Define *statistical significance, p-value, type I* and *type II errors,* and *effect size* and analyze these in an example

Null hypothesis significance testing

Methods for inferential statistics are routinely used to test hypotheses. While there are lots of different methods for hypothesis-testing that are suited to different circumstances, these methods generally involve the use of sample data to infer whether and to what extent the available evidence confirms or disconfirms competing hypotheses about the phenomenon. This is what happened when scientists at CERN rejected the possibility that the bump in their data was due simply to chance and instead posited a newly discovered particle, the Higgs boson, as the source of the data.

In this section, we'll focus on one prominent approach to statistical inference for testing hypotheses, namely null hypothesis significance testing (NHST).

In this approach, two competing statistical hypotheses are formulated. One is the ***null hypothesis***, which is a kind of default assumption; often, this just amounts to the hypothesis that nothing unusual is going on or that two variables are independent. For the scientists at CERN, this was the hypothesis that the bump in their data was just generated by background noise. The other hypothesis is the ***alternative hypothesis***, since it is posited as an alternative to the default assumption; this is a bold conjecture under investigation. For the scientists at CERN, this alternative was the hypothesis that a new particle, the Higgs boson, was responsible for the bump in their data.

The null hypothesis leads one to expect a certain range of data and is generally, but not always, just a negation of the alternative hypothesis—the bold conjecture. When the data collected are within that range, there's no grounds for questioning the null hypothesis. But when the data collected are sufficiently far outside the range of data expected from the null hypothesis, scientists reason that such data would be overwhelmingly unlikely if the null hypothesis were true. And so, in that case, they reason that the null hypothesis is likely to be false and should be rejected. The data instead support the alternative hypothesis—the bold conjecture.

This is basically a statistical version of the hypothetico-deductive (or H-D) method encountered in Chapter 6: expectations are derived from a hypothesis, and if observations don't match those expectations, the hypothesis is rejected (or, at least disconfirmed). In statistical hypothesis-testing, expectations for likely values of a random variable are determined from the null hypothesis. If a value is observed that falls far enough outside the expected range, the null hypothesis can be rejected, and, in its place, the alternative hypothesis can be tentatively accepted for further scrutiny. When scientists fail to reject the null hypothesis, their test results are deemed inconclusive. Technically this does not provide evidence in favor of the null hypothesis because the alternative hypothesis hasn't been ruled out by the data. Just like the H-D method, the value of NHST is when it identifies grounds to reject a null hypothesis.

In null hypothesis significance testing, as with estimation, inferential statistics is used to generate a probability distribution for potential outcomes based on the null hypothesis. This probability distribution represents the expectations if the null hypothesis is true. The scientists at CERN set out a protocol for statistical analysis in advance of gathering data, which enabled them to determine, for any bump in data observed, how probable that bump would be given the null hypothesis that no boson was responsible.

The probability distribution provides a statistical framework for assessing data that are collected. Data are evaluated for the degree to which they violate expectations based on the null hypothesis. For the unexpected bump in data observed at CERN, scientists determined that if the null hypothesis were true—that is, if no boson were responsible—then the probability of observing the bump in data would be extremely low: about one in three million.

So, probabilistic expectations are developed from the null hypothesis with the use of inferential statistics, and actual observations are compared with those expectations. Then, finally, scientists draw a conclusion from that comparison. This final step is always a judgment call. Scientists have to decide how unlikely the data would have to be, given the null hypothesis, to warrant them rejecting the null hypothesis. If the observations are not too far from what is expected from the null hypothesis, then scientists have no reason to reject the null hypothesis in favor of the alternative hypothesis. If, on the other hand, the observations do violate expectations, then this provides a reason to reject the null hypothesis in favor of the alternative hypothesis. This was the exciting scenario encountered at CERN: given the observation of a bump in data that would have been exceedingly unlikely without a boson responsible for it, the scientists rejected the null hypothesis and declared they had evidence for accepting the alternative hypothesis. That is, they declared they had discovered the long-sought Higgs boson, a new kind of fundamental particle.

The steps of NHST are summarized in Table 10.3, emphasizing how these steps conform to the basic ingredients of scientific methods—hypotheses, expectations, and observations—outlined in Chapter 2.

Developing probability distributions

How did the CERN scientists determine that the probability of the data they collected was so miniscule if the null hypothesis were true? That question is at the heart of null hypothesis significance testing. To answer it, let's turn our attention to a classic experiment on the tasting of tea, due to Ronald Fisher, a geneticist and one of the designers of NHST.

Imagine a friend asks you, when preparing tea for her, to add milk to the cup first and only then add the tea. She claims she can discriminate by taste the order in which the milk and the tea were poured into the cup—she thinks the tea tastes better when the milk is added first. Intrigued, you decide to test this claim. According to the steps

TABLE 10.3 Summary of steps in statistical hypothesis-testing

Step	Procedure
Hypothesis	Formulate alternative hypothesis (the bold conjecture) and corresponding null hypothesis (the default expectation)
Expectations	Determine what range of outcomes and probability distribution to expect if the null hypothesis were true
Observations	By experiment or observational study, determine one or more actual outcomes
Conclusion	Evaluate whether the actual outcome is unlikely enough given expectations from the null hypothesis to provide grounds for rejecting the null hypothesis

of NHST, summarized in Table 10.3, you should start by formulating the null and alternative hypotheses. In this circumstance, what would you choose for each of these? The bold and speculative conjecture is that your friend really can discern by taste the order in which milk and tea were poured into a cup. This would be surprising! The null hypothesis is simply that your friend cannot do this. This is, probably, your default assumption.

You prepare a cup of tea out of view, tossing a fair coin to determine whether you pour tea or milk first. (This randomizes the order, which is a way to control for extraneous variables; recall our discussion of randomization.) Then, your friend has a sip. She says that you added the tea first, and she's right. But this isn't terribly impressive; there was a .5 probability of her guessing correctly by accident, since either the tea was poured first in the cup or else the milk was. To see whether there is support for the alternative hypothesis, whether the evidence really supports your friend's claim, you need to have your friend make repeated guesses that are, on average, much more accurate than chance would allow.

You prepare eight new cups of tea at once, tossing a coin to determine milk-first or tea-first for each. You put the cups of tea in front of your friend and ask her to say of each whether the milk or the tea was added first. What does the null hypothesis lead us to expect? We can develop a probability distribution to let us know what to expect if your friend is just guessing at random. As with estimation, the probability distribution can be characterized by a mean and standard deviation.

If your friend is merely guessing, she is most likely to be right about four of the eight cups. This is the mean expected outcome, which you can calculate by multiplying the probability of success on each trial, which we said was .5, by the number of trials: mean = Pr(O=success) × # trials = .5 × 8 = 4.

In this context, the mean is the most likely outcome. If your friend were to make repeated guesses about the eight cups, and the null hypothesis were true, most of the time she'd be right about four cups. Since your friend is only making guesses about one series of eight cups, the mean indicates the most likely outcome. But this outcome is not guaranteed, even if the null hypothesis is true. By sheer luck, your friend might guess correctly more often or less, just as you might happen to get more or fewer heads than 50% in a series of coin tosses.

The standard deviation for the probability distribution can be calculated using this formula:

$$\sigma = \sqrt{[\text{mean} \times (1 - \text{Pr}(O=\text{success})]}$$

Notice that this is very different from the other standard deviation formulas we have encountered. Here, 1 is the total probability, and Pr(O=success) is the probability of success (in the tea experiment, guessing correctly) in a single trial. Multiplying that by the mean number of successes yields the variance; the square root of that number is, then, the standard deviation. For the tea experiment, the standard deviation is $\sqrt{[4 \times (1 - .5)]} = \sqrt{2} = 1.414$.

Figure 10.4 shows the probability distribution of the number of guesses your friend will get correct if she is randomly guessing, that is, if the null hypothesis is true. This

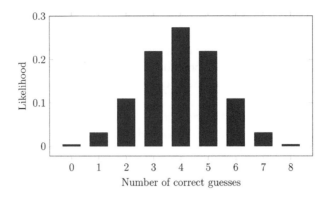

FIGURE 10.4 Probability distribution of the number of guesses your friend will get correct if she is randomly guessing

probability distribution can be estimated from the range, mean, and standard deviation that we have calculated.

The probability distribution can be used to establish the ***significance level*** required for rejecting the null hypothesis. This is a decision about how improbable, given the null hypothesis, an experimental result must be to warrant rejecting the null hypothesis. Later we'll discuss some further considerations for this choice. For now, we'll adopt .05 as the significance level, which is the most common choice.

Assessing statistical significance

With all of this preparatory statistical work completed, you are now ready to test your friend's tea-tasting skills. This is the third step NHST: collecting the data. You ask your friend to judge "tea-first" or "milk-first" for all eight cups of tea. She correctly judges all eight cups! Given this data, would you be tempted to conclude that you were wrong and your friend was right, that maybe she can perceptually discern something about tea and milk order?

There are two possibilities consistent with the data: (1) the null hypothesis is true, and our experimental result was an unlikely event of eight correct guesses purely by chance, or (2) the null hypothesis is false, and the alternative hypothesis that your friend can discriminate between milk-first and tea-first cups of tea is true. The goal of statistical hypothesis-testing is to do our best to decide whether (1) should be rejected in favor of (2). The probability distribution we've prepared can tell us exactly how unlikely our experimental result would be if (1) is the case, which should guide our decision of whether to reject (1).

From the outcome of the experiment, when your friend correctly judged all eight cups of tea as tea-first or milk first, we can calculate a ***p-value***. This is the probability of the observed data (or even more extreme outcomes) assuming the null hypothesis is true. The smaller the p-value, the more unlikely your observed data is if the null hypothesis is true. Notice that the p-value doesn't give us the probability that the null hypothesis is true or the probability that the alternative is true. It is just a yardstick we can use to answer the question: how likely is it to get data like you have observed if the bold conjecture you are interested in is not true?

We can visually estimate the p-value for the tea-tasting experiment from Figure 10.4: it's clear there is only a very small chance of guessing all eight cups correctly via random guessing. We can find the precise p-value with the multiplication rule; it's just the probability of the first guess being correct by chance multiplied by the probability of the second guess being correct by chance, and so on, for all eight guesses: $0.5 \times 0.5 \times 0.5 \times 0.5 \times 0.5 \times 0.5 \times 0.5 \times 0.5$, or 0.5^8, which is .0039. The probability of your friend guessing correctly on all eight cups by random guesswork is only .0039, or 0.39%. Put another way, if your friend tasted many series of eight cups of tea, she could get this outcome by guessing randomly only about one out of 256 series of eight cups. If she guessed all eight correctly by sheer luck, on her first try, she's really lucky!

Whether one should reject the null hypothesis is determined by comparing the p-value and the significance level. If the p-value is less than or equal to the significance level that had been selected, the researchers can reject the null hypothesis with reasonable confidence. When the p-value is greater than the significance level, we can't rule out the null hypothesis. The p-value of .0039 in our tea-tasting experiment is lower than our chosen significance level of .05, and so the outcome of this experiment is **statistically significant**. That means the outcome is unlikely enough if the null hypothesis is true that it provides grounds for rejecting the null hypothesis.

Most p-values can't be calculated so simply from probabilities as the one in our tea-tasting example. Another approach to evaluating statistical significance uses the mean, standard deviation, and 68–95–99.7 rule introduced in the previous section. By the 68–95–99.6 rule, outcomes that are two standard deviations away from the mean are the threshold for statistical significance at the significance level of .05. Two standard deviations in this case is 2.828 (1.414×2), so outcomes outside the range of 4 (the mean) +/− 2.828 are statistically significant. That range is 1.17 to 6.828. So, guessing eight cups correctly is statistically significant at the .05 level.

Notice that statistically significant results aren't necessarily important results. Here *significance* is a technical term that only means grounds for rejecting the null hypothesis. Many statistically significant results from NHST are theoretically uninteresting and practically irrelevant. Like the tea-tasting experiment, you could test whether people can reliably distinguish humans from dogs when passing them on the street. If your observations are statistically significant, thereby allowing you to reject the null hypothesis that people cannot distinguish between dogs and humans, you wouldn't be surprised. So, we should be wary of statistically significant findings that are nonetheless not worth our attention.

Another important feature of data that might be confused with statistical significance is *effect size*, which is a quantitative measure of the strength of a phenomenon, that is, of the magnitude of difference some variable makes. Statistical significance does not mean the effect size is large; the latter is a separate question. So, just as statistical significance doesn't guarantee an important finding, it also doesn't guarantee a strong influence. If you surveyed enough of the incoming students at a university, you might be able to conclude that their political views were more liberal or conservative, on average, than those of existing students—even if the difference were miniscule. Sometimes scientists represent the effect size along with the statistical significance of their findings, and this additional information can help put the finding into perspective.

Drawing conclusions

Comparing the *p*-value of the experimental outcome to the significance level provides a simple, objective criterion for deciding whether to reject the null hypothesis on the basis of observations. But there is a role for choice in what level of statistical significance to require. One can always ask whether this outcome is unlikely enough to reject the null hypothesis. This is a version of the more general decision we've seen elsewhere in this book regarding when there's sufficient evidence to believe a hypothesis.

In many fields, it's common to use a significance level of .05 as the dividing line. Observed results with a probability of less than .05 given the null hypothesis are said to be statistically significant at the .05 level. One can abbreviate this: $p < .05$. This is true of the outcome of our tea-tasting experiment, which is why we rejected the null hypothesis. Notice, though, that we could still be wrong: it's possible that our friend really was just extraordinarily lucky. But if we instead decided to play it safe and not reject the null hypothesis, we could be wrong about that as well. We might have then failed to detect our friend's tea-tasting superpower. By its very nature, statistical hypothesis-testing gives no guarantees.

The choice of significance level indicates the degree to which you're willing to accept the risk of erroneously rejecting the null hypothesis when it is true versus the risk of erroneously failing to reject the null hypothesis when it is false. These are the two different ways you could be wrong, and one or the other is always a risk. The risk of erroneously rejecting the null hypothesis when it is true is called a ***type I error***, or false positive. The risk of erroneously failing to reject the null hypothesis when it is false is called a ***type II error***, or false negative.

Scientists sometimes adjust the conventional .05 line for statistical significance considering whether type I or type II errors are riskier. Requiring a lower significance level means that you need stronger evidence to reject the null hypothesis; this decreases the chance of a type I error but increases the chance of a type II error. Requiring a higher significance level means that you need less evidence to reject the null hypothesis; this decreases the chance of a type II error but increases the chance of a type I error.

Imagine a new drug is being tested. If the drug is for a life-threatening illness with no treatment options otherwise—say, pancreatic cancer or Ebola—and experiments regarding the efficacy of the drug find it works better than a placebo with a *p*-value of .055, just missing the line for statistical significance at .05, researchers may still be inclined to bring the drug to market or at least continue testing. This result suggests the drug is very likely better than nothing! They want to avoid the type II, or false negative, error of accidentally rejecting a treatment that might be valuable. In contrast, if scientists are thinking about announcing a new particle, and they know their colleagues will scrutinize their findings, they may be especially careful about avoiding a type I or false positive error, and so will lower the significance level. The Higgs boson discovery was only announced after the probability of the experimental data was found to be just one in three million assuming the null hypothesis of no boson present. This significance level is so close to zero, it's difficult to even display numerically.

A related issue is that statistical tests vary in their power to detect an effect. The ***power of a statistical test*** is the probability that the test will enable the rejection of a null hypothesis. More powerful tests increase the chance of rejecting the null hypothesis, thus decreasing the chance of a type II or false negative error, where we fail to reject the null hypothesis when it is actually false. Power increases with sample size. In the tea-tasting experiment, we weren't able to reject the null hypothesis after one cup was guessed correctly, but we were able to after eight cups were guessed correctly. Increasing sample size to increase the power of a statistical test can be a good thing, but it also has a downside: this increases the chance of making a type I or false positive error. Studying a very large sample makes it relatively easy to uncover statistically significant findings, but this also makes it relatively easy to erroneously reject the null hypothesis—that is, to uncover findings that turn out to be false.

Statistical tests that increase power by using very large samples also enable findings that have very small effect sizes. For example, a certain gene has been linked with the chance of someone smoking cigarettes: if you have this gene, you are more likely to smoke cigarettes. Can researchers tell from your genes whether you have smoked, or will smoke? As the researchers acknowledged, absolutely not: there was only a very weak relationship. Finding factors with very small effect sizes has advantages and drawbacks. It can be useful to detect subtle statistical relationships, but weak statistical relationships are often uninteresting or unimportant. Further, it's not uncommon for people to take too seriously a statistical relationship with a very small effect size, especially if the finding fits with our expectations.

Box 10.1 Types of statistical hypothesis-testing

We have characterized null hypothesis significance testing as if it were a single technique, but there are actually multiple forms of statistical hypothesis-testing, with different names and different uses. Which type of test should be used depends on the kind of hypothesis under investigation, the type of data available, and other circumstances. The sort of statistical hypothesis-testing we've encountered in this section are *t* tests, where data about some group are used to test whether the group deviates systematically from what's typical. But *t* tests come in multiple varieties, depending on how many samples of data researchers collect and whether the alternative hypothesis predicts how the sample will deviate from the population mean. So, you might hear of one-sample or two-sample *t* tests, or one- or two-tailed *t* tests, for example. Another approach is analysis of variance, or ANOVA, which is used when there are more than two groups to compare. ANOVA also comes in multiple versions for different circumstances of statistical analysis. Pearson's r, in turn, is used to test whether a correlation in a sample provides a basis for concluding the variables in question are correlated in the population (the alternative hypothesis). All of these approaches to NHST share the same basic structure, but they differ mathematically and in how the results should be interpreted. There are also statistical techniques for when we can't assume a variable has a normal distribution. Calculations of any of these types of statistical tests are typically now carried out using software—so the hard part isn't the math but understanding and interpreting its significance.

EXERCISES

10.13 Recall: In your own words, define *statistical significance*, *p-value*, and *effect size*. Then, give an example of a result that is statistically significant but not important.

10.14 Apply: Consider the tea-tasting experiment again. Suppose that your friend correctly guessed six cups of tea instead of all eight. Decide whether this result is statistically significant at the .05 level by using the mean and standard deviation calculated previously and applying the 68–95–99.7 rule. Then, say whether this outcome is a basis for rejecting the null hypothesis.

10.15 Apply: It's estimated that 10% of the general population is left-handed. Imagine testing whether your group of friends contains an unusually large number of left-handed people. Let's say you have 75 friends, and 14 of them are lefties.
   a. Write out the null hypothesis and alternative hypothesis.
   b. Calculate the mean and standard deviation for how many of your group of friends would be expected to be lefties if the null hypothesis were true.
   c. From the information you calculated in (b), evaluate whether the data are significant at the .05 level.
   d. Decide whether to reject the null hypothesis, and justify your decision with statistical considerations.
   e. Is your decision at risk of a type I or type II error? Why?

10.16 Apply: Each of the following is a bold conjecture that can serve as an alternative hypothesis. For each, (i) formulate the null hypothesis, (ii) describe what a type I error would be and what a type II error would be, and (iii) say which kind of error would be more serious, and why.
   a. Adding water to toothpaste helps protect against cavities.
   b. This man is guilty of murder.
   c. The use of social media makes users depressed.
   d. The new drug is more effective than the old drug.
   e. The new drug is more dangerous than the old drug.
   f. Reading books promotes happiness.

10.17 Think: Classify each of the following statistical techniques as belonging to descriptive statistics, statistical estimation, or statistical hypothesis-testing. Give your rationale for each answer.

    a.   displaying a data set in a chart

    b.   surveying a group about their pizza preferences to decide if they have an unusual preference for anchovies

    c.   surveying a group about their pizza preferences in order to place an order

    d.   calculating the sample mean and sample standard deviation

    e.   surveying a group about their pizza preferences in order to guess what all Canadians' pizza preferences are

    f.   finding the mean level of preference for anchovies on pizza among a group and the standard deviation in that level of preference

    g.   rejecting a null hypothesis on the basis of data

    h.   finding the correlation coefficient of a data set

10.18 Think: Scientific journals tend to publish statistically significant results much more often than they publish findings of statistical insignificance. Why do you think this might be? Considering the earlier discussion about power, type I and II errors, and effect size, can you think of any concerns with this practice?

10.4 DIFFERENT APPROACHES TO STATISTICAL INFERENCE

After reading this section, you should be able to:

* Characterize three limitations of classical statistics and how Bayesian statistics solves each of them
* Describe how Bayesian statistics can be used (1) to establish the probability a hypothesis is true given an observation and (2) to compare how much an observation favors different hypotheses
* Describe two problems with Bayesian statistics

Problems with classical statistics

Null hypothesis significance testing, the approach to statistical hypothesis-testing described so far, is part of classical statistics, also called *frequentist statistics*. This approach is called "classical" because it is more or less standard—at least presently. But it's not the only game in town, and someday it may no longer be the standard way of doing statistics.

In this section, we'll describe three limitations of classical statistics. To set up the discussion, think back to the last step of the procedure of NHST. What results from this application of inferential statistics is a *p*-value: that is, the probability of the observation occurring given that the null hypothesis is true, which is then compared with a pre-established significance level to decide if the data are statistically significant. There are three oddities about this.

The first oddity is that scientists' primary interest in statistical hypothesis-testing is to figure out which hypotheses are true. But a *p*-value doesn't indicate how probable the hypothesis itself is, that is, how likely the hypothesis is to be true. It only indicates how probable observed data are if the null hypothesis is true. If the *p*-value is small enough, we can decide that the observations we in fact made are so unlikely given the truth of the null hypothesis that we should reject the null hypothesis. But we still don't know anything about the chance the null hypothesis is true.

A second limitation is that NHST doesn't allow us to take into account any prior information we might have in favor or against the truth of the null hypothesis. For example, in our tea-tasting example, it may be the right decision not to reject the null hypothesis even though your friend guesses correctly so often. What we know about how tasting works and about what properties a cup of tea can and can't have suggest this tasting feat should be impossible. Maybe your friend was just extraordinarily lucky, or maybe she had a way of cheating. In contrast, someone discerning two different types of wine in a blind taste test wouldn't really be that surprising. The same success rate may thus lead us to want to reject the null hypothesis of random guessing for wine tasting but not reject the null hypothesis of random guessing for tea tasting. As this illustrates, we often find different hypotheses more or less credible, and it seems this should influence our threshold for how much evidence we require to believe them. NHST does not allow us to take such prior information into account.

A third limitation of classical statistical techniques like NHST is that the statistical test doesn't directly relate to the alternative hypothesis at all. Again, all we are learning is the probability of the data we observed if the null hypothesis is true. And yet, the alternative hypothesis, the bold and speculative conjecture, is what scientists are truly interested in evaluating. How likely is the alternative hypothesis to be true? This is the million-dollar question in hypothesis-testing, but NHST gives us no way to answer it.

Bayesian statistics

The Bayesian approach to statistics solves all three of these problems with classical statistics. The Bayesian approach to statistical hypothesis-testing aims to determine when data count as evidence for one hypothesis and against a competing hypothesis and how data should change our degree of belief that each of these competing hypotheses is true. Bayesian statistics is based on Bayes's theorem and the subjectivist interpretation of probability, both introduced in Chapter 8. For Bayesians, an observation counts as evidence for a hypothesis when it raises our rational degree of belief in the hypothesis.

The simplest formulation of Bayes's theorem applied to a hypothesis and observation is:

$$\Pr(H|O) = \Pr(O|H)\Pr(H)/\Pr(O)$$

This formula states that the probability of a hypothesis H given an observation O—which is what we want to find out—is the probability of the observation given the hypothesis multiplied by the prior probability of the hypothesis, then divided by the probability of the observation.

Pr(H) is the prior probability of the hypothesis, while Pr(H|O) is the posterior probability of the hypothesis. The prior probability is our rational degree of belief before (prior to) making the observation, while the posterior probability is our rational degree of belief after (posterior to) making the observation. Considering the prior probability of a hypothesis is a way of holding implausible hypotheses to a higher standard of evidence than plausible hypotheses. So, we'd look for more evidence before agreeing that someone can tell milk-first or tea-first in a cup of tea than we would before agreeing that someone can tell the difference between different kinds of wine.

This formula takes three things as input: the prior probability of the hypothesis, Pr(H); the probability of the observation if the hypothesis is true, Pr(O|H); and the probability of the observation under all possible hypotheses, Pr(O). If all three of these probabilities are known—a major source of controversy for Bayesian statistics—we can use them to calculate the probability the hypothesis is true given the observation that has been made (or the data that have been gathered). And this is the main thing scientists want to discover from statistical hypothesis-testing.

Additionally, comparing prior and posterior probabilities shows us whether an observation confirms or disconfirms a hypothesis and by how much. If Pr(H|O) > Pr(H), then the observation O confirms hypothesis H. In other words, an observation confirms a hypothesis if the probability of the hypothesis—the rational degree of belief that the hypothesis is true—goes up once the observation has been made. A big increase in probability implies a large degree of confirmation, and a small increase implies a small degree of confirmation, while a big decrease in probability implies a large degree of disconfirmation, and a small decrease implies a small degree of disconfirmation.

Bayes's theorem also can be used to calculate the degree to which some observation or data set favors one hypothesis over another. Posterior probabilities can be calculated for any number of hypotheses from the same observation, considering the prior probabilities of the various hypotheses and the probability of the observation given each of the various hypotheses. These posterior probabilities can then be compared with one another to decide which hypothesis is more likely given the observation that's been made. Unlike classical statistics, this provides a comparative approach to hypothesis-testing.

There's also a shortcut to comparing how much an observation favors different hypotheses. It's possible to compare the likelihood of the observation given each hypothesis, or Pr(O|H_1) versus Pr(O|H_2), instead of posterior probabilities. These likelihoods are usually easier to find than posterior probabilities. An observation favors one hypothesis over another to the degree that the first hypothesis predicts the observation better than the other hypothesis does. This can be expressed numerically with the **Bayes factor**, which is the ratio of the probability of the observation given one hypothesis to the probability of the observation given another hypothesis, that is, Pr(O|H_1)/Pr(O|H_2).

Imagine that Lasha and Janine are interested in public opinion about the theory of evolution. Lasha believes that 70% of the public is convinced by the theory of evolution, while Janine believes that 60% of the public is convinced. They decide to poll 100 randomly selected people about their views on evolution to decide which of these is correct. Lasha's and Janine's different hypotheses lead them to have different expectations: Lasha predicts that about 70/100 will believe in the theory of evolution; Janine predicts it's about 60/100. (We can use tools of inferential statistics from earlier in this chapter to find the probability distribution each predicts for this random sample of 100 people.)

As it turns out, of the 100 people surveyed, 62 said they are convinced by the theory. According to the probability distribution based on Lasha's hypothesis of 70% belief in evolution, this observation has a probability of .02, that is, $Pr(O|H_1) = .02$. According to the probability distribution based on Janine's hypothesis of 60% belief in evolution, this observation has a probability of .08, that is, $Pr(O|H_2) = .08$. The Bayes factor is .08/.02, or 4. The survey result favors Janine's hypothesis over Lasha's by a factor of 4.

Problems with Bayesian statistics

We've seen a few of the benefits of Bayesian statistics over classical statistics, but Bayesian statistics faces its own problems. First, Bayesian statistics is often criticized for a lack of objectivity. There's often not enough information to have objective grounds for the prior probability—the probability of the hypothesis before data are gathered—but this prior probability is needed to use Bayes's theorem.

In some cases, we can calculate prior probabilities from established facts. For example, the probability that a 42-year-old woman has breast cancer can be estimated from the incidence of breast cancer in the general population. But such data is unavailable for many hypotheses. Recalling our earlier example, what prior probability should the scientists at CERN have assigned to the hypothesis that they'd detect a Higgs boson? It doesn't seem there's an objective way to assign a probability to this possibility before they ran their experiments. Without objective information to guide the selection of prior probabilities, individual biases and subjective values can find their way in. This is a problem because prior probabilities influence posterior probabilities in Bayesian statistics, and so biases and subjective values could then influence the conclusions drawn from experimental data. This possibility seems to undermine the objectivity of Bayesian reasoning and is perhaps the main challenge facing Bayesian statistics.

Some have responded to this challenge by suggesting rules for how prior probabilities should be established. Another response has been to suggest that exposing and taking into account people's different background beliefs is actually a good thing. Different choices of prior probabilities make it transparent how two scientists' judgments differ, so those differences can be justified by the scientists and assessed by others. In this respect, Bayesian statistics and classical statistics are similar. In NHST, researchers must decide on significance levels, sample size, and so forth, and these decisions are also judgment calls open to criticism. Nonetheless, the choice of Bayesian prior probabilities is a direct influence of background beliefs on degree of belief in hypotheses under investigation that many scientists are uncomfortable with.

A second problem for Bayesian statistics is that it's not obvious that posterior probabilities are always the best basis for updating one's beliefs. Some have suggested that abductive reasoning, discussed in Chapter 7, can be a better alternative. Recall that when people engage in abductive reasoning, they use explanatory considerations as evidence to support one hypothesis over others. You see cheese crumbs, small droppings, and some chewed up paper, and so you might reason that a mouse resides in your kitchen. It's not clear that these kinds of inferences follow Bayesian statistical analysis. The reasoning of paleontologists, for example, can be akin to "CSI"-style forensic work, what we called *methodological omnivory* in Chapter 4. They gather different kinds of evidence to assess the plausibility of hypotheses about the deep past. Statistics—Bayesian or classical—may not be involved at all.

EXERCISES

10.19 Recall: (a) Describe three limitations for classical statistics, especially as it's Watch Video 13 used in NHST. (b) For each of the three limitations for classical statistics you described, indicate how Bayesian statistics avoids it.

10.20 Recall: Write out Bayes's theorem; label prior probability, posterior probability, and likelihood; and describe what Bayes's theorem says when applied to a hypothesis and observation. Then, describe (a) how Bayes's theorem can be used to calculate the probability of a hypothesis given certain evidence, and (b) how a Bayes factor can be used to compare support for two competing hypotheses.

10.21 Apply: Suppose that your smartphone is being tested for a bug that affects about one device in a thousand. The bug could transfer all data on your phone to all your contacts. Your device has no apparent problem, and the test is accurate 90% of the time. This number means that if your smartphone actually has the bug, then the test result is positive with 90% probability, and if your smartphone does not actually have it, the test result is negative with 90% probability. After several anxious minutes, the test results come back: positive! Do you have good reason to be worried all your data will be exposed?
   a. Define the prior probability of the hypotheses that your device has the bug and that your device doesn't have the bug, the probability of the test result given the hypothesis that your device has the bug, and the probability of the test result given the hypothesis that your device doesn't have the bug.
   b. Explain how these probabilities can be used in Bayes's theorem to determine whether you should be worried.
   c. Think again about this situation, considering that, out of 1,000 devices, 100 will test positive for the bug. Would this consideration change any step in your reasoning?

10.22 Apply: You take an at-home covid test; it is negative. The false positive rate is less than 1%, while the false negative rate is about 20%.
   a. Write out the Bayes factor expression for this observation (negative test result) and the hypotheses that you do not have covid (H_1) and that you do have covid (H_2), then calculate the Bayes factor.
   b. What can you conclude from this Bayes factor? Can you conclude you do not have covid?

10.23 Think: Describe two problems for Bayesian statistics. For each, analyze how problematic you think the problem is, giving reasons for your opinion.

FURTHER READINGS

For a historically informed treatment of different approaches to statistical inference and their relationships, see Gigerenzer, G. (1993). The superego, the ego, and the id in statistical reasoning. In G. Keren & C. Lewis (Eds.), *A handbook for data analysis in the behavioral sciences: Methodological issues* (pp. 313–339). Erlbaum.

For a recent defense of the use of p-values and frequentism in statistical hypothesis-testing, see Mayo, D. G., & Cox, D. R. (2006). Frequentist statistics as a theory of inductive inference. In J. Rojo (Ed.), *Optimality: The second Erich L. Lehmann symposium* (pp. 77–97). Institute of Mathematical Statistics. And Mayo, D. G., & Hand, D. (2022). Statistical significance and its critics: Practicing damaging science, or damaging scientific practice? *Synthese, 200*(3), 220.

For short, critical surveys of problems with null hypothesis significance testing, see McCloskey, D., & Ziliak, S. (1996). The standard error of regressions. *Journal of Economic Literature, 34*, 97–114; and especially, Gigerenzer, G. (2004). Mindless statistics. *The Journal of Socio-Economics, 33*(5), 587–606.

For more on Bayesianism, see Howson, C., & Urbach, P. (2006). *Scientific reasoning: The Bayesian approach* (3rd ed.). Open Court.

For a compact comparison between frequentist and Bayesian approaches to statistical inference, see Sprenger, J. (2016). Bayesianism vs. frequentism in statistical inference. In *The Oxford handbook of probability and philosophy* (pp. 382–405). Oxford University Press.

CHAPTER 11

Causal reasoning

11.1 POVERTY AND THE CHARACTERISTICS OF CAUSAL REASONING

After reading this section, you should be able to:

- Describe the phenomenon of poverty and what the Niger experiment revealed about it
- List three characteristics of causal reasoning
- Give examples of how causal knowledge is of theoretical and practical importance

How can science help reduce poverty?

According to the United Nations' Sustainable Development Goals Report in 2022, more than 650 million people worldwide live in extreme poverty, surviving on less than $1.90 per day. This number may get higher in the coming years, especially if economic and social inequalities, war, and the environmental consequences of climate change worsen.

Poverty doesn't occur only in some remote regions of our planet. Poor people live in all countries and likely live in your neighborhood too. This is because poverty is not just about money. It is a complex phenomenon. Some of the salient aspects of poverty concern one's economic resources like assets and income, but other aspects are medical, psychological, and social, like malnutrition, psychiatric conditions, and lack of opportunity.

When people experience poverty, basic human needs go unsatisfied. The poor can suffer from hunger, malnutrition, ill health, deficient education, and insufficient access to technologies, infrastructure, social networks, and cultural opportunities. For individuals, poverty can cause or worsen social isolation, shame, anxiety, and depression. For communities and nations, poverty affects migration trends, public health, and political stability. These consequences partly depend on social norms preventing certain households, such as immigrant households, or individuals, such as women, from seizing new economic opportunities or benefitting from the support of a social network. Given all these complexities, how can scientists help to reduce poverty and prevent or mitigate its negative effects?

DOI: 10.4324/9781003300007-12

One promising idea comes from Niger. Niger is one of the world's poorest countries. In 2022, approximately half of its 25 million residents live in extreme poverty; in 2005, that number was 80%. In 2012, the government of Niger began giving small amounts of money to its poorest citizens. These small cash transfers had an obvious effect: they significantly improved the livelihoods of many of the poorest people. You may worry that people would use free cash in foolish ways. But several meta-analyses from developmental economics have shown otherwise. Rather than buying what are called "temptation goods," poor people typically use money received through a cash-transfer or microcredit program to improve their condition and the condition of their families.

Though cash transfers are helpful, they do not address the lack of opportunities associated with poverty. Researchers collaborated with Niger's government to determine whether the effects of its cash-transfer program would improve by adding other components. The researchers conducted a randomized controlled trial, randomly assigning 322 villages in Niger to a control group or one of three experimental groups. Households across all villages in the control and experimental groups received monthly cash transfers. Households in all three experimental groups also received training in business and administration. And then, the three different experimental groups provided (1) an extra cash grant, (2) life-skill training by social workers and a community film promoting communication, boosting aspirations, and noting women's lack of opportunities compared to men in the same village, or (3) the extra cash as well as life-skill training and the community film.

The researchers measured several variables related to the well-being of the participants in this experiment, especially variables related to women's empowerment, before

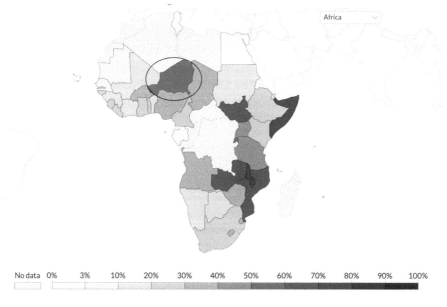

FIGURE 11.1 Share of population living in extreme poverty in African nations, 2019; Niger is circled

starting the experiment and at 6 and 18 months afterwards. They also calculated the cost-effectiveness of each intervention by comparing the monetary costs with the impact in each of the experimental conditions.

It turned out that the households in all conditions increased their consumption of food and domestic products. Households that also received psychosocial support showed improved levels of mental health, more positive expectations about the future, and increased social support, which in turn had further positive impacts on their economic situations over time. The researchers also discovered that integrating economic and psychosocial support to alleviate poverty had more lasting effects and low additional cost compared to simple cash transfers. This finding highlights that there is much more to poverty than a mere lack of money and that programs that target both monetary and psychosocial factors might be especially cost-effective.

Experiments like this can provide economists and policymakers with evidence about the causes and effects of poverty, as well as evidence about the relative efficacy of different interventions to reduce poverty. Economic concepts like poverty can seem abstract, but causal knowledge in economics plays key roles in the formulation of social policies and design of institutions. Such policies and institutions have many impacts on people's daily lives.

Three characteristics of causal reasoning

Causal knowledge is useful for goals such as alleviating poverty, mitigating global warming, preventing epidemics, and much more. But it can be tricky to uncover causal knowledge. Scientists have devised a number of methods to make reliable causal inferences in different experimental circumstances and for different theoretical and practical purposes. This chapter examines some of these methods for acquiring causal knowledge. Chapter 3's discussion of experimentation, Chapter 5's discussion of modeling, and Chapter 10's discussion of statistical hypothesis-testing are all relevant to using experiments and models to test causal hypotheses, and we will draw from those discussions in this chapter.

The Niger experiment investigating the causes and effects of poverty illustrate three general characteristics of causal reasoning that apply both in science and to everyday life. First, causal relationships can be uncovered from information about the timing, location, and frequency of events. If you get sleepy after lunch, perhaps eating lunch is a cause of your drowsiness. In the Niger example, researchers measured variables related to poverty before and after interventions to establish whether outcomes associated with reduced poverty were more frequent after some interventions than others.

Second, testing causal hypotheses often involves doing something in the world, such as performing an intervention. Intervening on a suspected cause while keeping other variables the same or ensuring they vary at random can provide more insight into causal relationships than just observing a correlation in two variables. In the Niger experiment, scientists tested the roles of different interventions using multiple experimental groups receiving different interventions. Because long-term economic prospects increased more, on average, in households receiving both monetary and psychosocial support, this suggests that poverty is not just about money and that policies should concentrate also on psychological and social factors.

Third, causal reasoning has great theoretical and practical significance. Causal information is an important variety of scientific knowledge and can be used to explain features of our world, as we shall see in Chapter 12. Besides theoretical knowledge, information about causes is also how we can make things happen—and prevent things from happening—in the world. Causal knowledge is not only crucial for effectively reducing poverty but also for treating medical conditions, improving environmental outcomes, and much more.

EXERCISES

11.1 Recall: Poverty is not merely a monetary deficit but a complex phenomenon. What are some of the causes of poverty? What are some of the negative effects that people deal with as a result of being poor?

11.2 Recall: The Niger experiment examined the effects on poverty of four experimental conditions. What were the four experimental conditions? What was the null hypothesis? What did the researchers find?

11.3 Think: Refresh your memory of external experimental validity, introduced in Chapter 3. Then, assess the external experimental validity of the Niger experiment described in this section. How confident can researchers be that these findings are accurate for residents of Niger? Should they be just as confident that the experimental results hold for poor households in other nations? Why or why not?

11.4 Recall: List the three characteristics of causal reasoning introduced in this section. Why is intervention useful for testing causal hypotheses? Why is causal knowledge more important than knowledge of correlations?

11.5 Apply: Provide your own example of (a) causal knowledge with theoretical importance and (b) causal knowledge with practical importance, and explain the importance of each.

11.6 Apply: Suppose you're a lead scientist following up on the Niger experiment to see if you can similarly reduce poverty in your local community. However, you yourself have very limited financial resources with which to conduct your study; for instance, you have no money for cash payouts. What interventions could you make in your follow-up study?

11.2 THE NATURE OF CAUSATION

After reading this section, you should be able to:

- Indicate how spatiotemporal contiguity and correlation each can be clues to causation but fall short of ensuring causation

- Define the *physical-process* and *difference-making* accounts of causation
- Assess the strength of a causal relationship and indicate how causal background is relevant

Clues to causation: spatiotemporal contiguity and correlation

Imagine that you are playing a game of billiards at your local pool hall. You hit the cue ball, which then rolls across the felt and strikes the 8-ball, which is itself then set in motion. Did the cue ball *cause* the 8-ball to move? The answer seems obvious. What else could have possibly made the 8-ball move?

In the 18th century, philosopher David Hume asked what our experience allows us to say about the nature of causation. He suggested there's no reason to say that causation is anything beyond regular associations between events. Hume's argument goes as follows. If you were to observe the cue ball hit the 8-ball, what you would have observed is just a series of events, one after another. You saw the cue ball moving towards the 8-ball, the cue ball touching the 8-ball, and then the 8-ball itself moving. There is no additional experience of causes and effects in all of that. We should then conclude that there is no ingredient to causation beyond just the series of observable events that are regularly associated. (You may recall from Chapter 7 that Hume was also skeptical about inductive reasoning.)

Hume's skepticism has motivated accounts of causation that go beyond mere regular association. Causal knowledge is important to science, but it's tricky to say what, exactly, causation amounts to. This skepticism can also motivate us to clarify the specifics of reliable causal reasoning. What is it, exactly, that you would look for to decide, for example, whether poverty causes social isolation, or whether routinely eating processed meats causes cancer?

Let's start by considering how spatiotemporal contiguity and correlation—together, Hume's regular associations between events—relate to causation. The perception of causal relationships was systematically investigated by psychologist Albert Michotte in the 1940s. Michotte's experiments showed that causal perception depends on spatial and temporal information. If two events—for example, pressing a piano key and hearing B-sharp—are spatially and temporally contiguous, that is, if they happen at the same time and place, then we perceive them as causally related. This is so even if we only experience these occurring together once. When there is a spatial or a temporal gap between two events, we are much less likely to perceive the one event as causing the other. Spatiotemporal contiguity can be a powerful indicator of a causal relationship.

On the other hand, spatiotemporal contiguity can also be misleading. Not all events that occur together are causally related, of course: some events occur together once or even several times by pure coincidence. So, spatiotemporal contiguity doesn't guarantee causation. Spatiotemporal contiguity is not necessary for causation, either. An effect can occur far away both in space and time from its cause. For instance, poverty in childhood can have effects much later in life, and poverty rates are influenced by geopolitical events that occur far away in time and space.

Indeed, many of the causal relationships investigated in science and important for everyday life are spatiotemporally separate to some degree. Sometimes the degree of separation is used to distinguish among an event's causes. ***Proximate causes*** are those occurring more closely in time and place to the caused event, while ***distal causes*** occurred further away in time or place. For example, when asked about the cause of a friend's poverty, you may describe her lack of social support and of a job at the moment, or you might note instead that many generations of her family have not accumulated much wealth and were denied access to social and economic opportunities. The former causes are proximate, while the latter distal. Proximate causes can be easier to identify, but distal causes are sometimes more important influences. This distinction between proximate and distal causes shows that events can have more than one cause. So, identifying a cause doesn't mean that you have identified the only cause or the most important cause, or that you have ruled out other causes. We have seen that poverty has many causes, for example.

Besides spatiotemporal contiguity, we also tend to use information about correlation between events to discern causal relations. Indeed, correlation can be used in combination with spatiotemporal contiguity as a guide to causation. As we know from Chapter 9, two variables are correlated when the value of one variable raises or lowers the probability of the other variable taking on some value. Different events can occur together a few times by chance, but if the values of two variables are correlated across many instances, this is some reason to think one variable may causally influence the other. For example, there is a clear correlation between childhood poverty and many negative life outcomes. It's thus worth investigating which (if any) of these life outcomes are caused by childhood poverty.

But correlation too is an imperfect guide to causation. For one thing, correlation is symmetric: if event *A* correlates with another event *B*, then *B* correlates with *A* as well. Causation isn't symmetric. If studying for an exam and getting a good grade on the exam are correlated, this is because studying causes getting a good grade and not the other way around. Yet a correlation alone doesn't provide information about which event is the cause and which event is the effect. Further, correlations between

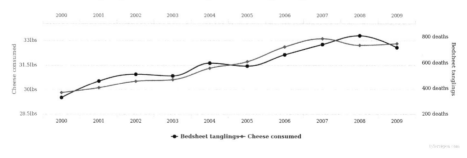

FIGURE 11.2 Per capita consumption of cheese correlates with the number of people who died from getting tangled in their bedsheets

events can exist even though neither event causes the other. Some events are correlated because they share a ***common cause***—a third event that causes both. Famously, ice cream consumption is correlated with drowning, but eating ice cream is not a cause of drowning. Instead, there is evidence that the occurrence of hot days increases both ice cream consumption as well as swimming, and so also drowning rates. Finally, some correlations are ***spurious***, that is, not causally related at all, such as the relationship between annual per capita cheese consumption and the number of people who died from becoming tangled in their bedsheets.

These are all ways correlations may fail to indicate causation. But the reverse can also happen: events can be causally related even when they are not correlated. Consider this example, due to philosopher Nancy Cartwright. Smoking cigarettes is well established as a cause of heart disease. Adequate exercise prevents heart disease. If smoking is strongly correlated with exercise, then this well-established cause of heart disease will also be strongly correlated with its prevention, and smoking and heart disease will not in general be correlated. This could be so even though smoking causes heart disease.

Box 11.1 Simpson's paradox

In August 2021, less than a year after Covid-19 vaccines first became available, headlines announced the findings of a new study published in the prestigious journal *Science*: "Nearly 60% of hospitalized COVID-19 patients in Israel fully vaccinated." The vaccine was supposed to help against Covid, but now more vaccinated people than unvaccinated were getting seriously ill! People worried that the vaccine's efficacy waned very rapidly, no longer offering protection after a few months. But no: the vaccine was still very effective at preventing severe illness from Covid-19. Instead, this was an instance of **Simpson's paradox**, where a correlation between two events disappears or is reversed when data are grouped in a different way. Israel had a very high vaccination rate: there were many more vaccinated than unvaccinated people. Further, age is correlated with the probability of vaccination: as in many other parts of the world, older Israelis were more likely to be vaccinated than younger Israelis. But older people are significantly more likely to suffer severe illness from Covid-19. The 40% of hospitalized Covid-19 patients who were unvaccinated was thus a much larger proportion at lower risk of severe illness than the 60% of hospitalized Covid-19 patients who were vaccinated. Accounting for the vaccination rate in the overall population and controlling for age revealed that Israelis over 50 were seven times more likely to be hospitalized if they were unvaccinated, while younger Israelis were 13 times more likely to be hospitalized if they were unvaccinated. There are many other real-life cases of Simpson's paradox. Understanding how and why the paradox emerges is important for drawing correct causal conclusions from statistical evidence.

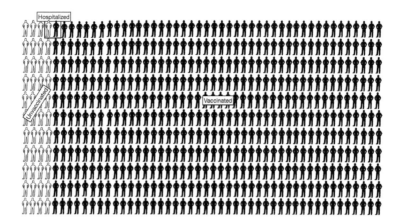

FIGURE 11.3 By 2021, nearly 92% of Israelis over 50 had been vaccinated for Covid-19. An unvaccinated Israeli over 50 had a nearly 1% chance of hospitalization from Covid, while a vaccinated Israeli over 50 had a .014% chance of hospitalization.

Accounts of causation: difference-making and physical processes

We have seen that spatiotemporal contiguity and correlation can be guides to causal relationships, but neither is a perfect guide. Despite Hume's skepticism, there must be more to causation than mere association between events. Here are two ideas about what that might be.

One idea is that causal relationships are relationships of difference-making. Put simply, the ***difference-making account of causation*** is the idea that if the occurrence of one event makes a difference to the occurrence of a second event, then the first event is a cause of the second event. If the billiard ball had not struck the 8-ball, then the 8-ball wouldn't have moved. If the billiard ball had struck the 8-ball in a different place or at a different speed, then the 8-ball would have moved in a different direction or at a different speed. The billiard ball's motion made a difference to the 8-ball's motion. Thus, the billiard ball's motion caused the 8-ball to move. This difference-making relationship goes beyond the mere correlation of events: a focus on intervention can help us see how. Recall that an *intervention* is a direct manipulation of the value of the independent variable. The idea of difference-making is that, for one variable to count as a cause of another, an intervention on this variable, while controlling all other variables, will change the value of the other variable. A difference-making account of causation uses the ideas of intervention and variable control to give an answer to what causation might be beyond mere correlation or association between events.

Box 11.2 Difference-making and counterfactual conditionals

According to the difference-making account of causation, causally relevant factors make a difference to whether an effect happens. The idea of difference-making relies on reasoning about counterfactual conditionals. Recall that a conditional is any if-then statement. Counterfactual conditionals have the following form: if C had occurred, then E would have occurred. For example, consider the conditional "if you had scored the penalty, then your team would have won the game." If this counterfactual conditional were true, this would be reason to say that you not scoring a goal caused your team to lose. Such conditionals are called *counterfactuals* because the antecedent of the conditional is contrary to fact. Recall from Chapter 6 that material conditionals are false only when the antecedent is true and the consequent is false. This isn't so for counterfactual conditionals. After all, the antecedent of a counterfactual conditional is always false. In our example, you didn't score a goal, but that doesn't guarantee your team would have won if you had—maybe that wouldn't have been enough to win. It's a matter of debate in philosophy how to assess whether a counterfactual conditional is true. It seems that to do so we need to consider how things would be in conditions unlike what has really occurred—in another "possible world," philosophers sometimes say. The difference-making account of causation makes use of the concept of an ideal intervention to help with this. If we imagine intervening on what really happened to make it so that you scored a goal, keeping everything else exactly the same, then would your team have won? If so (and if you didn't in fact win), then you not scoring a goal did cause your team to lose.

A second idea about the nature of causal relationships is that they are based on physical processes. On the ***physical process account of causation***, causation occurs when there is a continuous physical process connecting a cause to its effect, such that a cause transmits its effect with the transfer of mass, energy, momentum, charge, or other physical properties. When the billiard ball knocked into the 8-ball, some of its kinetic energy transferred to the 8-ball, which is why the 8-ball started moving. This is a physical process connecting the billiard ball's motion to the 8-ball's motion. Thus, the billiard ball's motion causes the 8-ball's motion. The physical process account gives a different answer than the difference-making account to how causation goes beyond mere association, pointing instead to the transmission of physical properties.

The physical process and difference-making accounts of causation may be compatible. But each account is more useful for thinking about causal reasoning in different circumstances. For some causal claims, physical processes are the benchmark; with others, they are difficult to track. Recall the research on alleviating poverty in Niger. How would you investigate energy transfer or other physical processes of the causes of poverty? You wouldn't. In contrast, it's clear how to think about increasing income with cash transfers, a suspected causal variable, and to test how those changes affect variables associated with poverty. This suggests a difference-making account is apt.

For other causal claims, the idea of difference-making doesn't neatly apply. The moon orbits around the sun because of the curvature of space-time. How could we intervene on this curvature to assess whether our intervention makes a difference to the orbit of the moon? It's a bit confusing; this seems like a feature of reality that can't be changed with an experimental intervention. And without even a way to conceive of such an intervention, we can't use difference-making to assess causal influence.

Watch Video 14

Strength of causation

Earlier in this section, we considered how correlation could provide clues to causal relationships. Correlation between variables occurs when the value of one variable raising or lowering the probability of the other variable takes on some value. These conditional probabilities are also useful in characterizing the strength of causal relationships.

Consider some cause, C, and its effect, E. To start at the extreme, if $Pr(E|C) = 1$, then the occurrence of a cause guarantees the effect. For example, having no income, family resources, or any other source of funds or necessities guarantees that someone will experience poverty. Having no source of funds or necessities is a ***sufficient cause*** of poverty, in the sense that this cause always brings about this effect. If $Pr(E|\text{not-}C) = 0$, then the occurrence of a cause is necessary for the effect to occur. For example, a motionless 8-ball will not roll across a flat pool table unless something strikes it. Another ball or other object striking an 8-ball is thus a ***necessary cause*** of a motionless 8-ball rolling across a flat pool table: a cause that must occur in order for the effect to come about. (See Chapter 6 for discussion of necessary and sufficient conditions, which is relevant to this use of necessity and sufficiency.) If both conditional probabilities hold, if both $Pr(E|C) = 1$ and $Pr(E|\text{not-}C) = 0$, then the cause is both necessary and sufficient for the effect. The cause occurring guarantees the effect will occur, and without the cause occurring, there's no way the effect will occur.

The strength of virtually all causal relationships falls short of these extremes. Most causes aren't necessary or sufficient for their effects to come about. For example, though having no income can often result in poverty, family or governmental support could interfere with that effect. Having no income isn't sufficient for poverty. Nor is it necessary: the *working poor* is a term for people who spend substantial time in paid employment and yet are experiencing poverty. A cause that raises the probability of an outcome occurring but does not guarantee the outcome is called a ***contributory cause,*** or ***partial cause,*** of that outcome. Most causes are contributory causes.

In general, you can judge the strength of a causal relationship with the following calculation:

$$S = Pr(E|C) - Pr(E|\text{not-}C)$$

This measures the degree to which the occurrence of the cause increases the probability of the effect occurring by taking into account the likelihood of the effect when the cause does and does not occur. If more people without an income experience poverty and fewer people with an income experience poverty in the US compared to the Netherlands, for example, then we could say that lack of income is a stronger influence on poverty in the US than in the Netherlands. It's important to notice that, although conditional probabilities can be used to characterize the strength of a causal relationship, they can't identify when one variable is a cause of another. Recall from our previous discussion of correlation that simply identifying a correlation between the conditional probability of variables doesn't mean one causes the other.

Most causes are merely contributory causes, rather than being necessary or sufficient for their effects, because causal relationships are often sensitive to other variables. The ***causal background*** consists in all the other factors that might causally influence an outcome, thereby also potentially affecting the causal relationship between the two events. The causal background for how lack of income causes poverty includes factors like family wealth, governmental support, savings, financial responsibilities, money management skills, and much more besides. This is one reason why poverty is a complex phenomenon. The conditional probabilities between a cause and its effect can change in different causal backgrounds or may only hold in some causal backgrounds.

Often, the causal background is ignored when causal claims are made. You do not generally consider that oxygen is a causally relevant factor when you explain how a defective stove caused an apartment fire, for example. But accounting for the relevant causal background is crucial in sound causal reasoning. For example, the value of cash transfers to mitigating poverty will depend on the causal background that shapes the extent to which lack of access to cash is responsible for experiencing poverty.

EXERCISES

11.7 **Recall:** How is spatiotemporal contiguity a clue to causation, and how does it fail to be a perfect guide to causation? How is correlation a clue to causation, and how does it fail to be a perfect guide to causation?

11.8 **Apply:** List three possible relationships that can result in a correlation between events *A* and *B* even though *A* does not cause *B*. Give a new example illustrating each.

11.9 **Recall:** Define the *difference-making* and *physical process* accounts of causation.

11.10 **Think:** What was Hume's skepticism about causation? Evaluate the merits of his concern, taking into account the discussion throughout this section.

11.11 **Apply:** Define *proximate causes* and *distal causes*. Then, for each of the following questions, give one explanation that cites a more proximate cause and one explanation that cites a more distal cause. You might need to invent some details about these causal relationships to answer this question; that's fine. Feel free to get creative.
a. Why did the Titanic sink?
b. Why did Ruth leave a tip after her meal at the restaurant?
c. Why did the hurricane happen?

11.12 Apply: Write down the formula that gives the strength of causal relationships. Then, considering that formula, order the following causal relationships from strongest to weakest, giving a brief rationale for the order of each item.

    a. Brushing your teeth, flossing, and visiting the dentist prevents cavities.

    b. Frequent smiling increases well-being.

    c. Eating pizza prevents getting the flu.

    d. Consuming anabolic steroids improves physical strength.

    e. Increase in the minimum wage produces higher attendance at football games.

    f. Warmer summers lead to longer periods of drought.

11.13 Think: For each of the causal relationships in 11.12, name one feature of the causal background that would make the causal relationship stronger and one feature of the causal background that would make the causal relationship weaker. It might help to consider the conditional probability relationship that gives the strength of causal relationships.

11.3 TESTING CAUSAL HYPOTHESES

After reading this section, you should be able to:

- Characterize how the difference-making account of causation relates to intervention
- Describe how intervention, direct and indirect variable control, random sampling, and statistical hypothesis-testing is each important to discerning causal relationships
- Articulate how statistical hypothesis-testing is used to test causal hypotheses

Experimental interventions and difference-making

Discovering causal relationships requires more than just passively observing what happens. The idea of causation as difference-making is useful for making sense of various methods scientists can use to acquire knowledge of causes. One such method is to run an experiment—ideally, a perfectly controlled experiment, as detailed in Chapter 3.

Suppose that you are a farmer interested in learning whether using a new fertilizer will increase your crop yield, that is, make a difference (positive difference) to the yield. This is a causal hypothesis. How would you test it? One way would simply be to try out the fertilizer on your crops and see what kind of a yield you get. But the causal background might vary from last year to this year in a way that affects crop yield: rainfall, temperatures, last frost date, and more. You wouldn't be able to distinguish that influence from the specific effect of the fertilizer on the yield.

A better approach would be to divide your field into different plots of equal size. You can then use the new fertilizer on some plots but not others and compare the crop yield from the fertilizer plots to the crop yield from the other plots. If the plots treated with the fertilizer produce, on average, a larger crop yield than the other plots, then the fertilizer made a difference; we could then say the new fertilizer *causes* increased crop yield. If the two groups of plots yielded about the same amount of crop, then the new fertilizer is probably ineffective. If the fertilizer plots do worse, the fertilizer makes a difference—but the wrong kind!

Applying concepts introduced in Chapter 3, we can say that the farmer has created an experimental group (plots to which the fertilizer is applied) and a control group (plots handled according to the farmer's past practices). The application of fertilizer to plots in the experimental group is an intervention. The farmer intervenes on a suspected cause to see whether this makes a difference to the suspected effect. The suspected cause is the independent variable, and the suspected effect is the dependent variable. The causal background consists of extraneous variables. Some extraneous variables are directly controlled by comparing yield in the same season, in the same field, with the same irrigation, and so on, while other extraneous variables are indirectly controlled by random assignment of plots of land to experimental group and control group.

Some experiments testing causal hypotheses aim to establish whether there is a causal relationship, and others aim to clarify the strength of a causal relationship. For example, some drug trials simply seek to establish safety: that a drug won't have negative effects. Others seek to establish efficacy: that a drug will have the expected positive effect. And still others aim to establish the relative strength of a causal relationship: whether some drug is more effective than another already on the market.

Experimental interventions introduce an external influence on a system, which can help disentangle causal relations. This is why the suspected cause is called an *independent* variable—the intervention independently determines its value, which, if all goes well, eliminates the possibility that the suspected cause is affected by the causal background. Other features of experimental design, such as using random sampling to select experimental participants and having a control group, are designed to minimize the chance that changes to the suspected effect are due to the causal background instead of the intervention. Altogether, these features help scientists test causal hypotheses, identifying whether a particular factor is a genuine difference-maker.

Statistical variation and testing causal hypotheses

In experimental tests of causal hypotheses, hypotheses about the existence, direction, and strength of causal relationships are used to develop specific expectations regarding how dependent variables will change in response to changes to independent variables. But testing causal hypotheses is complicated by extraneous variables and chance variation. The experimental techniques of intervention, direct and indirect variable control, random sampling, and statistical hypothesis-testing are motivated in large part by their ability to discern causal relationships from mere correlation, cause from effect, and causal influence from chance variation.

We just saw how intervention and variable control are useful. Let's now consider how statistical hypothesis-testing and random sampling help to test causal hypotheses in the face of statistical variation. In the fertilizer experiment we imagined earlier, we expected the new fertilizer to improve crop yield in the plots to which it is applied. Does this mean we should expect all fertilized plots to have better yield compared to all the standard plots? Surely not; varying causal background, and perhaps sheer random variation, will introduce variability among the plots. So, then, how many of the fertilized plots need to have a better yield to show that fertilizer makes a difference? To answer that question, we'd need to use inferential statistics.

As this illustrates, statistical hypothesis-testing is crucial for testing causal hypotheses. Inferential statistics can be used to generate specific expectations of a group's statistical properties, which is important for hypotheses that predict probabilistic causal influence. As we've seen, this is much more common than causes guaranteeing their effects. In null hypothesis significance tests, causal hypotheses play the part of the alternative hypothesis. The null hypothesis is usually that the posited cause does not actually influence the phenomenon of interest.

Box 11.3 Mill's methods

In the late 19th century, the English philosopher John Stuart Mill emphasized the role of observation and experimentation in discerning causal relationships. Mill identified five methods to evaluate causal hypotheses. These can still be useful, though statistical analysis and intervention are both important tools of causal hypothesis-testing not captured by Mill's methods.

1. **Method of concomitant variations**: Observing that the value of one variable changes in tandem with changes to the value of another variable. In modern terms, this just is using correlation between variables to infer that the variables might be causally related. Of course, mere correlation does not guarantee causation.
2. **Method of agreement**: Considering cases where an effect occurs to see what they have in common. If there is a prior event or condition common to all, then one may infer that this event or condition caused the effect.
3. **Method of difference**: Considering cases where an effect does not occur to see what distinguishes those from when the effect occurs. If an event or condition present when the effect occurred is absent when the effect does not occur, then one may infer it caused the effect.
4. **Joint method of agreement and difference**: Considering cases where the effect occurs to see what they have in common (method of agreement), as well as considering cases where the effect does not occur to see what was different (method of difference). If an event or condition common across cases where the effect occurs is absent when the effect does not occur, one may infer it causes the effect.
5. **Method of residues**: Comparing cases in which a set of causes brings about a set of effects to cases in which some of those causes bring about some of those effects and inferring, on that basis, that the absent cause(s) are responsible for the absent effect(s). Unlike the other methods, this is a way to reason about which of a set of causes are responsible for which effects.

So, for our farmer, the null hypothesis is that the fertilizer is causally inefficacious: the range of crop yield from the fertilized plots of land will only differ from the range of crop yield from the other plots by chance variation. The mean crop yield and crop-yield variation in the control group, together with the sample size, can be used to calculate the probability distribution for the experimental group's crop yield assuming the new fertilizer does not make a difference. If the measured crop yield is sufficiently unlikely, we can reject the null hypothesis and conclude the new fertilizer causes increased crop yield.

Another complication of expectations from a causal hypothesis due to causal background relates to external experimental validity, that is, the extent to which experimental results generalize from the experimental conditions to other conditions. For example, should we expect the new fertilizer to have the same effect as in our experiment in a rainier season, when the causal background is different? Due to considerations of external experimental validity, it's not sufficient just to ensure extraneous variables do not vary systematically between the experimental and control groups. Conclusions about a causal hypothesis may also go wrong when the causal background of an experiment is not sufficiently similar to the causal background in circumstances in which we seek to apply the causal knowledge.

Random sampling is helpful on this front. A sufficiently broad range of conditions that are randomly selected enables us to expect the range of experimental conditions to be similar to the real-world circumstances in which the causal knowledge would apply. This increases external experimental validity and, in turn, the relevance of our causal knowledge. Perhaps our farmer will have to test out the fertilizer over several seasons or on plots arranged to mimic expected variation such as in rainfall and temperature to know the new fertilizer causes increased crop yield across the range of growing conditions the farmer is likely to encounter.

EXERCISES

11.14 Recall: Describe the importance of intervention for testing causal hypotheses. Then, define the *difference-making account of causation* and characterize how it relates to intervention.   Watch Video 15

11.15 Apply: Headlines in popular media often misrepresent the scientific studies they discuss. In particular, many headlines suggest a causal relationship where the evidence provided by the scientific study only supports a correlation. Consider the following headlines. For each, (a) identify whether it makes either a causal or a correlational claim; (b) rewrite any headline using causal language so that it reads as a correlational study; and (c) suggest a possible explanation for each correlation that is not the posited or suspected causal relationship.

1. "Lack of Sleep May Shrink Your Brain" (CNN, September 2014)
2. "To Spoon or Not to Spoon? After-Sex Affection Boosts Sexual and Relationship Satisfaction" (*Science of Relationships*, May 2014)
3. "Daytime TV (Soap Operas) Tied to Poorer Mental Scores in Elderly" (Reuters, March 2006)

4. "Study Suggests Attending Religious Services Sharply Cuts Risk of Death" (*Medical Xpress*, November 2008)
5. "Facebook Users Get Worse Grades in College" (*Live Science*, April 2009)
6. "Texting Improves Language Skill" (BBC, February 2009)
7. "Study Suggests Southern Slavery Turns White People Into Republicans 150 Years Later" (*Think Progress*, September 2013)
8. "Dogs Walked by Men Are More Aggressive" (NBC News, November 2011)
9. "Want a Higher GPA? Go to a Private College" (*New York Times*, April 2010)
10. "Sexism Pays: Men Who Hold Traditional Views of Women Earn More Than Men Who Don't" (*Science Daily*, September 2008)

11.16 Apply: Choose two of the headlines from Exercise 11.15, and look up the text of both. Write a paragraph evaluating the strength of the evidence cited in the media report supporting the claim (causal or correlational) in the headline.

11.17 Recall: Describe why each of the following is important for discerning causal relationships: direct and indirect variable control, random sampling, and statistical hypothesis-testing.

11.18 Apply: How would you design an experiment to determine whether smoking marijuana causes schizophrenia?

a. Identify the intervention and describe how you would control extraneous variables.
b. Identify the expectations given the hypothesis, that is, what finding would enable you to conclude that smoking marijuana is a cause of schizophrenia.
c. Describe how statistical hypothesis-testing and random sampling are employed and why each is important.

11.4 CAUSAL MODELING

After reading this section, you should be able to:

- Describe how causal modeling can yield causal knowledge
- Define *causal Bayes nets* and describe how they are developed and applied
- Identify three assumptions of causal Bayes nets and discuss their significance

From correlation to causation

Besides controlled experiments and statistical hypothesis-testing, another important approach to gaining causal knowledge is causal modeling. Causal modeling involves the use of computational and statistical methods for representing, manipulating, and testing causal hypotheses. When experimental interventions cannot be conducted, causal modeling can be used to evaluate causal hypotheses. Even when interventions can be made, combining experiments with causal modeling can be used to evaluate whether correlational evidence supports the causal hypothesis and to help identify factors in the causal background that must be controlled during an experiment.

Like controlled experiments, causal modeling is also related to the difference-making account of causation, which defines causal relationships in terms of potential intervention and variable control. If an event C causes another event *E*, then an intervention on C influences the value of *E*. If C and *E* are merely correlated, share a common cause, or if *E* causes C, then interventions on C won't causally influence the value of *E*. In causal modeling, difference-making is identified from patterns of conditional probabilities between variables representing different features of a target phenomenon. Scientists can leverage these patterns of conditional probabilities to represent causal structure and reason about how variables influence one another.

The method of regression analysis, introduced in Chapter 9, is one of the oldest causal modeling procedures. It is used to estimate the correlation of two variables, conditional on all other measured variables. It's like drawing a best-fitting line for the relationship in the values of two variables based on data on a scatterplot. For a suspected causal relationship, regression analysis is used to estimate how a causal variable affects another variable. Merely running a regression analysis, however, cannot deliver causal knowledge when causal influence isn't yet known. We need additional evidence from meta-analyses, experiments, or non-experimental studies to supply that information, then regression analysis can be used to estimate the nature of the causal effect, including its direction and strength.

As in regression analysis, the building blocks of many causal modeling approaches are statistical correlations and probabilistic dependencies between variables that represent different features of a target phenomenon. In causal models, variables represent potential causes and effects. Here we will focus on Bayes networks, or "nets." A ***Bayes net*** is a kind of causal model that uses joint probability distributions to provide a compact, visual representation of causal relationships and the strength of those relationships. ***Nodes*** in the graph stand for variables of interest, and arrows connecting different nodes stand for hypothesized causal relationships between variables.

Conditional probabilities specify how each variable depends on its direct causes in a ***joint probability distribution***: the probability distribution for each of a set of variables, taking into account the probability of the other variables in the set. These causal models are called Bayes nets because they use Bayes's theorem to update the probabilities based on new information about the value of variables or their probabilistic relationships. (Bayes's theorem was introduced in Chapter 8 and Bayesian statistics discussed in Chapter 10.)

Developing Bayes nets

To better understand how scientists use Bayes nets to learn about causal relationships, consider this scenario from an introduction to the main concepts and applications of Bayesian networks:

> Suppose that a patient has been suffering from shortness of breath (called *dyspnea*) and visits the doctor, worried that he has lung cancer. The doctor knows

that other diseases, such as tuberculosis and bronchitis, are possible causes of this symptom, as well as lung cancer. She also knows that other relevant information includes whether the patient is a smoker (increasing the chances of cancer and bronchitis) and what sort of air pollution he has been exposed to. A positive x-ray would indicate either TB or lung cancer.

There's plenty of causal information here, but how that information relates to the case at hand is tricky to determine. Doctors can use causal Bayes nets to generate medical diagnoses. To construct such a model, the relevant variables first need to be identified. Each variable is represented with a node. While there's no uniquely right way of setting up the Bayes causal net, it helps to make choices about what nodes to include that enable us to represent the relevant aspects of the situation with enough detail to perform the desired reasoning. One possible modeling choice is shown in Table 11.1. In this case, the variables include *dyspnea, smoker, pollution exposure, age, x-ray result, tuberculosis,* and *lung cancer.*

The second step of constructing a causal Bayes net is to specify the system's causal structure by drawing arrows between the nodes. If one variable affects another, then the corresponding nodes should be connected with an arrow indicating the direction of the effect. Smoking and living in a polluted area are two factors affecting the patient's chance of having lung cancer. In turn, having lung cancer is a factor affecting the result of an x-ray and the patient's difficulty in breathing. If this is the structure of the situation, then we may draw the graph pictured in Figure 11.4.

Causal relationships represented in a causal Bayes net can take several forms. Causes can increase or decrease the probability of some variable taking on a given value, and there can be a feedback loop where two or more variables influence one another in a cyclical way. Most of the time, however, Bayes nets are assumed to be ***directed acyclic graphs*** (sometimes abbreviated DAG), which means that all the causal relationships are taken to go in one direction without feedback loops. This assumption means that earlier causes are not also later effects. You can see from Figure 11.4 that our graph makes this assumption; no arrows form circles like $X \rightarrow Y \rightarrow Z \rightarrow X$, and no arrow is bidirectional like $X \longleftrightarrow Y$.

TABLE 11.1 Possible values for variables in the dyspnea case

Variable	Values
dyspnea	{T, F}
smoker	{T, F}
pollution	{low, high}
x-ray	{positive, negative}
lung cancer	{T, F}
tuberculosis	{T, F}

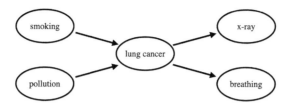

FIGURE 11.4 A causal graph for the dyspnea case

TABLE 11.2 Conditional probabilities of developing lung cancer given level of pollution exposure and whether one smokes

Pollution	Smoker	Pr(cancer = T | pollution, smoker)
high	T	0.050
high	F	0.020
low	T	0.030
low	F	0.001

After specifying the nodes and their structure, the strength of the relationships between connected nodes must be specified. To do so, one needs to define a probability distribution for each node, conditional on any node(s) that causally influences it. In the example of patient diagnosis, statistical information from previous medical studies or from observed frequencies are used to specify these probability distributions. For variables with no such information available, initial probabilities can be based on an intuition, guess, or estimation about the case. Recall from Chapter 10 that Bayesian statistics is supposed to make these initial guesses unimportant in the long run. Bayes nets allow us to infer conclusions even if we start off with imprecise or inaccurate initial probabilities.

Let's look at the node *cancer* in Figure 11.4. Its parent nodes are *pollution* and *smoker*, each of which can take two possible values for a total of four combinations of joint values: {<high, T>; <high, F>; <low, T>; <low, F>}. We can then specify the conditional probability of having cancer for each of these four cases. One way to represent these conditional probabilities is in a table, as in Table 11.2.

Reasoning with Bayes nets

Once all conditional probability distributions are determined, our causal Bayes net captures the relevant knowledge available. Now we can start to reason with it.

Reasoning with a Bayes net amounts to computing posterior probability distributions for one or more nodes of interest given the values of nodes that you have information about. These computations are governed by Bayes's theorem or other algorithms for computing approximations to posterior probabilities. Think of this as updating your beliefs about the value of a node based on changes to your beliefs about the values of other nodes. The arrows connecting nodes show the paths that probability distribution changes follow.

Belief updating can happen either from cause to effect or vice versa. For example, if we're certain that the patient has dyspnea, and her x-ray results are negative, then we can update our diagnosis about whether the patient has cancer, a causal influence on both dyspnea and x-ray results. In turn, updating our diagnosis about whether the patient has cancer will affect our beliefs about whether the patient smokes and lives in a high-pollution area, proceeding up the chain of causal influence. Or, if we are certain that the patient smokes, we can update our beliefs about her chance of having lung cancer accordingly, which is causally influenced by smoking status. This also influences our expectations of the x-ray result.

A different type of reasoning with causal Bayes nets regards the relationship between two causes that compete to explain an observed effect. In our case, *smoker* and *pollution* are two such causes. They compete to explain the value of the variable *cancer*, which they both influence. Suppose we learn that the patient has cancer. This new information raises the probability of both possible causes. Suppose that we learn further that the patient lives in a badly polluted city. Something interesting would now happen in our causal Bayes net. This new piece of information both explains the patient having cancer and lowers the probability that the patient is a smoker. Although the variables *smoker* and *pollution* are initially probabilistically independent, given that we know that the patient has cancer and lives in a highly polluted area, the probability that the patient is a smoker goes down. Knowing that the patient has been exposed to significant pollution accounts for the lung cancer and thus disrupts the probabilistic association between lung cancer and smoking. Now we needn't speculate that the patient was a smoker in order to explain the lung cancer.

In the simple case we've considered, a Bayes net is specified and then used to make causal inferences and predictions. In many other scientific applications, causal Bayes nets are incomplete in two ways. First, there are many variables that could be added that would precede, mediate, or follow the variables explicitly represented in the model. Second, information might be lacking about the causal relationships between the variables represented in the model. In that case, the structure of the network and the relevant probabilistic dependencies must be learned from data, since defining a complete Bayesian network would be too complex. For this, scientists rely on computational methods like machine learning algorithms that search correlational data for causal dependencies.

Cognitive neuroscientists, for example, are interested in the causal relationships between brain areas that support a cognitive capacity. To find out about these causal relationships, they often rely on brain imaging data, where subjects perform tasks that tap the cognitive capacity of interest while having their brain activity recorded. Neuroscientists use background knowledge about which brain regions might be involved in a task to focus their attention on activity in a few regions of interest, each one of which can be treated as a variable and represented as a node in a DAG. The challenge is then to discover the causal structure of these regions. Machine learning algorithms help neuroscientists to tackle this challenge. An algorithm searches the brain imaging data set to find the causal structure that best explains observed statistical dependencies between the variables of interest. Using DAGs, it is then possible to determine whether a causal structure is in principle identifiable from a probability distribution and to derive the probabilistic expression for this quantity.

Assumptions of Bayes nets

Reasoning with causal Bayes nets requires making several theoretical assumptions; these assumptions are needed to infer causal relationships from data about conditional probabilities. Three key assumptions are the common cause principle, modularity, and the causal Markov condition. These assumptions are not always satisfied by a data set, but when they are, causal Bayes nets are especially promising for learning about causal relationships between variables from their observed statistical features.

We encountered the idea of a common cause earlier in this chapter. The common cause principle says that every correlation between two variables is either due to a direct causal effect linking the correlated variables or is brought about by a third factor that causes both, that is, a common cause. This idea is of central importance in causal explanation and causal modeling. For example, suppose that two lamps in your room go out suddenly and simultaneously. You may look for whether the common power supply was interrupted. This interruption would be the common cause of the two simultaneous events and would explain this improbable coincidence.

Modularity is an assumption about how systems can be manipulated; it implies that interventions into causal relationships in a system should not change other causal relationships in the system. When a system is not modular, interventions on some causal relationship(s) change the nature of other causal relationships. If a system is modular, then dependencies between variables in a causal Bayes net model of the system that are not directly manipulated should not change. Modularity is a useful feature, as it allows for precise and focused predictions about what would happen in the target system if certain causal relationships were manipulated. When the modularity assumption is violated, Bayes nets won't correctly specify the state of a system after an intervention.

Closely associated with modularity is the causal Markov condition. One of the most important assumptions of causal Bayes nets, the **causal Markov condition** is the requirement that the probability of causal variables conditional on their parent causes are probabilistically independent of all their other ancestors. The basic idea is that remote causes are irrelevant to conditional probabilities, and thus to causal inference, when we know the immediate causes of an effect. In our medical diagnosis example, the value of *tuberculosis* is influenced by the value of *cancer*, but probabilistically independent of *pollution* and *smoker* conditional on *cancer*. This is because cancer and tuberculosis are causally related, whether the cancer was caused by smoking or by pollution.

The causal Markov condition is motivated by the idea that, when probabilistic dependencies are found between variables, these dependencies are due to one variable causing the other or to their sharing a common cause. The causal Markov condition specifies which variables will be probabilistically independent conditional on other variables in a set of variables under study. If the causal Markov condition holds of a set of variables associated with a system, then conditional independence between variables indicates the absence of a causal relationship. For example, a causal chain like $X \rightarrow Y \rightarrow Z$ implies that X, Y, and Z are all probabilistically dependent on one another but that X and Z should be probabilistically independent if Y takes a fixed value. In some cases, the Markov condition might fail if the set of variables included in the Bayes net omits common causes or includes variables that aren't relevant causes.

These are three key assumptions underlying reasoning with causal Bayes nets, though there are others as well. Understanding how different strategies for causal modeling work when some of their assumptions fail, and what kinds of errors they would yield, are two of the most important challenges of current causal modeling approaches.

EXERCISES

11.19 Recall: Explain what causal modeling is good for, pointing out its advantages and limitations.

11.20 Apply: For each of the following cases, (a) indicate the causal hypothesis involved, distinguishing causal variables from effects; (b) offer another plausible cause for the effect; and (c) explain whether the reasoning described in the case is good or bad with the help of a simple causal model (consider using a directed acyclic graph, or DAG).

    a. You have eaten your birthday dinner at your favorite pizzeria in town for the past 10 years. This year you got sick. This was also the first time your Uncle Sam was there. You conclude you got sick because Uncle Sam was there.

    b. Eryka normally goes to bed at midnight and gets up by 7:00 am each morning. She usually runs two kilometers after having some breakfast. This morning, however, she ran only half a kilometer and had to stop, as she was so tired. She recalled that she went to sleep unusually early the night before and concludes that too much sleep made her too tired to run.

    c. Phineas Gage's moral character changed dramatically after an explosion blew a tamping iron through his head. Gage was leading a railroad construction crew near Cavendish, Vermont, when the accident occurred. "Before the accident he had been a most capable and efficient foreman, one with a well-balanced mind, and who was looked on as a shrewd smart business man." After the accident, he became "fitful, irreverent, and grossly profane, showing little deference for his fellows. He was also impatient and obstinate, yet capricious and vacillating, unable to settle on any of the plans he devised for future action."

11.21 Recall: Describe (a) what a causal Bayes network is, (b) how a causal Bayes net is developed, and (c) how it is applied.

11.22 Apply: Consider the following story: A group of psychologists is interested in how intrinsic motivation of university students affects their exam results. They believe that intrinsic motivation affects both class attendance and home preparation (reading the textbooks, doing the assignments, etc.). They also believe that both class attendance and home preparation affect exam results. They do not believe that there are any further causal interactions. All relevant variables (intrinsic motivation, class attendance, home preparation, exam results) are considered to have two values: "High" and "Low" for intrinsic motivation, class attendance, home preparation; and "Pass" and "Fail" for exam results. The psychologists observe the following frequencies:

40% of all students have a high intrinsic motivation.

90% of all highly motivated students attend classes regularly, as opposed to 60% of all students with low motivation.

70% of all highly motivated students prepare well, as opposed to 20% of all students with low motivation.

80% of all students who prepare well and attend class regularly pass the exam.

60% of all students who prepare well and do not attend class regularly pass the exam.

45% of all students who do not prepare well and do attend class regularly pass the exam.

40% of all students who do not prepare well and do not attend class regularly pass the exam.

On the basis of this information, find the conditional probabilities and draw the causal Bayes net that corresponds to the story.

11.23 Recall: Describe the assumptions of common cause principle, modularity, and the causal Markov condition, and indicate why each is important for causal modeling.

11.24 Apply: Suppose that we measure the variables storm (S), barometer reading (B), and atmospheric pressure (A). You find that S and B are dependent, as are both B and A and S and A. Furthermore, you find that S and B given A are independent. From these constraints alone, what causal structure can you infer? Draw a simple DAG showing the causal relationship. Then, explain the role of the common cause principle and Markov condition in making reliable inferences about the causal relationships in the example.

FURTHER READINGS

For more on poverty, see Lister, R. (2021). *Poverty*. John Wiley & Sons. A more concise treatment is Wolff, J., Lamb, E., & Zur-Szpiro, E. (2015). *A philosophical review of poverty*. Joseph Rowntree Foundation. www.jrf.org.uk/report/philosophical-review-poverty

For more on normative and descriptive accounts of causation and their importance in scientific explanation and psychology, see Woodward, J. (2021). *Causation with a human face: Normative theory and descriptive psychology*. Oxford University Press.

For a pluralist view of the nature of causation and discussion of causal analysis, including causal Bayes nets, see Cartwright, N. (2007). *Hunting causes and using them: Approaches in philosophy and economics*. Cambridge University Press.

For a short introduction to causal modeling, with a focus on the epistemology of causation, see Eberhardt, F. (2009). Introduction to the epistemology of causation. *Philosophy Compass*, 4(6), 913–925.

For an advanced treatment of causal modeling, see Pearl, J. (2009). *Causality: Models, reasoning, and inference* (2nd ed.). Cambridge University Press.

CHAPTER 12

Explaining and theorizing

12.1 PSYCHIATRY AND SCIENTIFIC THEORIES

After reading this section, you should be able to:

- Summarize psychiatric knowledge bearing on depression and its treatment
- Describe the primary features of scientific theories and why they are important to science
- Indicate how scientific theories go beyond hypotheses

Understanding and treating depression

Depression is a mood disorder—not just mere sadness or some kind of pessimism. There is extensive variation in the patterns and severity of symptoms that clinically depressed patients experience. People afflicted by major depressive episodes often feel tired, irritable, and distracted; their thoughts and affect turn negative; activities they once found fun don't elicit the same level of joy or interest; their motivation to spend time with friends or family diminishes; and despair and hopelessness set in.

Approximately one in 20 adults (or 5%) worldwide experience a major depressive episode. Rates of clinical depression are rising steeply—especially amongst young adults in affluent countries. We know there are many causes of this disease, including social, psychological, and biological factors. People who have gone through traumatic events, for example, are more likely to develop depression. Disrupted sleep patterns, including disruptions caused by smartphone use, are a contributing factor to depression and other mood disorders. Obesity and depression can undergo feedback loops, where each condition makes the other condition worse. Dysfunctions in neuron growth and brain connectivity are salient in patients with clinical depression, too. And family and twin studies have highlighted various genetic factors associated with an increased risk of developing depression. These genetic, neurological, health, psychological, and social factors are among the known causes of this debilitating disorder.

Despite knowledge about some of the causal influences on depression, psychiatric understanding of depression remains limited. There's no clear-cut way to diagnose depression. Standard definitions of depression in diagnostic manuals are qualitative, listing clusters of possible symptoms. Like many other mental illnesses, depression

DOI: 10.4324/9781003300007-13

Symptoms and Warning Signs of Clinical Depression

persistent feelings
of sadness

loss of interest
in activities

physical aches
and maladies

difficulty sleeping
or oversleeping

changes in weight
or appetite

decreases in energy,
lethargy, and fatigue

difficulty thinking clearly
or efficiently

irritability, cynicism, or
frustration

FIGURE 12.1 Signs of depression

shows a high degree of comorbidity, which means that if a patient has depression, then that person is also likely to suffer from other conditions, such as anxiety, alcohol abuse, and bipolar disorder. Further, many of the associations that have been found between depression and genes, brain structures, and psychological and social events are weak or uncertain. These and other challenges mean that, currently, there is no general theory of depression—that is, no well-established conception of what it is or account of how it unfolds.

The medical treatment of depression is also a mixed bag. Several treatments work in many—but not all—cases. A number of antidepressant drugs are commonly prescribed as treatment, but it can be difficult to find the best medication or combination of medications for a specific patient, and effectiveness may change over time. Psychotherapy, such as cognitive behavioral therapy, is often an effective non-pharmacological approach to treating depression. But the efficacy of psychotherapy varies, and it's unclear when any one therapeutic approach will be particularly promising in treating depression. Psychedelic compounds and deep-brain electrical stimulation are sometimes used when medication and psychotherapy aren't effective. They show some promise, but research into the safety and efficacy of these treatments is ongoing.

Scientific theories

Chapter 1 discussed how science aims at the production of knowledge—an aim that is constitutive of the very meaning of the word *science*. We also saw that science aims to produce a distinctive kind of knowledge: scientific knowledge is explanatory knowledge of why or how the natural world is the way it is. Scientists aren't simply accumulating a list of confirmed hypotheses about our world and ourselves. The project is bolder: scientists are charged with helping us understand why, and how, things happen. Scientists are asked to furnish the tools for making sense of, predicting, and changing the world around us. To accomplish this, they develop ***scientific theories***, which are comprehensive accounts of phenomena, broader and more explanatory than individual hypotheses and models, and backed by more evidence. In studying depression, psychiatric researchers have discovered a variety of explanatory factors bearing on the mental illness, as well as a variety of approaches to treating it. But, while there is now much more knowledge about incidence and causes of depression and effective treatments, psychiatry is still far from a satisfactory understanding of the mental illness.

Theories often go beyond what is readily observable. The Darwinian theory of evolution by natural selection is a grand theory about the origins of all the diverse lifeforms on Earth, and Einstein's theory of relativity is a grand theory about the very nature of gravity, space, and time. Empirical evidence has been central to testing and confirming the content of these and other theories. But that content is more expansive than just what we're led to expect to observe in some particular circumstance. Evolutionary theory, for example, indicates what happened from the earliest years of life on Earth, and the various hypotheses comprising it have been confirmed by a wide variety of forms of converging evidence, from extensive fossil records and geophysical data to features and relationships among current lifeforms. As we saw in Chapter 3, relativity theory was dramatically confirmed by the light deflection observed during an eclipse, but the theory also tells us what would happen if we traveled at the speed of light and gives us a reason to believe that nothing but light will ever travel that fast.

In ordinary language, the term *theory* is sometimes used to mean that some thought is only a guess or that it hasn't been tested out. Scientific theories are not like that. Quite the opposite, scientific theories are major accomplishments, as both the Darwinian theory of evolution and Einstein's theory of relativity illustrate. Yet, because scientific theories have implications that are not immediately observable, they are never taken to be true beyond all doubt, no matter how much empirical data supports them. Scientists couldn't be more certain of the theories of evolution and relativity. Even so, it's possible that someday one or another of these theories, or another theory among our prized scientific achievements, will be replaced by a better theory. This possibility is due to the open and self-correcting nature of science.

Just as scientific theories go beyond the readily observable, theories also come about not simply by generalizing from observations. Often, when a new theory is formed, this is from a significant conceptual shift—some feat of imagination—that gives rise to a new way of thinking about observations. Darwin wondered whether the similar forms of life he observed across continents might be evidence that they dispersed from some ancient common ancestor. And he was inspired by an economist, Thomas Richard Malthus, about the pressures to survive created by population increases. Einstein was inspired by the puzzle of how to set clocks that are far apart to the exact same time, and by how observers' experiences vary depending on whether they are in motion, to reconsider the very nature of space and time. In both cases, extensive observations were subsequently obtained to empirically support the theories. But the initial idea was a kind of spark of insight, a different way of thinking about what we already know about the world.

Box 12.1 Reductionism, integration, and complex systems

Historically, many scientists and philosophers of science have expected that appealing to the fundamental laws of physics will ultimately explain all the features of the world, even in biological organisms and societies. This view is called **reductionism**. The idea behind this is that everything in the universe is composed of physical matter (sometimes called **physicalism**) and so their features are all explained by the laws governing physical matter. But reductionism has fallen out of favor. Even if everything in the world is made of physical matter and subject to physical law, our scientific knowledge of physics isn't likely to give us predictions and explanations of, say, how cognition works or how biological organisms have evolved. Over the past century or so, there have also been two trends in scientific research that seem to run counter to reductionism. First, many scientific questions have been productively explored by bringing together and integrating research from multiple fields. Our knowledge of depression, for example, is informed by research in neuroscience, molecular biology, cognitive science, and the social sciences. This pattern of integration contrasts with what we would expect if reductionism were true. Second, scientific research in several areas has productively advanced by recognizing patterns in behavior across systems with very different physical compositions. Complex systems research—across many fields of science and relying on a variety of methodologies, particularly from network theory and dynamical systems theory—productively explores how interactions shape phenomena and seeks patterns across very different types of systems.

EXERCISES

12.1 Recall: Summarize scientific knowledge about depression. Remember to consider what we know about its causes, its treatment, and related knowledge about human emotional regulation.

12.2 Think: What might a theory of depression offer that goes beyond the current state of scientific knowledge about depression? Motivate your response by referencing features of theories described in this section.

12.3 Recall: Define *scientific theory* and describe three primary features of scientific theories based on the discussion in this section.

12.4 Think: Briefly summarize the theory of evolution, and describe how the theory of evolution exhibits the three primary features of scientific theories. (You may need to research the theory to answer this question; make sure you use reputable sources, whose trustworthiness you can justify.)

12.5 Recall: What do scientific theories add to science, beyond the processes of hypothesis-testing we have mostly focused on in this book? How does theorizing relate to hypothesis-testing?

12.6 Apply: Review the discussion of the Higgs boson discovery in Chapter 10. This discovery provided additional confirmation of the so-called standard model in physics. Investigate this theory, and then answer the following questions about it as best you can.
   a.   What is the theory a theory of—that is, what phenomena is it supposed to be about?
   b.   What are the central parts of the theory?
   c.   What kind of evidence has been used to support the theory?
   d.   Does the standard model make claims about things that we can't directly observe? About what kinds of things?
   e.   What are some of the considerations that sparked the development of the theory?
   f.   Has the theory undergone any changes over time? What change(s) and why?

12.2   USING SCIENCE TO EXPLAIN

After reading this section, you should be able to:

*   Describe the scientific goal of providing explanation and say why it is important
*   Define the nomological, pattern-based, and causal accounts of explanation
*   List one advantage and two problems with each of these three accounts of explanation

Scientific explanation

We said that scientific theories are important for producing explanatory knowledge. The explanatory knowledge produced in science is a special kind, explicitly supported by evidence through the use of methods discussed in this book. But there's significant overlap between scientific and everyday forms of explanation, too. All of us sometimes notice things that cry out for explanation. We routinely ask questions like "Why did you kiss him?" "How much does exercise improve my health?" "Why did the economic crisis happen?" "How did the dinosaurs go extinct?" Even children regularly engage in this pursuit of explanatory knowledge.

   Many children—and adults too—have wondered why the sky is blue. A parent might quickly answer that the sky is blue because it looks that way to us, or because that's just the way the sky is. Such answers don't explain why the sky is blue; they offer no insight into why or how the phenomenon is the way it is. A satisfying explanation of why the sky is blue relies on some sophisticated scientific theorizing: sunlight travels in straight lines unless some obstruction either reflects it, like a mirror; bends it, like a prism; or scatters it, like the molecules of gas in the Earth's atmosphere. Because blue light has shorter wavelengths, it is scattered more than other colors in the spectrum. That's why we normally see a blue sky. In contrast to most parents' quick answers to this question, this explanation appeals to other facts about the world and scientific laws or theories to give a deeper understanding of the phenomenon.

Generating explanations plays important cognitive roles. As we discussed in Chapter 7 with abductive reasoning—also known as *inference to the best explanation*—explanatory considerations can be used as evidence in support of a hypothesis, making the hypothesis more credible. Generating explanations to oneself or to others also routinely facilitates the integration of new information into existing bodies of knowledge and can lead to deeper understanding; this is called the ***self-explanation effect***. Performance on a variety of reasoning tasks can be improved when one is asked to explain. That's why explaining ideas and responding to explanatory questions is such a good way to learn new material. And instructors and tutors learn material faster and with more depth of insight in virtue of explaining it to others.

Perhaps most important among its cognitive roles, explanation produces understanding. Understanding the world around and within us is a supreme achievement that is central to science. ***Understanding*** is a kind of insight, grasping why or how something came about or is the way it is. This makes it possible for us to intervene in the world and to anticipate what will happen next. When we understand how a system works—say, the tidal system of the San Francisco Bay, an example from Chapter 5—we can anticipate how changes in some features of the system (like the Reber plan) will lead to changes in some other features (tides, salinity, and so forth).

Because explanations generate understanding, they satisfy our curiosity. To satisfy our curiosity and have that experience of "Aha! Now I get it!" can feel really good. Psychologist Alison Gopnik once likened understanding to orgasm. Sex evolved to feel good because it leads to babies, which is needed for a species to continue. Similarly, Gopnik reasoned, understanding is enjoyable because explanations help people navigate their environment. And so, the desire to satisfy our curiosity has led humans to ever more sophisticated and accurate theories about our world.

The satisfaction of curiosity is no guarantee of a good explanation, though. Sometimes people have a sense that they understand something without genuinely understanding it—explanations can be wrong. People also can fall prey to an ***illusion of explanatory depth***, believing they understand the world more clearly and in greater detail than they actually do. We all regularly overestimate our competence and depth of knowledge; recall our discussion in Chapter 1 of the cognitive errors, like confirmation bias, that science is designed to correct for.

An illustration of how one can be dangerously misled by the feeling they understand something is the public reception of climate change research. As you may recall, climate change was originally called "global warming." But this terminology misled many people about what they should expect to experience. When a season was not warmer than usual in some location, some people were tempted to doubt the reality of climate change, for it seemed to them as if things weren't getting warmer after all. But climate change does not produce warmer temperatures in every location at every point in time. Instead, it produces a global increase in average temperatures over time, as well as increasingly extreme weather—both hot and cold—and storms.

Another example of the illusion of explanatory depth concerns public reception of neuroscience. Experimental data suggest that people are often misled into judging bad psychological explanations as better than they really are when those bad explanations are accompanied by completely irrelevant neuroscience information. The allure of neuroscience information, which sounds technical and cites biological details, might interfere with people's ability to critically evaluate the overall quality of an explanation. This interference can have negative practical effects when, for example, it is exploited by advertisements for "brain training" that promise cognitive enhancement. This is the opposite of the climate change case. Instead of scientific expertise being disregarded because of personal experience, scientific credibility is misapplied to get people to believe something there's not actually sufficient evidence for.

Given the centrality of explanation to science and the potential for all people, including scientists, to feel like they understand something even when they do not, it's an important task to clarify the nature of scientific explanation. If we can say what features good explanations must have, then we will be better able to judge whether something counts as an adequate explanation.

Philosophers of science have thought long and hard about what features scientific explanations should have. Some have suggested that explanations should cite laws of

FIGURE 12.2 Oklahoma Senator James Inhofe denying the climate crisis in the US Congress while brandishing a snowball (February 26, 2015)

nature in order to account for phenomena. Another idea is that explanations should show how phenomena fit into patterns. Still others have suggested that explaining is a kind of causal reasoning, and that explanations should state what causes a phenomenon.

Explaining with laws and patterns

The idea of successful explanations appealing to scientific laws is at the heart of the **nomological account** of explanation (from the Greek *nomos*, meaning law). According to this view, phenomena are explained only when explanations reference laws that can account for them.

The nomological account of explanation was developed most fully by the philosopher of science Carl Hempel. Hempel proposed that explanations are logical arguments that appeal to general scientific laws to derive statements about the occurrence of the phenomena we want to explain. Hempel thought that some nomological explanations were valid deductive arguments, while others were strong inductive arguments. In either case, the premises must include at least one scientific law, which enables us to derive the phenomenon we want to explain. Hempel believed that knowing the scientific laws and background conditions that guarantee or make very likely the occurrence of the phenomenon would lead people to realize the phenomenon in question was to be expected. By rendering phenomena expectable, scientific explanations reveal our world to be ordered, proceeding in accordance with general laws.

Many scientific explanations fit this nomological account of explanation. Consider how scientists might explain the increase in the average global temperature of Earth's atmosphere. Energy from stars that reaches a planet is reflected off planetary surfaces. Greenhouse gases—carbon dioxide, water vapor, methane, and others—have molecular structures that absorb energy and then release it. Greater concentration of greenhouse gases in an atmosphere thus leads to increased temperature at the surface of a planet. These are law-like generalizations that describe these regularities in nature. Next, note that the concentration of greenhouse gases in Earth's atmosphere has increased. This is a background condition, a fact about current circumstances here on Earth. Together, these claims deductively imply the conclusion that the Earth's average temperature has increased. This argument is deductively valid with all true premises, yielding a simple nomological explanation of global warming.

Box 12.2 Scientific laws

Historically, many scientists have taken the purpose of science to be discovering laws of nature, which can in turn be used to provide explanations and make predictions. Examples include Newton's law of universal gravitation in physics; the three laws of thermodynamics, central to physics and chemistry; and the laws of supply and demand in economics. A scientific law can be thought of as a set of rules for inferring what must follow from some set of conditions.

The law of supply, for example, is a way to infer the relative price of goods from the quantity of goods supplied. Newton's law of universal gravitation is a way to infer the force between two bodies based on their masses and the distance between them. And yet, many useful scientific insights do not provide exceptionless rules for how things must occur. Indeed, the law of supply in economics is at best an approximation for how supply influences price. Even Newton's law of universal gravitation fails to take into account the influence of other forces, such as magnetic charge. Further, many scientific advances do not even attempt to capture laws of nature at all. Philosophers of science thus debate what is required for something to count as a law, as well as the extent to which scientific inquiry involves discovering laws as opposed to, say, causal mechanisms.

Just as phenomena can be explained by laws, scientific laws themselves can be explained by appealing to other, more comprehensive laws. For example, consider Galileo's law that bodies fall toward Earth at a constant acceleration. This law can be deductively derived from the Newtonian law of gravitation. The Newtonian force of gravity explains the constant acceleration of bodies falling toward Earth. Newtonian laws, in turn, can be explained by appealing to the more comprehensive general theory of relativity developed by Einstein. The Earth's gravity is explained as a distortion of space caused by the Earth's mass. Objects speed up as they fall toward Earth, just as a ball rolling from the edge to the center of a bowl speeds up.

The idea of explaining scientific laws with reference to other, more general laws introduces a second account of explanation. According to the **pattern account**, explanations fit particular facts into a more general framework of laws and principles. This also has been called a *unification* account of explanation, since the number of assumptions required to explain phenomena decreases when an explanation is provided. Descriptions of phenomena, and scientific laws as well, are unified by uncovering the basic patterns that govern them.

Although we previously gave a nomological explanation of global warming, this can also be construed as a pattern explanation. The phenomenon of anthropogenic global warming on Earth due to human activities is explained as one instance of a general pattern of atmospheric effects on planetary climate, a pattern that also includes Earth before human activity drastically increased greenhouse gas concentration in the atmosphere and other planets with different atmospheres entirely.

One advantage of the pattern account over the nomological account is that it doesn't require scientific explanations to cite laws. Pattern explanations can cite regularities that may not qualify as laws. In place of the law requirement, there's an emphasis on descriptions that fit the phenomenon to be explained into a wider pattern, to see the phenomenon as one instance of a more general regularity. Our earlier discussion of depression illustrates the advantage of this. Scientific knowledge of depression is not developed enough to provide scientific laws that convey the relations between properties of this disorder and that govern the occurrence of depression or effectiveness of treatments. Indeed, the phenomenon of depression may not be governed by general laws at all. But there are still descriptions of patterns that can be explanatory.

For example, to explain why selective serotonin reuptake inhibitors (or SSRIs), the most commonly prescribed category of depression medication, are effective, one can point out that low serotonin levels are associated with depression, and taking SSRIs increases serotonin concentration in the brain by blocking its reabsorption by neurons. There is no law that decreasing serotonin leads to depression or that increasing it will effectively treat depression. But there is a broad pattern between serotonin levels and depression that this explanation, and the treatment itself, rely upon.

Another example comes from evolutionary theory, which explains many facts about the traits of organisms and the relationships among them with reference to a simple pattern that plays out in a multitude of ways. The pattern at the heart of evolutionary theory is that natural selection acts on variation among organisms to produce cumulative change in species. The theory of evolution is not a single general law of nature. It recognizes many different influences on evolution and that evolution proceeds in different ways depending on several factors. Evolutionary explanations are thus not productively construed as nomological explanations. But they do conform to the pattern account rather well.

The idea behind the nomological and pattern accounts of explanation—that explanations make phenomena less surprising by referencing laws or by showing how they fit into a wider pattern—captures important features of many scientific explanations. But both also face significant objections.

We've already encountered one problem with the nomological account: many good explanations don't appeal to any scientific laws. We saw this with evolutionary explanations and with the example of explaining why SSRIs treat depression.

The pattern account faces a related difficulty. It focuses on explanations that show how to fit a phenomenon into a wider pattern, but some explanations are highly specific. Consider the explanation for humans' large brains. As early hominids (ancestral humans) faced environmental changes and developed more complex social exchanges, this led to the selection for and eventual evolution of increasingly large and complex brains, eventually tripling human brain size from its ancestral size. Brain size increased especially rapidly between about 800,000 and 200,000 years ago. This explanation is compelling, and there's significant evidence backing it up. But it is also highly specific. It appeals to the general pattern of evolutionary theory, again, selection acting on variation to produce cumulative change; but much of the explanation is highly specific to human's unique evolutionary past.

A second problem for both nomological and pattern accounts is that they neglect a key feature of explanation: asymmetry. If one thing explains another, then this explanatory relation does not usually hold in reverse. Consider the following example. Your mobile phone sends you a weather alert, and you explain this with reference to the fact that a storm is approaching and the generalization that weather alerts are sent out when storms are approaching. But, it seems, this explanation doesn't work in reverse. You can't explain the approaching storm by citing the weather alert you received and

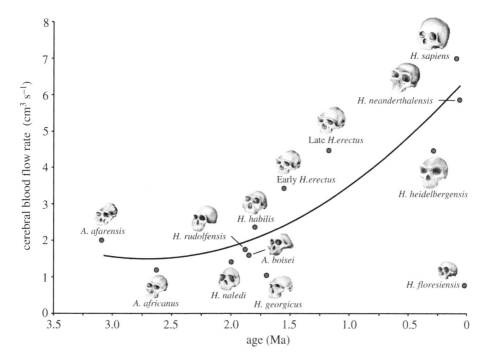

FIGURE 12.3 Evolution of human brain size

the generalization that weather alerts are sent out when storms are approaching. This mixes things up: the storm isn't approaching because you received the weather alert. The weather alert gives you evidence of the storm, but it isn't an explanation for why the storm is happening.

But the nomological and pattern accounts of explanation don't illuminate this asymmetry. According to the requirements for explanation on those views, a generalization about weather alerts being sent out when storms approaching and the background condition of receiving a weather alert should explain the storm occurring! This seems strange to say the least. An approaching storm can be a good explanation for why you receive a weather alert, but the weather alert is no explanation for why the storm is approaching. You can do a lot with your mobile phone, but you can't usually bring about a storm.

Causal explanation

Another account of explanation is inspired by this asymmetry of explanation that the nomological and pattern accounts fail to illuminate. According to the *causal account* of explanation, explanations appeal to causes that bring about the phenomenon to be explained. On this view, appealing to the approaching storm explains the weather alert on your phone because the storm causes weather alerts to be sent out. But, as we just noted, your phone's weather alert doesn't cause the storm—so on the causal account of explanation, it doesn't explain the storm either.

The causal account seems to apply well to many explanations in science, including in fields that have not developed scientific laws. As we emphasized in Chapter 11, knowledge of causes enables prediction and manipulation of phenomena, via intervention on causal factors. It's plausible that identifying causal factors is also central to explanation.

In one variety of causal explanation, the focus concerns how causal factors regularly combine into systems that produce the target phenomenon. Some call this variety of explanation *mechanistic*. The search for causal mechanisms seems to play an especially important role in some parts of the social and life sciences. For example, the scientific explanation of an organism's regulation of blood sugar appeals to how pancreatic endocrine hormones maintain blood sugar within a range of about 70–110 mg per 100 ml blood. If glucose (blood sugar) decreases below this range, pancreatic alpha cells secrete glucagon, which causes the liver to release stored glucose. If blood sugar increases above the range, pancreatic beta cells secrete insulin, which causes adipose tissue to absorb glucose from the blood. This explanation is also part of the explanation of diabetes, which is a disorder characterized by the pancreas producing insufficient amounts of insulin. This explanation displays how an interrelationship among pancreatic hormones, liver tissue, and blood sugar ordinarily work together in a complex way to maintain blood sugar levels within a narrow range.

We've seen that the causal account of explanation accommodates the asymmetry of explanation. It can also address the other concern raised earlier with the nomological and pattern accounts of explanation, that not all explanations seem to feature scientific laws or to describe broad patterns. Some causal relationships follow very general patterns, or are law-like in nature, but others are not. If you heat ice, it will melt or evaporate. There are virtually no exceptions to this. If you heat chocolate, it will usually melt—but if you heat it too quickly, it gets thick and lumpy instead. This is a general pattern, but it has some exceptions. In contrast, perhaps the evolution of large brains will only happen once in history. Perhaps background conditions had to be just right for such a trait to evolve. But all of these, from the law-like to the highly particular, are still cause-effect relationships. And on the causal account, all are successful explanations.

However, the causal account of explanation faces its own difficulties. First, just as it seems some explanations may not cite laws or patterns, it also seems that other explanations don't cite causes. For example, we said previously that Einstein's general theory of relativity explains why Newtonian law of gravitation holds, which in turn explains Galileo's law of constant acceleration. But it would be strange to say that general relativity causes the Newtonian law of gravitation, or that the law of gravitation causes constant acceleration. A related concern is that, if it is right that general patterns and scientific laws can help us understand the world, at least sometimes, then the causal account of explanation is lacking because it doesn't identify that explanatory value.

A second difficulty with the causal account of explanation stems from the observation that many phenomena have many causes. For this reason, causal explanations may come too easily. Causal explanations may cite only one or a few causal influences, when we know there are many causal influences on the phenomenon that's explained. How is this enough to explain the phenomenon? Some respond to this challenge by saying that the more causal information you can give, the better explanation you have. Others seek another principle to decide what causal information belongs in an explanation. Yet another difficulty is that, as discussed in Chapter 11, there are different conceptions of what causation is, but an account of causal explanation requires a successful definition of *causation*.

So far, we have talked about these three accounts of explanation as if one is right and the others wrong. But it's possible that each account captures certain elements of what helps us understand the world. One initial reason to think this might be so is that each of these accounts of explanation aptly characterize some, but not all, of the examples of successful explanation we have discussed. Perhaps laws, patterns, and causes all can contribute to our understanding, and so any of these can be an ingredient of explanation, even if none is involved in every single explanation in science.

EXERCISES

12.7 **Recall:** What are three ways explanation is useful for our thinking? How are these valuable to science?

12.8 **Think:** In your own words, describe why explanation is important to science.

12.9 **Recall:** Define *illusion of explanatory depth*. Why is this risky in science?

12.10 **Recall:** Define the nomological, pattern-based, and causal accounts of explanation. List one advantage and two problems with each of these three accounts of explanation.

12.11 **Apply:** Choose one account of explanation: nomological, pattern, or causal. Find a novel example of a scientific explanation that seems to conform to the account you chose. Describe the example, making clear how it succeeds as an explanation. Then describe why this example should be seen as conforming to the account of explanation you chose.

12.12 **Think:** Construct a chart or table listing the strengths and weaknesses of each of the three accounts of explanation discussed in this section. Decide which account(s) of explanation is the most promising, and support your answer with an argument.

12.3 SCIENTIFIC BREAKTHROUGHS AND THEORY CHANGE

After reading this section, you should be able to:

- Define and give an example of a scientific breakthrough
- Outline Kuhn's view of the four stages of science
- Describe how the discovery of oxygen led to changes to the field of chemistry

Scientific breakthroughs

We said in section 12.1 that, when a new theory is formed, this sometimes involves a significant conceptual shift or feat of imagination that gives rise to new ways of thinking about observations. No scientific theory is accepted beyond any doubt, and theories are sometimes replaced by successors. The differences between a theory and its successor can be minor, while in other cases there is a radical shift. The most significant *scientific breakthroughs*, when there has been a radical shift in theory in some field of science, have been changes in worldview: they involve comprehensive revision to how background assumptions, data, and ideas are combined, and thus to which scientific theory is supported.

Consider again theories of our universe and the bodies within it. The worldview that arose with Aristotle had great scope and logical coherence. The Aristotelian theory of falling bodies claims that heavy bodies fall faster than light ones, and its geocentric conception of the universe placed Earth at the center, which fits with most common observations of how the world is. But, over time, observations were made that the Aristotelian worldview couldn't easily accommodate. Eventually it was replaced by a Copernican, and then Newtonian, conception of the universe, with the Earth not a fixed center, but a planet in motion around the Sun.

Because of the dramatic change in worldview, astronomers from the 4th century BCE and the 17th century would have agreed about the positions of the stars in the sky, but they would have radically different interpretations of those observations. Similar observations provided clues to constructing the theories, but the differences between those theories were vast. This is the Scientific Revolution of the 16th and 17th centuries, discussed also in Chapter 2.

Additional radical shifts followed on the heels of the rejection of the Aristotelean worldview, and with these changes came radical revisions to accepted ideas about the position and movement of Earth, the shape of orbits, and the nature of universal forces. In general, each new theory accounted for some body of evidence better than its predecessor. Still, most of the changes were rather radical changes in perspective. The same is true of the later replacement of Newtonian mechanics with Einstein's theory of relativity, when universal forces were replaced by non-Euclidean geometries of space-time.

Scientific breakthroughs have occurred in other fields of science as well. This is as you'd expect if scientists are genuinely open to revising or replacing any theory when doing so is warranted by the evidence. And breakthroughs seem rewarding and significant: there's a sense that, after a scientific breakthrough, we more clearly understand our world. An initial spark of insight leads to a conceptual shift or feat of imagination that results in reinterpreting existing data to support a new theory, and then more data are discovered that confirm this new theory.

From another perspective, though, the idea of scientific breakthroughs is also troubling. What happened to our scientific knowledge from before the breakthrough—were scientists just wrong? How do we know that our current best theories won't suffer the same fate and also be rejected for new and better theories? Can we trust our current scientific theories at all, then? These are deep and troubling questions that strike right at the heart of science. But let's postpone that discussion until later in this chapter, after we have a fuller picture of what scientific breakthroughs are like, and how and why they occur.

Kuhn's scientific revolutions

The series of scientific breakthroughs in the 17th century just described suggests we might think about scientific breakthroughs in terms of revolution. Revolutions are pretty dramatic; think of political revolutions like the French revolution at the end of the 18th century, the fall of the Soviet Union two hundred years later, or the Arab Spring in the early 2010s. A *scientific revolution* is a radical change of a reigning theory being overturned or abandoned in favor of a new theory, often involving an alternative worldview. Scientific revolutions don't just change which scientific theories are accepted; they also influence the trajectory of science itself, including how to interpret evidence, which scientific procedures are accepted, and often the social and institutional structure of science, such as who is accepted as a scientific authority.

The sociologist, historian, and philosopher Thomas Kuhn wrote a famous book, *The Structure of Scientific Revolutions*, first published in 1962. In this book, Kuhn advanced an influential model of scientific change based on the notion of revolution. He suggested that scientific revolutions have occurred and will continue to occur periodically as an important part of science. In his view, this prevents science from proceeding in a straight line by accumulating an increasing body of knowledge and an expanding store of explanations. Kuhn argued that science instead proceeds in phases.

Kuhn called the earliest phase of science *pre-paradigm*. This is characterized by the existence of different schools of thought that debate very basic assumptions, including research methods and the nature and significance of data. Data collection is unsystematic, and it's easy for theories to accommodate new observations because the theories are inchoate, or undeveloped. Such theories can easily be adapted in different ways to accommodate new observations. There are many puzzles and problems but not very many solutions.

Kuhn's second phase is the normalization of scientific research. One school of thought begins to solve puzzles and problems in ways that seems successful enough to draw adherents away from other approaches. Kuhn called this period *normal science*, because scientific researchers' widespread agreement about basic assumptions allows the research to become stable. Scientific practices become organized. Laboratories or other workspaces may be set up, research techniques and methods become widely accepted, and agreed-upon measurement devices are improved.

During normal science, scientific developments are driven by adherence to what Kuhn called a *paradigm*. Broadly conceived, a paradigm is just a way of practicing science. It supplies scientists with a stock of concepts, symbols, and assumptions about the world that they can use to communicate more effectively. A paradigm also involves methods for gathering and analyzing data, as well as habits of scientific research and reasoning more generally. For example, in 20th-century psychology and psychiatry, the behaviorist overthrow of the introspectionist and psychoanalytic paradigms weren't just a change in theory; rather, this ushered in new experimental tools, methods of operant conditioning, and ways of thinking about how to perform science. A paradigm stems from elements of the reigning school of thought, a risky conjecture that has been sufficiently confirmed by evidence to be taken on board and developed into a scientific theory.

Kuhn thought that, during a period of normal science, each field of science is governed by a single paradigm. But, scientists in the grip of some paradigm have often ended up with observations that are at odds with it, or that lead to worrying puzzles called **anomalies**: deviations from expectations that resist explanation by the reigning theory. Usually, anomalies are just noted and set aside for future research. But anomalies can accumulate, and this creates a kind of increasing tension for the accepted scientific theory. Some scientists begin to worry that the theory might not be right after all.

The accumulation of anomalies sets science up for a crisis. A crisis occurs when scientists finally lose confidence in the reigning theory in the face of mounting anomalies. For Kuhn, a paradigm is only rejected if a critical mass of anomalies has led to crisis and there's also a rival paradigm to replace it. Another theory has been developed by some renegade scientists, and the problems with the existing paradigm mean that this new theory—together with its background assumptions, methods, and so forth—can finally get attention. If this is so, a crisis might be followed by a scientific revolution. In this phase of science, all the elements of the accepted paradigm are up for negotiation. Data, interpretations of data, background assumptions, methods and technical apparatus, and so on—any and all might be rejected, replaced, or reinterpreted from the perspective of the new paradigm. This four-stage view of scientific change is summarized in Table 12.1.

The chemical revolution

The Scientific Revolution began when geocentrism was replaced with heliocentrism in astronomy: the Earth was no longer seen as the central heavenly body but instead taken to revolve around the Sun. According to Kuhn, this perfectly fit his ideal description of

TABLE 12.1 Kuhn's four stage account of scientific change

Stage	Features
1. Pre-paradigmatic science	Different schools of thought debate basic assumptions
2. Normal science	A paradigm is accepted, and efforts are devoted to basic research and puzzle-solving
3. Crisis	Scientists lose confidence in the dominant paradigm upon the accumulation of anomalies
4. Revolution	The old paradigm is rejected in favor of a new one

scientific revolution. Another abrupt revolutionary change in science that Kuhn recognized as a scientific revolution involved sweeping changes in the field of chemistry in the 18th century. Two of the protagonists of this scientific revolution were the chemists Antoine-Laurent and Marie-Anne Paulze Lavoisier. When they began their work, scientific understanding of matter and its transformations was still grounded in the Aristotelian worldview. Aristotle had believed that all earthly materials are composed of the elements air, earth, fire, and water. This theory of the four elements had been slowly modified by the medieval alchemists, who aimed to turn base metals into gold and to produce an elixir of immortality. By the 18th century, alchemists believed all things were compounds of sulfur, mercury, and salt.

In the early 18th century, one pressing scientific question was: what happens when something burns? Alchemists thought that materials lose sulfur when they change into slag, rust, or ash by heating. The physician and chemist Georg Ernst Stahl modified this idea, developing the theory that every combustible material contains a universal fire-like substance, which he named *phlogiston* (from Greek, meaning flammable). Combustible materials like wood tend to lose weight when burned, and Stahl explained this change in terms of the release of phlogiston from the combustible material to the air. When the air becomes saturated with phlogiston or when a combustible material releases all its phlogiston, the burning stops. Stahl believed that the residual substance left behind after a metal burns is the true substance of the original metal, which lost its phlogiston during combustion. This residue, which was called "metal calx" (what we would now call an *oxide*), has the form of a fine powder. Both metal calx and the gases produced during combustion could be captured, measured, and experimentally manipulated.

Unlike gases and calx, though, phlogiston was an utter mystery. Nobody had isolated it, and nobody had found a way to manipulate it experimentally. In fact, phlogiston seemed to have properties that were inconsistent with Stahl's theory. Stahl believed phlogiston had a positive weight. When you burn a piece of wood, the remaining ash loses phlogiston, and it weighs less than the original log. But in other cases, for example, when magnesium or phosphorus burn, the residue left behind weighs more than the original material. If phlogiston is released during the burning process, why was there a gain in weight in these cases?

Intrigued by this anomaly, the Lavoisiers experimented with various metals and gases to investigate why and how things burn. They observed that lead calx releases air when it is heated. This suggested that combustion and air were, somehow, linked. Explaining the link was a difficult task, however, because at that time little was known about the composition and chemistry of air. Meeting Joseph Priestley helped. Priestley had discovered a gas he called "dephlogisticated air," which was released by heated mercury calx. This gas was thought to greatly facilitate combustion because, being free from phlogiston, it could absorb a greater amount of the phlogiston released by burning materials. Candles burning in a container with "dephlogisticated air" would burn for much longer, for example. This gas, Priestley observed, facilitated respiration too: mice in containers with dephlogisticated air lived longer than mice placed in containers without this gas.

FIGURE 12.4 A depiction of one of the Lavoisiers' experiments from the late 18th century

The Lavoisiers tried to replicate Priestley's experiments; and, based on their own results and observations, they elaborated a new theory of combustion. The central idea was that combustion was the reaction of a metal or other material with the "eminently respirable" part of air. Believing (incorrectly) that this kind of air was necessary to form all sour-tasting substances, or acids, Lavoisier called it *oxygène*, from the Greek words for acid generator. According to this new theory, combustion did not involve the removal of phlogiston from the burning material but, rather, the addition of oxygen.

This emerging rival paradigm set the basis for the revolution from which modern chemistry emerged. In the 1780s, the Lavoisiers and other scientists adopted the idea of a chemical element, and of chemical compositions of simpler elements. This new system of chemistry was set out by Antoine-Laurent Lavoisier in a textbook in 1789. This book didn't just describe the theory, but also explained the effects of heat on chemical reactions, the nature of gases, and how acids and bases react to form salts. It also described the technological instruments used to perform chemical experiments. And it contains a "Table of Simple Substances"—the first modern listing of chemical elements.

After the publication of this textbook, most young chemists adopted Lavoisier's theory and abandoned phlogiston. "All young chemists," he wrote in 1791, "adopt the theory, and from that I conclude that the revolution in chemistry has come to pass." From a Kuhnian perspective, the next phase of normal science had begun.

EXERCISES

12.13 Recall: Define *scientific breakthrough* and give an example.

12.14 Think: How are scientific breakthroughs valuable for scientific knowledge? How are they challenging for scientific knowledge?

12.15 Recall: Describe the features of each of Kuhn's four stages of science: pre-paradigm science, normal science, crisis, and scientific revolution. Illustrate each stage by describing how it applies to the chemical revolution.

12.16 Recall: The case of phlogiston presented an anomaly for Stahl's theory. How so? Describe the extensive changes that the discovery of oxygen led to for the field of chemistry.

12.17 Apply: The advent and rapid development of computers was a major technological advance. Given your background knowledge, would you describe the development of computers as a Kuhnian revolution? Why or why not?

12.4 SCIENTIFIC PROGRESS AND THE SECURITY OF SCIENTIFIC KNOWLEDGE

After reading this section, you should be able to:

- Give two examples of incremental theory change in science and describe their differences from Kuhnian paradigm shifts
- Identify two challenges for scientific knowledge from scientific breakthroughs and at least one response to each challenge
- Describe three reasons to think scientific knowledge is safe, regardless of the possibility of radical theory change

Incremental theory change

Kuhn's notion of scientific revolution seems to accurately characterize some episodes in the history of science the times of especially radical transformation in accepted scientific knowledge. But, this is a particularly extreme form of scientific change. Most other episodes of scientific change seem to be less dramatic, and there's also a question about whether ordinary scientific activity fits Kuhn's characterization of normal science.

Incremental changes in science are more common and less abrupt than Kuhn's account suggests. Consider, for example, the Darwinian revolution in the 19th century. Darwin's theory of evolution has had a dramatic and lasting impact on our understanding of the nature of lifeforms, the relationships among different species, and how species have changed over time, and—like the change from geocentrism to heliocentrism—the establishment of Darwinian theory was arguably a scientific revolution if anything is. But Darwin did not produce the first evolutionary theory, nor has evolutionary theory remained entirely the same as what Darwin first described. Changes in the field of biology, both before and after Darwin's revolutionary breakthrough, have been more gradual than a Kuhnian paradigm shift.

The general idea of evolution is that whole species—not just individuals—can change over time, and this idea is many centuries old. The nature of biological change as a scientific research program can be traced to the work of several naturalists over 50 years before the publication of Darwin's *Origin of Species* in 1859. Even Darwin's specific ideas about evolution were significantly influenced by other scientific work; we've already mentioned the influence of the political economist Thomas Richard Malthus. And another scientist working at the same time as Darwin, Alfred Russel Wallace, was independently developing a theory of evolution by natural selection strikingly similar to Darwin's. So, although Darwin's ideas were a tremendous breakthrough, they built upon existing scientific work, and they were inspired by and related to concurrent scientific work by others.

Furthermore, the science of biology since the Darwinian revolution has not simply consisted in the application of Darwin's ideas, as Kuhn would have us expect for a period of normal science. Rather, our understanding of evolution has been in continual development. The so-called *modern synthesis* in the early 20th century integrated the existing knowledge of genetics and Darwinian evolutionary theory, which had previously been construed as competing theories. Other elements of evolutionary theory have been revised since, like the recognition of nongenetic influences on traits and how significantly organisms shape their environment.

Another point in support of incremental rather than revolutionary scientific change is that theory change doesn't always involve the rejection of existing theories. Sometimes it comes from the joining of theories, as in the modern synthesis just mentioned, and other times it can come from new methods. Biologist James Watson and physicist Francis Crick, for example, reached their groundbreaking conclusion that the DNA molecule exists in the form of a double helix by applying a new modeling approach to data gathered by Rosalind Franklin. Using cardboard cutouts to represent the chemical components of DNA, Watson and Crick tried to make different arrangements, as though they were solving a jigsaw puzzle. Through this concrete model-building, the double helix structure of DNA was identified. This had enormous consequences for subsequent biological research.

Nonscientific pursuits, such as mathematics and philosophy, can drive scientific theory change too. The development of a new kind of geometry—non-Euclidean Riemannian geometry—paved the way for Einstein's theory of relativity. Einstein's theory adopted this purely mathematical geometry as a description of physical spacetime. One basic difference between Euclidean and non-Euclidean geometry concerns the nature of parallel lines. In Euclidean geometry, there's only one line through a given point that is parallel to another line. In some non-Euclidean geometries, there are infinitely many lines through a point that are parallel to another line and, in others, there are no parallel lines. This development in mathematics made it possible for Einstein to wonder whether the geometry of our own universe might be non-Euclidean. Around this same time, the development of new logical systems of predication, quantification, and operators helped move far beyond the syllogistic logic that had dominated for the past two millennia. These developments in turn affected, and were affected by, these upheavals in mathematics and led to the generation of set theory, mathematical logic, and computer science.

In summary, scientific theories undoubtedly change. But these changes may be influenced by changes in method, data, mathematics, or other relevant ideas or practices. And the changes tend to be incremental and cumulative more often than radical and transformative.

Box 12.3 Merton's priority rule

According to sociologist of science Robert Merton, the first scientist making a novel discovery is rewarded with prestige and recognition. The second discoverers, runners-up, get nothing. Priority in discovery matters for patents and copyrights in applied science. It also matters more generally for recognition and prestige, which are the common currencies of scientific credit. Prestige in science comes in different forms. It can include promotion, prizes like the Fields Medal or Nobel Prize, and eponymy—that is, having a discovery named after oneself, like Parkinson's disease. If the first scientist to publish a new discovery receives recognition and the seconds get nothing, then priority is a strong motivator for scientists, which then shapes the reward structure of their field. Scientists often worry about their work being scooped, and priority disputes occur regularly in science. Famous cases include Isaac Newton versus Gottfried Leibniz over the invention of calculus, Charles Darwin versus Alfred Wallace over the discovery of natural selection and evolution, and more recently, researchers at MIT versus UC Berkeley over the discovery of CRISPR-Cas9, a very significant genome-editing technology.

Scientific progress

Earlier, we raised worries about how scientific breakthroughs may undermine our confidence in scientific theories. If some theories that were seemingly well supported by evidence are eventually rejected, who's to say our current theories won't also be rejected? And do such scientific breakthroughs make it so that science isn't really making progress at all? Let's consider these questions in a bit more depth and isolate a few important considerations.

When scientific theories change, do we have reason to think that the new theory is an improvement on the last one and that science is progressing towards truth? This question is complicated by two features of theories and theory change. First, theories often appeal to phenomena that cannot be directly observed. Examples of this we have encountered in this book include the Higgs boson, the first moments of the universe's existence after the Big Bang, and the original common ancestor of all life on Earth.

How can we ever be sure we are right about these and other phenomena like them? Second, at least some instances of theory change have been radical: scientists eventually rejected phlogiston, decided they were wrong about the placement of Earth in the universe, and learned that psychiatric conditions such as depression are not imbalances of bodily humors. How can we ever be sure that our scientific findings are on a path to truth, when the next radical revision could be right around the corner?

There's at least one influential argument suggesting that, despite all this, we have reason to believe that our best mature scientific theories are approximately true. This argument is an abductive inference, or inference to the best explanation, from the success of science. It begins with the observation that our best scientific theories are extraordinarily successful: they enable scientists to make empirical predictions, to explain phenomena, to design and build powerful technologies. What could explain this success? One possible explanation is that our best scientific theories are approximately true. If these theories were not approximately true, then the fact that they are so successful would be astonishing. So, it seems, the best explanation for the success of science is that our best theories are true—or, at least on the path to true and getting closer. This is sometimes called the ***no miracles argument***, suggesting that all of the scientific successes we know about would be utterly mysterious—that is, would have no explanation—unless scientific theories are approximately true.

Yet, some believe that this conclusion is overly optimistic. Here's an inductive argument for the opposite conclusion. If we examine the history of scientific theories in any given field, we find plenty examples of older and successful theories rejected in favor of newer, more successful theories. So, most past theories, which were also predictively successful and empirically adequate, turned out to be false. Therefore, by generalizing from these cases, our current scientific theories are likely to be false as well, standing a good chance of eventually being replaced themselves by new theories. This is sometimes called the ***pessimistic meta-induction***, since it draws a pessimistic conclusion about the eventual fate of today's scientific theories based on inductive reasoning from the rejection of many past scientific theories. The upshot of this argument is that we do not have a strong reason to think our current best scientific theories are approximately true.

Why scientific knowledge is safe

The pessimistic meta-induction raises important questions about how certain we can be about our current scientific theories. What it doesn't do is undermine the fact that science is the single most successful project of generating knowledge that humans have ever embarked on. There are reasons to think scientific knowledge is trustworthy and safe, even if we can't be certain that our current scientific theories are the final say on how the world is.

First, consider the point we made earlier about incremental theory change. Some theoretical breakthroughs may resemble Kuhnian paradigm shifts, but many more developments in scientific theory are small and incremental, suggesting that previous successes may persist—and be further built upon—as scientific theories advance. This seems so for the development of both psychiatry theories of depression and evolutionary theory. And, even when new research has shown limitations in earlier theories, often the abilities and insights based on those theories persist. Sometimes, it's even the case that scientists return to earlier, rejected theories for inspiration when current theoretical frameworks come up short. For example, recall from Chapter 2 how limitations of the germ theory of disease inspired a restored focus in public health to social determinates of disease.

Second, scientific theories are only one part of the overall scientific endeavor. Even if some of the theoretical claims at the heart of a scientific theory are eventually shown to be wrong or only approximately true, that can leave intact the value of many ideas, observations, and technological advances associated with the theory. Those ideas, observations, and advances qualify as factors in the production of scientific knowledge in their own right. For example, although phlogiston is no longer posited as a chemical created from combustion, the observable features of combustion that motivated phlogiston theory, such as the residues produced and their volumes, are still counted as part of our body of knowledge of combustion.

Third, scientific knowledge is safe despite the possibility of future breakthroughs because science is not just a collection of theories. At its heart, science is a collection of methods, a set of recipes for investigating our world featuring common ingredients—most prominently, hypotheses, expectations, and observations. Parts of these recipes, like empirical observation and attempts to explain, are common to people of all ages all around the world. Other parts, like experimentation and mathematical tools, have developed in several cultures at different times. And some aspects, like a social institution balancing trust and skepticism, developed with contemporary institutionalized science. This set of recipes for science has persisted and been improved through continual refinement for centuries. It will continue to evolve, but it is unlikely ever to be supplanted, even if individual scientific theories are sometimes abandoned. Over the long arc of history societies have much to gain, and little to lose, by relying on the best science of their day.

EXERCISES

Watch Video 16

12.18 Recall: Describe why Darwinian evolutionary theory is better seen as incremental theory change. How does this instance of theory change differ from Kuhnian paradigm shifts?

12.19 Recall: Define the *no miracles argument* and the *pessimistic meta-induction*. Explicitly state the conclusion of each argument.

12.20 Think: Consider again the *no miracles argument* and the *pessimistic meta-induction*. Assess each argument, one at a time, writing at least one paragraph evaluating each. Then, for each, say whether you are ultimately convinced of its conclusion.

12.21 Think: How does the existence of scientific breakthroughs, or revolutions, challenge the ideas of scientific truth and scientific progress? Motivate the concern as well as you can. Then, evaluate the merits of the concern, thinking back to all you've read in this book.

12.22 Recall: Describe three reasons to think scientific knowledge is safe, regardless of the possibility of radical theory change.

12.23 Think: How convinced are you by these three reasons to think scientific knowledge is safe? Why? Can you think of additional grounds for trusting scientific knowledge? What lingering concerns, if any, do you still have about science's trustworthiness?

FURTHER READINGS

For more on depression and psychiatry, see these volumes in Oxford's *A very short introduction* series: Scott, J., & Tacchi, M. J. (2017). *Depression*. Oxford University Press and Burns, T. (2018). *Psychiatry*. Oxford University Press.

For an introduction to scientific explanation, see McCain, K. (2022). *Understanding how science explains the world*. Cambridge University Press.

For Kuhn's view of scientific revolutions, see Kuhn, T. (1962). *The structure of scientific revolutions*. University of Chicago Press.

For more on scientific realism, see Anjan Chakravartty's "Scientific realism." In E. N. Zalta (Ed.), *The Stanford encyclopedia of philosophy* (Summer 2017 ed.). https://plato.stanford.edu/archives/sum2017/entries/scientific-realism/.

CHAPTER 13

Science in society

13.1 ANIMAL BEHAVIOR AND SCIENCE'S SOCIAL CONTEXT

After reading this section, you should be able to:

- Summarize sexual selection theory and give examples of a showy trait and an aggressive trait thought to evolve through sexual selection
- List three concerns with classic sexual selection theory and suggestions for how the theory might be expanded or rethought on the basis of these concerns
- Describe three ways in which science can be influenced by its social and historical context

Sex and reproductive strategies in animals

Some animal traits have clear value to the animal. Predators like the leopard have eyes on the front of their head to make it easier to track their prey, while prey like the antelope have eyes on the sides of their head so they can keep an eye out to ensure their safety. Eye placement evolved differently in predator lineages and prey lineages because each placement has different advantages.

Other traits are less obvious in their function. Since Charles Darwin first developed the scientific theory of evolution by natural selection in the 19th century, scientists have puzzled over the emergence and biological function of traits like the peacock's long, colorful tail feathers, also called a *train*. What's puzzling about the peacock's showy train is that it makes it easier for predators to spot the peacock and more difficult for the peacock to move quickly. And it's not the kind of trait that would just occur by accident!

The widely accepted explanation for the peacock's colorful train is that the showy tail feathers don't help with survival but, instead, with reproduction. This trait—and many others, across many species of animals—is thought to evolve through *sexual selection*, when a trait is valuable simply because it appeals to potential mating partners. If the trait is inherited by offspring, then sexual selection can explain how the trait can become more pronounced and widespread over time. So, the peacock,

DOI: 10.4324/9781003300007-14

FIGURE 13.1 The peacock's showy tail feathers

over time, is thought to have evolved a longer, showier train because of this process of sexual selection.

Sexual selection theory traditionally emphasizes how female mate choice can lead males to evolve showy traits—the peacock's train, the cardinal's red coloration, the lion's mane—and how male competition for mating opportunities with females can lead males to evolve aggressive traits—the elephant seal's huge size and aggressive behavior, the deer's large antlers, the rhinoceros beetle's horns. But some scientists have criticized this focus. Among them are anthropologist and primatologist Sarah Blaffer Hrdy and biologist Joan Roughgarden.

Hrdy researched primate behavior in the second half of the 20th century. She criticized the assumption behind sexual selection theory that the norm across animal species is for choosy females to select among many promiscuous males who are eager for mating opportunities. Darwin presented this as a matter of the "coy" female engaging with the "eager" male. The problem with this idea is that, across many species, females are actually quite eager to engage sexually. So, a basic assumption of sexual selection theory turns out often to be inaccurate. Beyond that, Hrdy emphasized that sexual behavior is a very small portion of all animals' lives—also relevant is time spent finding food, resting, fleeing dangers, rearing their young, playing, building homes, socializing, and so on.

Joan Roughgarden is a biologist who researched genetics and, later in her career in the early 21st century, turned to research on the variety of sexual, reproductive, and parental strategies across animal species. Roughgarden's book *Evolution's Rainbow* catalogs that variety. Animals engage in a wide variety of mating strategies, such as bluegill sunfish, which have three distinct male types with dramatically different body sizes, appearances, and mating strategies. Roughgarden suggests these different types should be thought of as distinct genders. She also emphasized the wide variety of factors that determine the sex of animals, that is, whether they are male or female. In some species, including humans, genes play a central role in determining whether an animal is female or male. In others, like frogs, environmental factors such as temperature determine sex, and in still other species, sex of an individual changes over their lifespan.

Animals also vary widely in their parental strategies. Some species leave their offspring to fend for themselves immediately, while other species invest significantly in caring for and protecting their young. Sexual selection theory has emphasized female contribution to caring for offspring. But, in species that care for their young, this is sometimes a two-parent project; sometimes a male project, as with seahorses, whose males carry the young around in a pouch much like kangaroos; and sometimes a group collaboration, where larger social groups collectively rear their young, as in several species of birds.

After cataloging the extensive variation in animals' sexual, reproductive, and parental strategies, Roughgarden began to reconsider how those strategies may have evolved. She criticized sexual selection theory for focusing exclusively on reproductive sexual encounters, ignoring how sexual activities can play nonreproductive social roles like resolving conflicts, and how pairs and communities can cooperate to successfully raise offspring. She also pointed out that animals engage in a wide variety of social encounters, with members of the opposite sex and same sex alike, and that these can also have evolutionary significance.

Roughgarden has developed several competing hypotheses for traits targeted by sexual selection theory. These hypotheses emphasize extensive social exchanges and cooperation as potential reasons for the evolution of traits typically explained as the result of sexual selection. While most biologists still think of sexual selection theory as a cornerstone of evolutionary theory, many biologists' understanding of sexual selection theory has at least been updated to acknowledge more variation in how reproductive pressure influences evolution. Sometimes females compete for reproductive access to males, and this has led, in some cases, to females having showy traits or aggressive traits. Sometimes males invest more in offspring care than females do. And sometimes broader social groups are relevant to the evolution of sexual and reproductive behaviors.

If Roughgarden and Hrdy are right, then more attention should be paid to the evolutionary impacts of how animals engage with one another—sexually, cooperatively, and competitively—outside of reproduction. This might lead some traits that have been explained as instances of sexual selection to be reclassified as the evolutionary result of the influence of social dynamics.

Returning to the example of the peacock's colorful tail feathers, this is still broadly accepted as a primary example of a showy trait resulting from sexual selection. But more than a decade ago, researchers in Japan published a paper showing that, in local feral populations of peafowl, more elaborate tail feathers weren't associated with mating success. Male displays involving the colorful train are certainly involved in mating behavior. But other researchers have found that displaying the tail feathers by fanning them out in mating rituals serves to amplify peacocks' verbal calls, which raises the question of whether the trains themselves are important to mating or just their role in amplifying verbal cues. Furthermore, peacocks also display their tail feathers when faced with a predator, making themselves seem larger and distracting the predator, as a bite to the colorful train simply leads to the loss of some feathers rather than bodily injury. Sexual selection theory is meant to account for traits that wouldn't evolve by natural selection alone, but it seems like the peacock's colorful tail feathers may have direct contributions to survival as well.

Science in a social context

Beyond her primatology research, Sarah Blaffer Hrdy also asked *why* primatologists began to take issue with standard sexual selection theory in the 1970s, first beginning to notice the active sexual lives of female primates. She suggests that, especially when studying primates, scientists tend to identify with and observe more closely animals of the same sex as themselves. Humans are, after all, one species of primate. So, in Hrdy's view, an increasing number of women in the field of primatology brought with it increasing attention to female primates and their behaviors, instead of the previous focus on male primates with females more in the background.

Joan Roughgarden is also reflective about how her social identity influenced her research. In the preface to her book about variability in sex and reproduction in animals, she reflects on attending her first Pride Parade in San Francisco and transitioning shortly after to confirm her identity as a transgender woman. These experiences led her to wonder about all the diversity across the animal kingdom, and how evolution and development lead to such diversity. Roughgarden associates this with her growing realization that sexual selection theory was overly limited, as well as with the insight that "kindness and cooperation are basic to biological nature."

Way back in Chapter 2, we introduced the idea that science—including both its aims and its methods—is influenced by *social values*, or priorities and moral principles accepted in some community. In the scientific investigation of sex and reproduction in the animal kingdom, the influence is felt in the question of which sex is focused on (male or female), on whether reproductive variability beyond biological sexes—what Roughgarden urges us to think of as different genders—is investigated, on which traits and patterns are emphasized, and perhaps even on whether the focus is on the cooperative or competitive value of traits.

Because scientific reasoning is a fundamentally human endeavor, it always occurs in specific social, institutional, and historical contexts. Scientists make observations, develop theories, make discoveries, and interact with one another all within specific interpersonal, institutional, and cultural circumstances. Further, science is embedded in institutional structures like universities, labs, museums, journals, and publishing companies. These social contexts and institutions in which science occurs are influenced by history, by who is part of them, and by the social identities scientists bring to their work.

Further, social and historical context influences the nature of science. Even as science aspires to produce knowledge that is not limited by specific perspective, scientific theories are in some ways creatures of the times, places, and people who created them. Some have suggested that Darwin's statement of sexual selection theory was strongly influenced by Victorian moral sentiment in its assumption that, throughout the animal kingdom, male animals are promiscuous and aggressive and female animals "coy" and selective. This looks suspiciously like gender norms in Victorian England, Darwin's cultural setting. While Darwinian evolutionary theory was certainly a tremendous step forward for biology, it was also influenced by the time and place of its creation, and perhaps by the interests and personal values of the individuals who created it.

So, science seems to be shaped by its social, institutional, and historical context. Science also is regularly used to promote particular social aims, either explicitly or implicitly. Specific scientific aims can support different social aims. Consider how Roughgarden is explicit about her motivation to explore variation in sex and reproductive strategies in the animal kingdom in order to support social aims of inclusion. Even if some scientific research does not relate directly to social aims, a scientist's interests and values may still influence how the aims of their scientific research connect with social concerns.

The difficult truth is that, throughout history, science has often been used to promote objectionable social aims and, at times, has even been pursued in ways that incorporate morally horrendous views like eugenics and scientific racism. Science has been used to expand power over others, to invent nuclear and chemical weapons for the purpose of mass destruction, and to amass wealth for the few, as with research for the fossil fuel industry. Science has also been used to promote misinformation, as when the British doctor Andrew Wakefield fraudulently claimed that the combined measles, mumps, and rubella (MMR) vaccine was linked to autism or when tobacco corporations paid scientists to present cancer research in a way calculated to mislead the public.

Science has also been used to directly harm and oppress marginalized groups, as when the Nazis ran deeply cruel experiments on the prisoners of concentration camps and when the US Public Health Service ran the Tuskegee syphilis experiment. In this clinical study, researchers withheld treatment from 399 impoverished, rural, Black men who had syphilis. They never informed the participants that they had syphilis or that there was a cure for the disease. Scientific research has also indirectly supported racism and sexism by focusing on aims like identifying a genetic basis for racial differences in intelligence or neurological differences between men's and women's brains.

All of this suggests that an important aspect of investigating science is learning about science's relationship to society. We need to understand how science is shaped by scientific and historical context, ways in which science can be influenced by social values, how to uncover problematic values or problematic roles for values in science, and how science can be used to promote positive social aims.

EXERCISES

13.1 Recall: Summarize sexual selection theory and give one example of a showy trait and one example of an aggressive trait thought to evolve through sexual selection.

13.2 Apply: Choose an example of a showy or aggressive trait, either from the text or a new one. Using Google Scholar, a library catalogue, or a similar search tool, find 3–5 research articles focused on that trait as an instance of sexual selection. Summarize the main finding of each article in one sentence; you can probably do this just by reading their titles and abstracts. Do these articles together provide adequate support for the hypothesis that the trait is an instance of sexual selection? Why or why not?

13.3 Recall: Describe (a) Hrdy's criticism of the assumption that female animals are "coy" while males are "eager," (b) Roughgarden's point about variety of sexual, reproductive, and parental strategies, and (c) Roughgarden's idea about how cooperation might be important instead of competition. For each, say how it challenges sexual selection theory.

13.4 Recall: How did Hrdy think research into sexual selection was influenced by social factors? How did Roughgarden think research into sexual selection was influenced by social factors?

13.5 Think: In your own words, describe how you think historical and social factors have influenced research into the evolution of sex, reproduction, and parental strategies. Give one example of an influence that you think was problematic for scientific or ethical reasons and one example of an influence that you think was acceptable (scientifically and ethically).

13.6 Apply: Choose an example of a scientific theory or finding discussed anywhere in this book. Describe how you think historical and social factors may have influenced the research bearing on that theory or finding. Does the influence by historical and social factors you identified call the theory or finding into question, or not?

13.2 PARTICIPATION IN SCIENCE

After reading this section, you should be able to:

- Describe how people with some social identities have been historically excluded from or marginalized in science
- List three ways in which diversity of scientists is beneficial to science
- Indicate how members of the public can be included in scientific research and how this can affect scientific research

Exclusion from science

Just as we must acknowledge the historic contribution of science to immoral social aims and objectionable values, including classism, racism, and sexism, we also must acknowledge that science has never been—and still is not—fully inclusive. This means that many people—because of their geographical location, institutional affiliation, language, sex, race, or creed—do not receive a fair chance to participate in and contribute to scientific inquiry. Historically, the institutions that house today's science disproportionately developed in Europe and, later, the United States and predominantly included wealthy, White men. Women, people of color, people without wealth, and people in other nations have always contributed to science. But people from these and other marginalized social groups have historically had very limited access to resources to make scientific contributions and very little recognition afforded to them for their scientific contributions.

Polymath Alan Turing did groundbreaking research in computer science, logic, mathematics, cryptography, and morphology in Great Britain. During World War II, he helped crack the code used by the Nazis to protect their military communication, an achievement that many historians believe was the single greatest contribution to the Allied victory. Turing was also a visionary of artificial intelligence. You may have heard of the Turing machine and Turing test that he invented; he anticipated that human intelligence would one day be matched by machines. Turing was also gay, and at the time this was illegal in Britain. Despite his groundbreaking contributions to the computer, or "digital," revolution, Turing was arrested and chemically castrated by the British government. Humiliated and resentful, he committed suicide at the age of 41.

If being outed as gay in mid-20th century Great Britain was awful, matters were no brighter for women in science for most of history. Cecilia Payne-Gaposchkin's groundbreaking dissertation *Stellar Atmospheres* in 1925 became a cornerstone of modern astrophysics, but she couldn't get a professorship, so she had to do low-paying adjunct teaching for the next 20 years. Rosalind Franklin was a chemist and x-ray crystallographer, who we mentioned in Chapter 5 for her important contributions to the discovery of DNA's structure in the mid-20th century. Franklin was responsible for an x-ray diffraction image that was shared with Watson and Crick without her knowledge (pictured in Figure 13.2). After seeing that image, Watson and Crick developed their physical model of DNA. They went down in history as having discovered DNA's double helix structure, eventually winning the Nobel Prize for this work. In contrast, Franklin's enormous contributions to the discovery were recognized only after her death.

A similar story is that of neuroscientist Kathleen Montagu, who published her discovery of the neurochemical dopamine in the human brain in 1957. Her work was largely overshadowed by a very similar discovery three months later by Swedish neuropharmacologist and Nobel Prize winner Arvid Carlsson and colleagues. This is a common enough phenomenon that is has been named. The **Matilda effect** is the bias against recognizing women scientists' achievements, whose work is often uncredited or else attributed to their male colleagues instead.

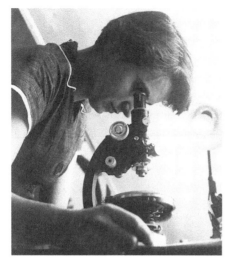 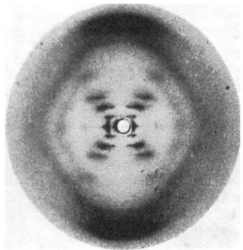

FIGURE 13.2 a. Rosalind Franklin (left); b. Franklin's x-ray diffraction image that famously inspired Watson and Crick's double helix model of DNA

Societal prejudice and structural exclusion have made it more difficult for not only women but also racial and ethnic minorities, sexual and gender minorities, people with disabilities, first-generation college students, people from low socioeconomic backgrounds, residents of the Global South (Latin America, Asia, Africa, and Oceania), and other marginalized groups to be supported in their scientific work and even to become scientists in the first place. Even today, opportunities and rewards in science disproportionately go to men from families with financial resources and college educations and who live in wealthy nations. That said, the inclusion of people from social groups underrepresented in science is increasing, as more attention is focused on making science more inclusive and equitable.

Diversity in science

When only certain kinds of people participate in science, science suffers. Systemic exclusion and marginalization result in science squandering or losing out entirely on the contributions of the people who were excluded or marginalized. A second way in which science suffers is that there are fewer role models for aspiring scientists. When scientists like Turing or Franklin are dishonored or not acknowledged, younger people do not have the opportunity to look up to them. When groups of people are systematically underrepresented and marginalized in science, young people may not see themselves as potential participants in science.

These are reasons for the scientific establishment to prioritize inclusion, that is, to take steps to ensure people with a full range of backgrounds have a fair opportunity to become scientists and to gain recognition for their achievements. We should also look back to the history of science, revisiting standard accounts of who discovered what, who qualified as scientists, and which societies influenced the development of science as an institution. We have tried to keep that in mind as we selected examples and scientists to feature in this book, though we acknowledge we did not fully succeed. There is a broad project needed, and to some extent underway, to update our collective understanding of science to more fully credit the full range of individuals and societies that shaped it.

Diversity in science has important effects beyond simply not losing out on scientific discoveries or future scientists. Who participates in science and who is excluded affects the nature of scientific knowledge and its trustworthiness. The features of scientists—nationality, gender and sexuality, socioeconomic background, race and ethnicity, religious belief, political affiliation—all help determine what questions scientists are interested to investigate, what bold conjectures they come up with, and perhaps in some cases which methods they tend to use. When a group of scientists in some field is sufficiently diverse, all differences among them can contribute to the range of questions, richness of ideas, and ultimately the quality of inquiry. If, instead, only certain kinds of people participate in science or in some particular field of science, then the kinds of questions posed, ideas generated, and interpretations of data may reflect the limited perspectives of those scientists. For example, recall Hrdy's observation of the changes to primatology research that occurred when a critical mass of women became primatologists. And, back to our point about role models, the influx of women to primatology might well have been influenced by the publicity received by Jane Goodall's research on chimpanzees in the mid-20th century.

Charles Henry Turner was an African American zoologist who pioneered the study of animal cognition, particularly in insects. Born in the US just two years after the end of slavery, Turner was the first Black student to graduate with a master's degree from the University of Cincinnati and the first Black student to be awarded a PhD at the University of Chicago. Turner was a victim of racism and could not find work in a university. He worked in an all-Black high school, while continuing his research on animal behavior. Turner rejected the prevailing ideas of the time that animals like birds, bees, and ants do not have sophisticated abilities of perception or cognition. His research demonstrated that these animals possess impressive powers of memory, learning, and problem-solving, drawing attention to sophisticated behaviors like mound building in ants and hunting habits of wasps. Although Turner could not work in a university and did not have access to adequate laboratory equipment, he still managed to publish the results of his studies in prestigious scientific journals. His research was ahead of its time, predating more recent cognitivist approaches to analyzing and explaining animal behavior.

The value of diversity of science, then, is more than just who is recognized for what discovery, how breakthroughs are received, and who gets to be a scientist. People with different social identities can bring different styles of reasoning to the table, and scientific progress often demands creativity and seeing things anew. For these reasons, the conscious inclusion and encouragement of people from diverse social groups and with

FIGURE 13.3 Charles Henry Turner, African American zoologist working in the late 19th and early 20th centuries

diverse social identities to participate in science doesn't just benefit those individuals and benefit society; it also makes science more successful at achieving its paramount goal of generating knowledge.

Public participation in science

Science welcoming new scientists from diverse social groups and with diverse social identities is one way to gain new perspectives in science. Another way, which is also gaining popularity, is to include members of the public as collaborators in scientific research. *Participatory research* is any scientific research in which members of the public who aren't trained scientists participate in doing the research. Participatory research goes by many names; you may have heard it called citizen science, community-based research, or action research. We encountered this in Chapter 7's discussion of how scientific research helped uncover the Flint water crisis. Community members in Flint reached out to environmental scientists for help, and a local pediatrician led research into pediatric lead exposure.

Box 13.1 Open science

The term *open science* indicates several practices that support the free sharing of scientific research and calls for broadening the demographic composition and theoretical approaches of scientific research. The overarching aim of open science is to enhance the quality of research, accelerate the rate of discovery, and make science more inclusive. Open science has several dimensions. One is open access to research products, where scholarly outputs such as software, scientific articles, and books are disseminated open source — free to read, use, and share by anybody. An example is open-access repositories of electronic preprints of scientific papers such as *arXiv*. Another more controversial example is *Sci-Hub*, a shadow library website created by Kazakhstani computer programmer Alexandra Elbakyan, providing free access to millions of published research papers without regard to copyright and paywalls. Another dimension of open science is pedagogical. The aim here is to make a diverse range of teaching and learning materials freely available and accessible through schools, libraries, museums, and even theaters. A third dimension of open science is preregistration and open data: specifying one's research plans in advance of a study and making all data available and accessible in public repositories. These all aim to improve the transparency, accountability, and replicability of research. Finally, another dimension are calls to identify the social, economic, and political dynamics that have historically shaped and still shape participation in science and to work to empower marginalized scientists, striving to make scientific institutions fairer.

Scientists sometimes choose to include the public in research efforts in order to increase the amount of data that can be collected. For example, the Cornell Lab of Ornithology in the United States has involved members of the public in collecting data about bird breeding, courtship, habitats, and colors for decades, which has enabled more data and in a broader range of locations. But that's not the only benefit from public participation in the Cornell Lab's research. Participation in gathering data about birds is also an opportunity for students and interested adults to learn more about birds and about how scientific research occurs. And it is an important opportunity for education about conservation and participation in conservation efforts.

Some participatory research even involves members of the public in shaping the research aims and methods. When research focuses on topics of community interest, such as health outcomes, local environmental conservation, and community access to services, involving members of the relevant communities early in the process of designing the research enables scientists to benefit from their inside perspective. This can help the research target issues of genuine community concern, in ways the community will value, and it can help legitimize the research in community members' eyes, which can help with recruiting participants into the study as subjects (to respond to surveys, be interviewed, or donate genetic samples, for example). In Chapter 2, we mentioned how local, indigenous knowledge is increasingly recognized as valuable for certain kinds of scientific research, for example, on how to adapt to local effects of climate change. That is one pattern participatory research can follow.

Because of science's history of exclusion and its contributions to injustice, members of some communities hesitate to participate in scientific research. If participatory research is pursued with true concern for community member's priorities and is developed to empower public participants and their communities, then participatory research can perhaps help ameliorate this problem. But it's worth pointing out that participatory research can also be pursued in problematic ways. This can occur, for instance, if researchers do not empower public participants to influence the research, do not heed community members' recommendations, or do not follow through in their commitments to the community related to the research. Indeed, science has developed a bad reputation for so-called helicopter or parachute research, where scientists from wealthy nations visit poorer countries to study underserved communities and local ecosystems, perhaps collecting artifacts or biological samples, and publishing results in scientific journals with no involvement from local scientists and without concern for whether and how the community is impacted or whether community members have access to the knowledge generated.

EXERCISES

Watch Video 17 **13.7** **Recall:** Describe how people with some social identities have been historically excluded from or marginalized in science. List three social identities affected by this exclusion and marginalization.

13.8 **Apply:** Choose one social identity underrepresented in science (such as a racial or ethnic minority, a sexual or gender minority, people with disabilities, first-generation college students, people from low socioeconomic backgrounds, or residents of the Global South). Use the internet to identify three prominent scientists with the social identity you have selected. For each, provide their name, the approximate time period of their career, their research field(s), and a brief summary of their main scientific contributions.

13.9 **Recall:** List three ways in which diversity of scientists is beneficial to science.

13.10 **Think:** Describe three steps that you think could be taken to increase diversity in science. Then, indicate any concerns or downsides you can think of for each of the steps. Which of the steps you described do you think should be implemented, and why?

13.11 **Recall:** Define *participatory research*. Describe three ways in which participatory research can improve scientific research.

13.12 **Apply:** Go to the website www.zooniverse.org; this is a major platform for participatory research projects. Look at some of the projects on the website and find one that's particularly interesting to you. Describe the project, what the aim of the research is, and why you think it is designed to include public participation. What is the value for the scientific research of public participation? What is the value for the public of participating in this research? Make sure you provide the name and link of the project you chose.

13.3 SOCIAL VALUES AND SCIENCE

After reading this section, you should be able to:

- Define the *value-free ideal* and describe what about it is correct and what is incorrect
- Give an example of when values have influenced science in a problematic way
- List five ways in which values influence science in legitimate ways and give an example of each

The value-free ideal

In Chapter 1, we emphasized how the institution of science has been developed to control for and overcome human limitations in reasoning like confirmation bias. One feature of this is that science is supposed to provide a way to subject our pre-existing ideas to rigorous test. Wanting something to be true isn't good grounds for believing it is true, and science provides methods of hypothesis-testing, reasoning, and theorizing to evaluate the grounds for our beliefs. This idea inspires what has been called the **value-free ideal** for science, which is the thesis that scientific reasoning should proceed free from the influence of our values—such as moral, social, or political beliefs.

There is something right about the value-free ideal. Whether or not we *want* something to be true is irrelevant to whether it is in fact true, and science aims at the production of genuine knowledge. So, in science, hypotheses and theories should be judged not for their desirability but instead for the grounds for thinking they are true. Some scientific questions—about the reality of climate change, humans' evolutionary history, and gender differences in cognition, for example—may evoke emotional reactions, but those emotions aren't relevant to how scientific research bears on the questions. Instead, these questions can only be answered by conducting experiments or studies, constructing models, evaluating evidence, applying statistical reasoning—by applying the recipes for science.

On the other hand, it is clear that science regularly is influenced by values. For example, we mentioned earlier how Darwin's evolutionary theory encodes Victorian morality and how Hrdy's and Roughgarden's research was inspired by values each of these scientists held. Positing the value-free ideal as an ideal enables us to acknowledge that sometimes values have in fact played a role in science, while insisting that they should not. Still, if science regularly does incorporate values, it's worth asking whether being value-free should even be considered an ideal for science to aspire to.

Indeed, we discussed in Chapter 2 how social values can influence both the aims and methods of science. Is discovering a vaccine for the Zika virus, which is easily transmitted to humans by mosquito bites and leads to serious birth defects when pregnant women are infected, more important than discovering new facts about quasars, pulsars, supernovas, or other astral phenomena? Deciding this requires consideration of social values. It's clear that the US Public Health Service should not have withheld syphilis treatment from 399 impoverished Black men without their knowledge in the Tuskegee syphilis experiment; saying why requires consideration of values. The value-free ideal is not only inaccurate of how science has in fact played out; it's also undesirable as an ideal for science to aspire to. Values should, at least in some ways, influence scientific reasoning.

How values shape science

The value-free ideal suggests that science should simply be a source of objective facts about the world, immune to influence by human values. On the other extreme, some people believe science only serves preexisting political or economic values. The right view of how values should influence science is somewhere between these two extremes. Science is not and should not be value-free, but there are ways in which science should limit the influence of values.

Scientists are human beings with different moral, political, and religious values, and we have seen that the social context for science and who participates in science can both influence scientific findings. And yet, the recipes for science are designed to limit the kinds of influence social and individual values have on science. When scientists' values lead them to violate the recipes for science—acceptable approaches to data collection and modeling, to hypothesis-testing and abductive reasoning, to statistical analysis, and so forth—this is illegitimate. When scientists use values to supplement, inform, or guide the use of the recipes for science, this can be legitimate.

In his book, *A Tapestry of Values*, philosopher Kevin Elliott divides the legitimate roles values can play in science into five categories, as helping to answer five different questions. These questions are summarized in Table 13.1. Answers to these questions are needed in order to know which scientific methods to employ, on which phenomena, and to what end. These roles for values thus align with our suggestion that legitimate uses of values supplement or guide the methods of science but do not violate those methods.

To begin, scientists' individual values and societies' collectively held values help answer the first question about what to study. Individually, a researcher's interests and values surely shapes what field of science they pursue, which lab they work in, and what problems they tackle. Collectively, we choose what kinds of scientific research to support when funding agencies, including tax-supported governmental agencies, designate the areas of research they fund and which specific research projects in those areas to fund.

Beyond what to study, values also inform decisions about how phenomena should be studied. Scientists can bring different questions, methods, and background assumptions to bear on any given topic, and these choices in how research is pursued reflect

TABLE 13.1 Five questions that our values help answer when doing science

Question	Example
1. What should we study?	what kind of research is prioritized for funding
2. How should we study it?	how the initial hypothesis and assumptions about the causal background both guide experimental design
3. What are we trying to accomplish?	getting the most accurate information vs. accurate-enough information quickly enough to guide policy
4. What if we are uncertain?	how much evidence to require before accepting or rejecting a given hypothesis
5. How should we talk about the results?	the level of certainty conveyed to the public about some scientific finding

Source: Adapted from Elliott (2017)

researchers' and society's values. One instance of this influence is how the initial hypothesis and assumptions about the causal background both guide experimental design, including the nature of the intervention and which extraneous variables to control. As with the initial choice of what to study, funding agencies can also influence how phenomena are studied. For example, research into depression can focus on the efficacy of cognitive therapy; the role of sleep, diet, and exercise; or the efficacy of pharmaceutical intervention. The choice to strongly prioritize pharmaceutical intervention to the exclusion of other focuses reflects the outsized influence drug companies have had on medical science.

The third question that a focus on values helps to answer is what, exactly, scientists are trying to accomplish in studying some phenomenon. This is an even more fine-grained decision than just what to study and how to study it. In climate research, for example, scientists might prioritize getting information about climate trends available quickly so that it can guide policy, or they might prioritize getting as accurate of information as possible, no matter how long it takes. This decision about the aim of research is influenced by values, including views about the social role the scientific research is expected to have.

Fourth, values influence how scientists proceed in the face of uncertainty. Scientific results are never free from uncertainty. There's the basic problem of measurement error. We've also seen that, if observations don't match expectations, it could be the fault of the hypothesis, or it could be the fault of some auxiliary hypothesis. In an experiment, no matter how perfectly controlled, there's always the chance that an unexpected confounding variable has interfered with the finding. In statistical hypothesis-testing, scientists choose whether to reject the null hypothesis or not, and either choice could be wrong, resulting in a type I error (false positive) or type II error (false negative). These are just a few examples of the unavoidable uncertainty in science and the need to choose how to proceed in the face of **inductive risk**, that is, the risk of wrongly accepting or rejecting a scientific hypothesis.

These kinds of uncertainty are all forms of underdetermination. Recall from Chapter 3 that underdetermination is when the evidence available to scientists is insufficient to determine which of multiple theories or hypotheses is true. Some believe that there is even permanent underdetermination in science: that there will never be enough evidence to conclusively decide in favor of one hypothesis or theory and against all possible alternatives. When scientists face underdetermination, they must choose what to believe or whether to suspend judgment.

Because of this unavoidable uncertainty, scientists must decide how much evidence to require before endorsing a theory or hypothesis (or before rejecting a theory or hypothesis). Safety is very important to us, whether for drinking water or medications, so toxicology tests must have a high bar for success. There is a lower bar for deciding whether a new drug is more effective than an already available drug. Scientists also must decide how to represent scientific uncertainty to the public. In 1988, climate scientist James Hansen declared that climate change, global warming, was occurring. He described that as a decision based on weighing "the costs of being wrong to the costs of not talking." There was already enough evidence for Hansen to be relatively confident in his choice to speak up. Decades later, of course, there is now incontrovertible evidence of climate change.

This introduces a fifth category of values' legitimate influence on science, regarding how scientists, journalists, and others who communicate scientific findings to the broader public should talk about those findings. As Elliott stresses, this isn't just a decision to be accurate. Scientific findings also can be discussed in their relationship to previous findings, their potential social effects, or—picking up on the point just above—their level of certainty. This framing influences whether and how the public will engage with research, and this is a choice not dictated by scientific methods but by scientists' and society's priorities.

So, what scientists should study, how they should study it, what they aim to accomplish, how much evidential support should be required, and how scientists should communicate their results all depend on moral considerations—on values. These are legitimate influences of values on science. Recognizing these roles for values in science is crucial. This enables us, as a society, to critically assess what values are employed at each of these junctures. The influence of values on science can be problematic or even nefarious if the wrong values are employed at any one of these stages. Figuring out the right and wrong values is tricky, and it is not a matter for scientists alone to decide. Instead, this is an issue that needs to be engaged with broadly in our society.

Examples of problematic values influencing science are, unfortunately, very easy to come by. Here's one. In 2017, the US president Donald Trump proposed that NASA resources should be dedicated to exploring the solar system instead of to climate change research. This research priority—a decision about what to study—reflects the values endorsed by a small but vocal contingent of politicians in the US. Choosing not to fund climate change research amounts to deciding that knowledge about the rate and impact of climate change is relatively unimportant. Space exploration is important! It, too, should be funded. But because climate change is already having disastrous effects on populations, the environment, and economies across the world, and because the long-term costs of ignoring it will be disastrous, pulling funding from NASA's earth science division in order to avoid investigating climate change and its effects upheld the wrong priority.

We have suggested that science doesn't have to be free from values to be trustworthy and objective. What matters is that values influence science in the right ways and that science effectively resists the influence of harmful, unethical values. Values, even good values, shouldn't play the wrong roles in science; we should never decide a theory is true simply because we wish it were true. Further, the wrong values shouldn't influence science, even through the proper channels. To better understand how science earns its trust and objectivity, it's important to acknowledge the many roles of social and moral values in scientific reasoning and to critically examine the values that influence our science. By doing these things, we can clarify what values should influence science, and in what ways.

EXERCISES

13.13 Recall: Define the *value-free ideal* for science and describe what is correct about it and what is incorrect about it.

13.14 Think: Describe an example of when values have influenced science in an illegitimate way. (Your instructor might ask you to take an example from this section or to come up with a different example.) Then diagnose what went wrong: what was wrong about the values or the nature of their influence, and what was the detrimental effect to science?

13.15 Recall: List the five ways discussed in this section in which values legitimately influence science. Give an example of each.

13.16 Think: Describe the value-free ideal of science. Then, summarize the view of how values can legitimately factor into science. In your view, does that view of values' influence violate the value-free ideal, or is it consistent with that ideal? Give an argument in support of your answer.

13.17 Apply: Suppose you are working for an NGO (nongovernmental organization) on the task of measuring poverty levels across countries. For each of the following decisions, describe at least two ways to proceed, and say how values are relevant to making the decision.
  a. Which countries will you include in the study?
  b. How will you define and measure poverty?
  c. What extraneous variables will you take into account?
  d. How will you make comparisons across countries?
  e. How will your results be publicized?

13.18 Apply: Look back at the case of climate change discussed in Chapter 1. Identify at least five ways in which values are likely to have affected that research and describe how those values have impeded or promoted scientific knowledge of climate change.

13.4 CHANGES IN SCIENCE AND NEW CHALLENGES

After reading this section, you should be able to:

- Describe the move toward ABC research in the 21st century and what remains unchanged
- List three challenges to science's capacity for self-correction related to incentive structures and sources of funding
- Characterize the current challenges of diversity and public communication in science

Changes in 21st century science

Antireductionist, big, and corporate. This is the distinctive new "ABC" of scientific research in the 21st century. ABC research investigates complex phenomena at multiple scales (antireductionist), capitalizes on big collaborations between many scientists with diverse expertise and big data (big), and involves business partnerships with private corporations (corporate). An example is proteomics, which is the study of how proteins influence organisms. Proteins, the building blocks of organisms, are complex molecules that play many vital biological functions. These functions depend on proteins' physical shapes. Proteins acquire their distinctive shapes when the long chains of hundreds or thousands of smaller molecules composing them, called amino acids, fold into a wide variety of three-dimensional physical structures.

Since the discovery of the structure of DNA in 1950s, one of biology's grandest challenges has been the problem of protein folding. This is the problem of determining the 3-D shape of the proteins in an organism from proteins' amino-acid sequence. Unlocking the mystery of how proteins fold is key to preventing and treating many diseases, like cancer; to understanding how life emerged on Earth and how organisms develop; and to many bioengineering applications. This is not just a very important problem; it is also a very hard problem. For any organism, the number of possible amino acid sequences and proteins is very large, and any given sequence of amino acids might fold into an huge number of different 3-D structures. So, it has been difficult, if not impossible, to tackle this problem with traditional experimental methods.

A machine learning tool called *AlphaFold*, first developed in 2018, has been a game-changer in the field of proteomics. Developed by the Google-owned firm DeepMind, AlphaFold leverages deep-learning artificial neural networks trained on big data sets about hundreds of thousands of experimentally determined protein structures to predict the shape of thousands of other proteins, in organisms from bacteria to humans, from their genetic sequence. Its predictions are surprisingly accurate. AlphaFold has some limitations—for example, it is not designed to predict the effects of mutations on a protein's structure—but it is already changing the field of proteomics. It saves biologists a lot of time-consuming manual labor, so it accelerates research. It can also open new avenues of research, as it makes available new biological hypotheses and interpretations.

Research relying on AlphaFold exemplifies ABC research in its antireductionist approach to studying protein folding in all its variety in an open-ended, exploratory way; by involving many scientists with diverse expertise, as well as big data methods; and by relying on corporate partnership in producing needed technology. Of course, not all contemporary scientific research fits this pattern, and it's more relevant to some fields than others. But ABC research is a good characterization of recent trends in scientific research nonetheless. Let's look at each of these features.

Reductionist methods attempt to account for a system by considering the behavior of its fundamental components, explaining even complex features as the result of component parts. It's increasingly clear that this approach is inadequate to account for complex, possibly chaotic, systems such as the Earth's weather and climate; the economic, cultural, and political behaviors of societies; and the workings of the brain. The same goes for the intricate dynamics of protein folding. Antireductionist approaches grapple directly with complexity, employing a variety of methods and approaches from multiple disciplines, often including big data techniques, to probe how networks might self-organize, what surprising patterns exist, and how systems flexibly adapt to different circumstances. Proteomics and other "omics" disciplines, like genomics, exemplify this anti-reductionism. And even in fields that fit this model less closely, some techniques employ antireductionist methods.

ABC research is also big. Scientific research has become highly collaborative and ever hungrier for data and computing power. For example, the scientific article first

FIGURE 13.4 Sequencing data of genome analysis and particle molecular structure

introducing AlphaFold had more than 30 authors. Each of these authors brought specialized expertise to the research, from machine learning and software engineering to molecular and system biology. Studies in fields like particle physics, genetics, and medicine can feature even larger collaborations, among thousands of scientists who together co-author a single peer-reviewed paper. Running a study in these fields can be like running a Hollywood movie. It requires specialized competences and well-defined roles in a team pursuing an overarching research goal. Large collaborations also typically involve sophisticated equipment and large data sets. And, as we just saw, big data and computational approaches also are made necessary by embracing antireductionist approaches.

Box 13.2 Co-authorship in large collaborations and with AI

Pick any recent issue of a scientific journal. Most articles you'll find will list several co-authors, possibly hundreds or even thousands. The co-authors of a scientific paper are often from different countries and based in different institutions. They have different disciplinary backgrounds, so different co-authors may not fully understand other co-authors' contributions. The research sometimes involves complex methods or instruments, whose workings may not be transparent to all collaborators. Increasingly often, scientific authors also rely on AI tools such as large language models, for example ChatGPT, for generating and editing text. In a few cases, AI chatbots have also been credited as co-authors of published research. But what if a large collaborative project, where expertise and responsibility are decentralized and distributed among several individuals and even AI tools, produces a published research article introducing a patentable new discovery? Or containing a mistake? Who deserves deserve credit for the discovery or who should be held accountable for the mistake? And can an AI be a genuine co-author? These questions are important—and difficult—to answer. Co-authors are those who make a significant scholarly contribution to an article and who publicly agree to be legally, epistemically, and morally accountable for the published research. This is clear enough. But questions about co-authorship, credit and blame, and accountability become murkier with large collaborations, especially with the advent of AI tools, as there is no single, universal norm for how co-authorship should be determined in collaborative work.

Finally, ABC research is corporate. Large-scale collaborations geared towards understanding complex phenomena can also benefit from involving private corporations. This isn't inevitable. Most scientific research is still conducted by researchers based at universities and funded by governments and other public institutions. But, there is a trend toward scientific research led by teams of scientists based at corporate research labs like Google DeepMind. With the emergence of machine learning and artificial intelligence, an increasing number of scientists have been recruited from universities to big companies like Amazon, Meta, Google, Tesla, and Uber. The same trend applies to smaller firms and start-ups, especially in the biotech and food sectors.

The declared mission of corporate science is sometimes basic research, sometimes innovation, sometimes profit. The complicated interplay of these aims has contributed to the discovery of life-saving drugs, smartphones, and electronic vehicles. It can also boost research in an entire field by making valuable data sets and methodologies openly available to all, as in the case of AlphaFold, where the researchers at DeepMind have made openly available their data sets and computer code. But the relationship between science and corporations is complicated. This foregrounds challenges like the corruption of science from a motive of profit rather than pursuit of knowledge and how patentable data and methodologies should be shared.

Again, not all contemporary scientific research fits this model of antireductionist, big, and corporate. Reductionist approaches remain valuable in numerous research projects; not all scientific research is or should be collaborative; and much research proceeds without influence from corporations or clear application to industry. Note also that, even in ABC research, big data and machine learning tools like AlphaFold are not likely to replace human scientists. Science's power consists, at least in part, in bringing to bear new ideas in creative ways, making judgment calls about how research should proceed, and looking beyond correlations to seek explanation. In many instances, those aspects of science are best conducted by human intellects.

Challenges facing science

Scientific knowledge is worthy of trust due to its characteristics outlined in Chapter 1—especially its capacity for self-correction based on evidentialism and social structure, and the many methods and patterns of reasoning used to find evidential support for hypotheses and theories. We saw in Chapter 1 that science's capacity for self-correction requires scientists' openness to criticism and dissent, their sincere and transparent communication of their results and uncertainties, and scientific communities' welcoming of diverse perspectives. Trustworthy scientific knowledge is produced when scientists' judgments are critically and openly assessed in light of other data, competing interpretations, and alternative possibilities.

Yet this process of collaboration and criticism, and thus science's capacity for self-correction, currently faces significant challenges. Some of these challenges relate to the incentive structure in science and how it shapes scientific findings in ways that can undermine trustworthiness. Facing up to these challenges requires us to think carefully about the scientific process, the role of incentives in shaping that process, and what values are thus finding their way into science.

First, scientists face pressures that make them more likely to focus only on research that will lead to surprising, new results. As we have seen, science is a social practice that occurs in institutions like universities and national research centers. Scientists are professionals who get paid for teaching and for doing research. But university salaries are, in most cases, not enough to fund scientific research. Scientists need extra money to pay for scientific instruments and lab equipment, for experimental participants, and for their assistants. This extra money is generally awarded by public agencies like the ERC (European Research Council) in Europe and the NSF (National Science Foundation) and NIH (National Institute of Health) in the US. The competition is fierce: every year the number of applicants for funding grows, while, partly due to budget cuts, the number of available awards shrinks.

Scientists' ability to secure grants determines their career prospects. And their chance of securing grants depends on their quantity and quality of publications, frequency of citations, previously awarded grants, and the public attention they are able to attract. "Publish or perish" is the phrase coined to capture the increasing pressure in science to rapidly and continually publish research articles in order to sustain one's career. The competition for space in prestigious journals is also fierce; many journals reject over 90% of the articles submitted. And journal editors usually prefer to publish novel results that support an exciting hypothesis rather than negative results, even if they are robust and well documented, or studies replicating earlier findings. The tendency to publish surprising, new results more often than negative results, replication studies, and exploratory work is called *publication bias*.

Publication bias is common in all scientific journals but especially strong in the most prestigious journals. Consequently, scientists have a strong incentive to produce surprising, positive results. Their careers literally depend upon it! So, this is where scientists tend to put more of their attention, compared to writing up negative findings, replicating or assessing previous results, or even conducting preliminary, exploratory investigations.

Publication bias, combined with the scarcity of resources and employment opportunities, generates a challenge to science's capacity for self-correction. For one thing, science's openness to criticism depends on researchers publishing negative findings and attempting to replicate existing studies to see if the evidence holds up. Replication of previously published studies can increase the credibility of scientific claims when the supporting evidence for these claims is reproduced. When findings are not replicated, this can improve experimental designs or data analysis or even give scientists reason to rethink accepted theories. So, a disincentive from publishing negative findings and conducting replication studies undermines self-correction and the accumulation of trustworthy knowledge across science.

The scientific research associating specific foods with cancer risk, for example, may be seriously distorted by a lack of attention to replication. In Chapter 9, we discussed how statistically significant associations with cancer risk have been claimed for most food ingredients, from beef to tea. Careful analysis of this literature highlights that many published studies report implausibly large effects, even when the actual evidence is weak and effect sizes are small. Publishing negative results and greater emphasis on replicability would improve the internal and external validity of research about how different foods influence cancer risk.

Second, the incentive structure of current science also negatively impacts scientists' communication of their results. The emphasis on producing exciting, publishable results may lead scientists to cut corners in how experiments are designed and how data are analyzed and presented. For example, whether a conscious decision or not, scientists may not randomize group assignment in their experiment or may not control for some known confounding factors. Another common shortcut is ***data dredging***, where data mining techniques are used to uncover patterns in the data that support a hypothesis not under investigation. Data dredging increases the chance of a type I error (see Chapter 10), making it more likely that the endorsed hypothesis is wrong. Relatively few studies report effect sizes and measures of uncertainty in a transparent way, so it can be hard for others to assess the quality of a study.

Third, science's capacity for self-correction and thus its trustworthiness can be compromised by the financial interests funding scientific research. Fierce competition in academic science also is leading scientists to abandon academic research for jobs in industry. IT, AI, and pharmaceutical, chemical, and agricultural industries are hiring more and more scientists, often with better pay and greater job security. This raises another worry about sincere and transparent communication of scientific results. Being funded by a private company to carry out research may pose conflicts of interest, introducing funding bias. ***Funding bias*** is when researchers distort the results or modify conclusions of their study due to pressure from the study's funders. This can happen through the ways in which values influence science—what to study and how, what the aim is, how to handle uncertainty, and how to present the findings. This can also happen via intended or unintentional improper influence on data or methods. Regardless, funding bias leads to corporations having outsized influence on the nature of our scientific knowledge and, in some cases, even misleading the public with bad information.

Another challenge to science's trustworthiness relates to diversity of ideas and of practicing scientists. Diversity in scientific approaches fosters science's capacity for self-correction. But the institutional structure of contemporary scientific inquiry has reduced incentives for and freedom to pursue research challenging existing scientific ideas or moving in new directions. This has fostered increasing specialization in the sciences, and it has also shielded popular theories and methods from being challenged by competing, and perhaps better, theories and methods. Further, as we have seen earlier in this chapter, diversity among practicing scientists—in their social identities, worldviews, and more—is also important for science's capacity for self-correction. Yet the diversity of professional scientists remains limited.

Another challenge for science's trustworthiness relates to public communication. Too much science is inaccessible to the general public. Scientific studies get published by for-profit journals, and these journals typically put articles behind pricey paywalls. Academic institutions pay for their faculty and students to have journal access, but those without institutional affiliation are left without easy access. By the time science is reported in newspapers and popular magazines, it is often characterized inaccurately and misleadingly, full of hype and exaggeration. This too is due to an incentive structure, this time for journalism: media outlets are rewarded for splash and clicks, not for careful accuracy. This can fuel serious misunderstanding of scientific findings.

In this book, we have painted a picture of science as fallible, but with a formidable set of tools to help in the discovery of knowledge. Some scientific knowledge has had dramatic practical importance—just think about the outstanding progress of the medical sciences and of computer science and AI. Other knowledge regards fascinating aspects of the faraway universe and the strange behavior of microparticles. The ability of science to produce trustworthy knowledge requires openness to criticism and dissent, the pursuit of meaningful questions, and the communication of results in a sincere and transparent way. Only by ensuring these features are maintained and strengthened can science be self-correcting, generate objective knowledge, and thus deserve our trust.

EXERCISES

13.19 Recall: Describe three ways in which science has changed in the 21st century. Then, describe three important ways in which science is the same, even with these changes. For the second part of this question, you may want to consider topics from earlier in the book.

13.20 Think: Describe how big data and machine learning approaches are used in science. How transformative do you think these approaches will be for scientific methods and accomplishments? Provide justification for your response in 2–3 sentences.

13.21 Recall: List five current challenges for science described in this chapter. For each, indicate how it threatens science's objectivity and/or public trust in science.

13.22 Apply: List five potential ethical problems arising from scientific research funded by the pharmaceutical industry. For at least three of these problems, describe a concrete action to address that ethical problem that could be taken by government, pharmaceutical companies, or some other party.

13.23 Apply: Choose a current or recent example of societally important scientific research. Describe the nature of scientific uncertainty and what kinds of action had to be taken even with this uncertainty. Describe how social, economic, and moral considerations might have factored into the decision of how to proceed.

13.24 Apply: In recent years, there have been several retractions of published scientific articles that have captured the world's attention. In 2015, it was the retraction of a paper about gay marriage that was initially published in the prominent scientific journal *Science*. Read the description of this case on *Retraction Watch* <https://retractionwatch.com>, then answer the following questions.

   a. What risks do people who report misconduct in science (whistleblowers) face?

   b. Were human subjects "harmed" in this case, and if so, how?

   c. Describe how data management issues influenced this case.

   d. Describe how authorship issues influenced this case.

   e. Does this case raise any conflict of interest?

   f. What issues does the case raise about collaborating with others?

   g. Describe how replication issues influenced this case.

FURTHER READING

For a readable account of the wide variation of sex and reproductive strategies in the animal kingdom, see Roughgarden, J. (2004). *Evolution's rainbow*. University of California Press (reprint 2013).

On participation in science and relationship to the public, see Potochnik, A. (2024). *Science and the public*. Elements in Philosophy of Science, Cambridge University Press.

For an introduction to the roles values play in science, see Elliott, K. (2017). *A tapestry of values: An introduction to values in science*. Oxford University Press and (2022). *Science and values*. Elements in Philosophy of Science, Cambridge University Press.

On complexity, see Mitchell, M. (2000). *Complexity: A guided tour*. Oxford University Press.

Glossary

68–95–99.7 rule: the percentages of values that lie within one, two, and three standard deviations around the mean of a normal distribution

abductive inference: a type of nondeductive inference that attributes special status to explanatory considerations; also sometimes called *inference to the best explanation*

absolute risk: the number of individuals experiencing some condition in relation to the relevant population

abstraction: omitting or ignoring known features of a system from a representation or account of it

accuracy: the extent to which a model represents the actual features of its target system

addition rule: the probability that any of a number of outcomes will occur is the sum of their individual probabilities, subtracting the probability of multiple outcomes occurring

affirming the antecedent: a valid pattern of deductive inference in which a conditional statement and its antecedent are used as premises for concluding the consequent is true; also, *modus ponens*

affirming the consequent: an invalid pattern of deductive inference in which a conditional statement and its consequent are used as premises for concluding the antecedent is true

algorithm: a procedure for obtaining some outcome that halts in a finite number of steps

alternative hypothesis: in statistical hypothesis-testing, the alternative to the null hypothesis; the bold conjecture under investigation

ampliative inference: an inference in which the conclusion expresses content that surpasses what is present in the premises

analogical model: representations with features similar to focal features of a target system

anomaly: deviation from expectations that resist explanation by the reigning scientific theory; on Kuhn's view, motivation for scientific revolution and paradigm shifts

antecedent: in a conditional claim, a condition that guarantees some consequence; logically prior to the consequent

appeal to ignorance: concluding that a certain statement is true because there is no evidence proving that it is not true

appeal to irrelevant authority: appealing to the views of an individual who has no expertise in a field as evidence for some claim

applied research: scientific research used to develop some tangible output like techniques, software, drugs, or new materials; often, a core motivation is to generate products for profit

argument: a set of statements in which some (premises) are intended to provide rational support or empirical evidence in favor of another (the conclusion)

assumption: a specification that a target system must satisfy for a given model to be similar to it in the expected way

auxiliary assumptions: assumptions about how the world works that often go unnoticed but that are needed for a hypothesis or theory to have the expected implications

axiom: a statement accepted as a self-evident truth about some domain, used as a basis for deductively inferring other truths about the domain (see *theorem*)

background conditions: the physical, technological, and social aspects of an experiment or study

bar chart:visual representation of statistical distribution in which bars of different heights show amount for different values of a variable

base-rate fallacy:neglecting the base rate or prior probability of some occurrence and focusing only on the individual information at hand

basic research: scientific research that aims at knowledge for its own sake

Bayes factor: the ratio of the probability of the observation given one hypothesis to the probability of the observation given another hypothesis, i.e. $Pr(O|H_1)/Pr(O|H_2)$

Bayes net: a kind of causal model that uses joint probability distributions to provide a compact, visual representation of causal relationships and the strength of those relationships

Bayes's theorem: $Pr(H|O) = Pr(O|H)Pr(H)/Pr(O)$, also $Pr(H|O) = Pr(O|H)P(H)/Pr(O|H)P(H) + Pr(O|\text{not-}H)Pr(\text{not-}H)$ for opposed hypotheses; a mathematical result at the heart of Bayesian statistics

bell curve: see *normal distribution*

biased variable: a random variable for which some outcomes are more likely than others (unfair)

big data: very large data sets that cannot be easily stored, processed, analyzed, or visualized with traditional methods

bimodal distribution: two values in a range are the most common; in a histogram, there are two peaks

biological sex: the categories male, female, or intersex, which, in humans, are determined by X and Y chromosomes, hormones, reproductive organs, and other physical traits

blinding: when researchers or subjects are temporarily kept unaware of group assignment, hypotheses under test, or other experiment details; also called *masking*

calibration: comparing the measurements of one instrument with those of another to check the instrument's accuracy and adjusting it if needed

case study: a detailed examination of a single individual, group, system, or situation in a real-life context

causal account: the view that explanation involves appealing to causes that brought about the phenomenon to be explained

causal background: all the other factors that might causally influence an outcome, thereby also potentially affecting the causal relationship between the two events

causal Markov condition: the requirement that the probability of causal variables conditional on their parent causes are probabilistically independent of all their other ancestors; an assumption of causal Bayes nets

central limit theorem: the statistical claim that samples with a large enough size will have a mean approximating the mean of the population

cluster indicators: markers or features of some variable that allow researchers to more precisely define it without oversimplification

cohort study: a study in which researchers select a group of subjects sharing some defining trait and study them over time, in comparison to another group of subjects that is as similar as possible except without this trait

collectively exhaustive outcomes: when at least one outcome of a set of outcomes must occur at any given time

common cause: when neither of two covarying types of events causes the other but a third event causes both

computer simulation: computer program developed from data about a phenomenon to simulate its behavior; also *computer model*

conclusion: in an argument, a statement that is supported by the premises

conditional probability: the probability of an event's occurrence given that some other event has occurred; expressed $Pr(X|Y)$ where X is conditional on Y

conditional statement: a statement in which one circumstance, the antecedent, is given as a condition for another circumstance, the consequent; the antecedent guarantees the occurrence of the consequent

confidence interval: an interval within which the value of the variable should lie for a given percentage of possible samples

confirmation: a possible consequence of the H-D method; the observation matches the expectation based on the hypothesis, providing evidence in favor of the hypothesis

confirmation bias: the tendency to look for, interpret, and recall evidence in ways that confirm and do not challenge existing beliefs

conflict of interest: financial or personal gains that have the potential to inappropriately influence scientific research, results, or publication

confounding variables: extraneous variables that are not controlled and affect the relationship between the independent and dependent variables

conjunction fallacy: the error in judgment when we judge a conjunction of two events to be more likely than either one of the events on their own

consequent: the condition that arises from, or is guaranteed by, the antecedent

contributory cause: a cause that raises the probability of an outcome occurring but does not guarantee the outcome; also called a *partial cause*

control group: a group similar to the experimental group but that does not receive the intervention and experiences other value(s) of the independent variable

correlated variables: the value of one variable raises or lowers the probability of the other variable taking on some value

correlation coefficient: describes the direction and strength of correlation; a positive or negative sign indicates positive or negative correlation, and a number between 0 and 1 indicates the strength of the correlation

counterexample: a situation, real or imagined, in which the premises of an argument are true but the conclusion false; shows that a deductive argument is invalid

crucial experiment: an experiment that decisively adjudicates between two hypotheses

curve fitting: extrapolating from a data set to expected data by fitting a continuous line through a data plot; there are always multiple different lines consistent with the data

data: public records produced by observation, sensory experience, or some measuring device; allow observations to be recorded and compared

data cleansing: identifying and correcting errors in a data set by deciding which data are questionable and should be eliminated

data dredging: using data mining techniques to uncover patterns in the data that support a hypothesis not under investigation

data model: a regimented representation of data, often with the aim of discerning whether the data count as evidence for a given hypothesis

deception: when researchers actively misinform participants about some aspect of an experiment or study

deductive argument: an argument in which the truth of the premises is supposed to guarantee the truth of the conclusion; in a valid deductive argument this is so, in an invalid deductive argument it is not

degree of belief: amount of confidence in the truth of a given hypothesis

denying the antecedent: an invalid pattern of deductive inference in which a conditional statement and the denial of its antecedent are used as premises for concluding the consequent is also false

denying the consequent: a valid pattern of deductive inference in which a conditional statement and the denial of its consequent are used as premises for concluding the antecedent is also false; also *modus tollens*

dependent variable: a variable measured for changes after an intervention to an independent variable. The value of the dependent variable is anticipated to depend on, or be affected by, the independent variable

descriptive claim: a statement about how things are without making any value judgments

descriptive statistics:tools for summarizing, describing, and displaying data in a meaningful way

difference-making account of causation: the idea that if the occurrence of one event makes a difference to the occurrence of a second event, the first is a cause of the second event

direct variable control: when extraneous variables are all held at constant values during an intervention

directed acyclic graphs: abbreviated DAG; graphs in which all the causal relationships are one-directional (none of a cause's effects are also among its causes) and do not move in a circle (following a series of cause-effect relationships will not lead you back to an earlier cause as a later effect)

distal causes: causes that occurred further back in time from the effect, and perhaps further away as well

double-blind: an experiment or study in which both scientists and subjects are unaware of which subjects are in the experimental or control groups

Duhem-Quine problem: the idea that scientific hypotheses can never be tested in isolation but only against the background of auxiliary assumptions

ecological validity: the degree to which experiment circumstances are representative of real-world circumstances

effect size: a quantitative measure of the strength of a phenomenon

empirical investigation: inquiries that ground the justification for beliefs about the world in sensory information and observations

estimation: predicting properties of a population on the basis of a sample

eugenics: the idea that a human population can be improved by controlling breeding; historically linked to racist and classist science that threatened human liberties and dignity

evidence: anything that plays the role of supporting belief

evidentialism: the thesis that a belief's justification is determined by how well it is supported by evidence

exemplar: a model that is one of the target systems it is used to represent

expectation: a conjecture about observable phenomena, which are based on a hypothesis and which should be true if the hypothesis is true

experiment: a type of empirical investigation where researchers perform an intervention that changes some feature of a system and observe the effects, with the aim of understanding how the system works or why a certain outcome occurs

experimental group: a group that receives the intervention to the independent variable, or experiences the intended value of the independent variable

explanatory knowledge: sufficiently justified truths about how things work and why they are the way they are

exploratory experiment: an experiment that may not rely on existing theory and is not aimed to test a specific hypothesis; used to gather data to suggest novel hypotheses or to assess whether a poorly understood phenomenon actually exists

external experimental validity: the extent to which experimental results generalize from the experimental conditions to other conditions—especially to the phenomena the experiment is supposed to yield knowledge about

extraneous variables: all other variables besides the independent variable that may influence the value of the dependent variable; if uncontrolled, these may become confounding variables

fair: a random variable that has independent outcomes that are all equally likely

falsifiable: when it is possible to describe what kind of evidence would, if found, show a claim to be false; without being falsifiable, a claim would be unscientific

falsificationism: the thesis that scientific reasoning proceeds by attempting to disprove claims rather than to prove them right

field experiments: an experiment conducted outside of a laboratory, in the participants' everyday environment

frequency: how often some outcome occurs in a data set

frequency distribution: how often a variable takes on each value in some data set

frequency interpretation: the idea that the probability of an outcome is the limit of its relative frequency

funding bias: when researchers distort results or modify conclusions of their study due to pressure from the study's funders

gambler's fallacy: inferring from past variation from the expected frequency of outcomes that there will be future variation from the expected frequency in the opposite direction; for example, "the coin has landed on heads a lot; I bet it will land on tails next time."

gender: behaviors, social roles, appearances, and identities of individuals traditionally associated with the expression of masculinity, femininity, or non-binary features

generality: a model's applicability to a greater number of target systems

germ theory of disease: the theory that microbes like bacteria, viruses, and fungi, or other "germs" cause illnesses

histogram: a graphical display of data that uses bars of different heights to represent a frequency distribution

historical sciences: fields of science that investigate past events, such as archaeology, paleontology, and cosmology

hypothesis: a conjectural statement about what some aspect of the world is like, which is not (yet) backed by sufficient, or perhaps any, evidence

hypothetico-deductive method: to test a hypothesis, an expectation is deductively inferred from the hypothesis, then compared with an observation; violation of the expectation deductively refutes the hypothesis, while a match with the expectation nondeductively boosts support for the hypothesis

idealization: an assumption made without regard for whether it is true, often with full knowledge it is false

illusion of explanatory depth: believing that one understands the world more clearly and in greater detail than actually is the case

illusion of understanding: failure to appreciate the depth of one's ignorance about a topic when one lacks genuine understanding

independent outcomes: the probability of each outcome does not affect the probability of the other outcomes

independent variable: a variable that is changed or observed at different values in order to investigate the effect of the change

indigenous knowledge: true claims based on the observations, practices, and ideas developed about some geographic region by people native to the area, typically outside the institution of science

indirect variable control: extraneous variables are allowed to vary but researchers ensure the variation is independent from the value of the independent variable

inductive inference: an inferential relationship from premises to conclusion that is one of probability rather than necessity

inductive generalization: an inference to a general conclusion about a class of objects based on the observation of some number of objects in that class

inductive projection: an inference to a conclusion about the feature of some object that has not been observed based on the observation that some objects of the same kind have that feature

inductive risk: the risk of wrongly accepting or rejecting a scientific hypothesis

inference: a logical transition from one thought to another

inferential statistics: using statistical reasoning to draw broader conclusions on the basis of limited data

instruments: technological tools or other kinds of apparatus used in experiments

intelligent design: the idea that lifeforms are so complex that they couldn't possibly have come about without the help of an intelligent designer (such as the Judeo-Christian God)

internal experimental validity: the extent to which researchers can infer accurate conclusions about the relationship between the independent and dependent variables

intervention: a direct manipulation of the value of the independent variable

Islamic Golden Age: an important period in the development of science prior to the Scientific Revolution from roughly the 8th through 13th centuries, stretching from Central Asia to the Iberian Peninsula

joint probability distribution: the probability distribution for each of a set of variables, taking into account the probability of the other variables in the set

laboratory experiment: an experiment conducted in a laboratory, giving scientists control over interventions performed and direct and indirect control of many extraneous variables

life history: (biology) the traits and circumstances that affected survival and reproduction of members of a species

likelihood: the probabilities of getting the observed data given the truth of some specific hypothesis

logic: the study of the rules and patterns of inference

longitudinal study: an observational study in which observations are made of the same variables over time, in many cases over a long period of time

machine learning algorithms: step-by-step procedures run on powerful computers that enable scientists to mine large data sets for patterns or to perform other tasks

material conditional: a conditional statement (with an antecedent and consequent) that is false only if the antecedent can be true while the consequent is false

mathematical model: mathematical equations that use variables, parameters, and constants to represent one or more target systems

Matilda effect: the bias against recognizing women scientists' achievements, whose work is often uncredited or else attributed to their male colleagues instead

mean: a measure of the central tendency of a data set; the sum of all values for that variable in the data set divided by the number of instances

measurement standards: rules or norms for regulating the use of quantity concepts, which help create a meaningful scale to apply across instruments

mechanistic model: a model that represents the component parts and operations constituting some recurring process

median: a measure of the central tendency of a data set; the middle value in a distribution when the values are arranged from the lowest to the highest

meta-analysis: technique to combine multiple experiments or observational studies of the same hypothesis to strengthen the conclusions that can be drawn

metascience: using scientific techniques to study science itself

methodological omnivory: the use of multiple methods and specially tailored tools to generate evidence about specific targets

mode: a measure of the central tendency of a data set; the most frequent or numerous value in a data set

models of phenomena: models that represent target systems and are used to investigate those systems

modus ponens: See *affirming the antecedent*

modus tollens: See *denying the consequent*

monotonic inference: an inference that cannot be invalidated by the addition of new information

multiplication rule: the probability that two independent events both occur is the result of multiplying their individual probabilities

mutually exclusive: a set of values that a variable cannot take on simultaneously

natural experiment: an intervention on an independent variable that occurs naturally without experimenters influencing the system

natural phenomena: objects, events, or processes that are sufficiently uniform so as to be susceptible to systematic study

naturalism: the thesis that science provides natural explanations of natural phenomena

nature of science: the orientation, values, and methods that are specific to science and allow it to generate knowledge in the ways that it does

necessary cause: a cause that must occur in order for the effect to come about

necessary condition: a condition that must be satisfied for the occurrence of the specified outcome

negatively correlated: when greater values for one variable are related with smaller values for a second variable

neglect of probability: a bias in judgment or decision under uncertainty that comes from disregarding probabilistic information

no miracles argument: an abductive argument that the best explanation for the success of science is that mature scientific theories are approximately true

nodes: used to represent variables in causal Bayes nets

nomological account: the idea that explanation involves inferring statements of phenomena from scientific laws and statements of initial conditions

normal distribution: a unimodal distribution with the most common values clustered around the middle, with decreasingly common outcomes as the values get higher or lower

normal science: the most common phase of science, according to Kuhn; scientific research is stable, based on widespread agreement about basic assumptions; this follows either pre-paradigm science or scientific revolution

normative claim: a statement about how things ought to be, which might or might not correspond to how they in fact are

null hypothesis: a kind of default assumption; in statistical hypothesis-testing, often this just amounts to the hypothesis that nothing unusual is going on or that two variables are independent

observational study: collecting and analyzing data without performing interventions and sometimes without aiming to control extraneous variables

observation: any information gained from the senses—not only what one sees, but also what one hears, smells, touches, or any other way one experiences the world

observer-expectancy effect: the influence of a scientist's expectations on the behavior of experimental subjects

openness to falsification: the requirement that a claim be willingly abandoned when the preponderance of evidence indicates that it's false

operational definition: a specification of the conditions when some term applies, enabling measurement

outcome space: the set of all values a random variable can take on; also called *sample space*

outliers: measured values for a variable that are notably different from the other values in the data set

p-**value**: the probability of the observed data assuming the null hypothesis is true

paradigm: according to Kuhn, a way of practicing science; provides scientists with a stock of assumptions about the world, concepts and symbols for effective communication, methods for gathering and analyzing data, and other habits of research and reasoning

parameter: a quantity whose value can change in different applications of a mathematical equation but that only has a single value in any one application of the equation

participatory research: scientific research in which members of the public who aren't trained scientists participate; sometimes called *citizen science, community-based research,* or *action research*

pattern account: the view that explanation involves fitting statements of phenomena into a more general framework of laws and principles

perfectly controlled experiment: an experiment in which all variables are controlled except for the independent variable, an intervention is performed on the independent variable, and the effects on the dependent variable are measured; no confounding variables are possible

pessimistic meta-induction: reasoning inductively from the rejection of many past scientific theories to the conclusion that many of our current best scientific theories are also not true

phenomenological analysis: description and analysis of what some experience is like for a particular individual

phenomenon: an object, event, regularity, or process that exists or occurs; plural *phenomena*

philosophy of science: the investigation of science, focused especially on questions of what science should be like in order to be a trustworthy route to knowledge and to achieve the other ends we want it to have, such as usefulness to society

physicalism: the view that everything in the universe is composed of physical matter

physical constant: a quantity believed to be universal and unchanging over time

physical process account of causation: the idea that causation occurs when there is a continuous, physical process connecting a cause to its effect, such that a cause transmits its effect with the transfer of mass, energy, momentum, charge, or other physical properties

pie chart: visual representation of statistical distribution in which a circle is divided into different-sized slices to depict how much of the outcome space for a variable falls into different categories of values

placebo effect: the phenomenon of an experimental subject's expectations leading to the outcome the subject expects; this can be an extraneous variable

plagiarism: the fraudulent theft of someone else's ideas, scientific results, or words, which are subsequently presented as one's own work without giving proper credit

population: a large collection of entities that share some characteristic

population validity: the degree to which experimental entities are representative of the broader class of entities or population of interest

population variance: a measure of the variability of a data set; the average of the squared differences of values from the mean

positively correlated: when greater values for one variable are related with greater values for a second variable

posterior probability: the probability of a hypothesis conditional on an observation that has been made; Bayes's theorem can be used to calculate this

power of a statistical test: the probability that the test will enable the rejection of the null hypothesis

pre-paradigm: the earliest phase of science according to Kuhn; characterized by the existence of different schools of thought that debate very basic assumptions, including research methods and the nature and significance of data

precision: the extent to which a model finely specifies features of a target system

premise: a statement that provides support for some conclusion; the starting point for an inference

prior probability: the probability of an outcome without conditionalizing on known information (also: base rate)

probability: how often some outcome is expected to occur

probability distribution: how often a variable is expected to take on each of a range of values

probability theory: a mathematical theory developed to quantify uncertainty and to reason about random variables, or outcomes that are individually unpredictable but that behave in predictable ways over many occurrences

problem of induction: the concern that inductive inference cannot be logically justified, since any possible justification would need to employ inductive reasoning and would thus be circular

proximate cause: a cause that occurs closely in time and perhaps in space to its effect

pseudoscience: a nonscientific activity designed to look enough like science to deceive people into thinking it has scientific legitimacy

publication bias: the tendency to publish surprising, new results more often than negative results, replication studies, and exploratory work

qualitative data: information that is non-numerical and without some other standard that makes it easily comparable, such as diary accounts, unstructured interviews, and observations of animal behavior

qualitative variable: a variable with values that are not numerical but descriptive, such as the variable sport, with values basketball, hockey, and so on

quantitative data: data that is easily comparable, often in numerical form, such as numbers, vectors, or indices

quantitative variable: a variable with numerical values, such as height or percent correct on an exam

random sampling: the individuals composing the sample are selected randomly from the population

random variable: a variable whose values are individually unpredictable but predictable in the aggregate

randomization: the use of arbitrariness or some chance procedure like a lottery to assign experimental entities to experimental and control groups

rangethe difference between the smallest and largest values of variables in a data set.

rational degree of belief: the extent to which a rational agent should believe some claim

reasoning: a cognitive process of drawing inferences in support of some conclusion

reductionism: the view that appealing to the fundamental laws of physics will ultimately explain all the features of the world

refutation: a possible consequence of the H-D method; the observation contradicts the expectation deductively inferred from the hypothesis, so the hypothesis is deductively proven to be false

regression analysis: finding the best-fitting line through the points on a scatterplot

regression to the mean: a statistical phenomenon where, if a sample of values is extremely higher or lower than the mean, the next sample will tend to be closer to the mean

relative frequency distribution: a frequency distribution that records proportions of occurrences of each value of a variable rather than absolute numbers of occurrences

relative risk: a comparison between the incidence of a condition in two groups, generally expressed as a ratio or percentage

reliability: the extent to which an instrument accurately and consistently measures what it is supposed to measure

replication: performing an experiment or study again to check whether the result remains the same

representative sample: the experimental entities studied do not vary in any systematic way from the general population

retrograde motion: the historic astronomical observation of planets seeming to stop in their orbit, reverse course back across the sky, stop again, and reverse yet again to continue on their original path

robustness: a measure of insensitivity to features that differ between a model and the target system

robustness analysis: analyzing multiple models or different versions of a model to determine whether and to what extent their results are consistent

sample: a subset of the population of interest to be studied

sample mean: the most likely average value of the trait of interest in a population

sample size: the number of individual sources of data in a study, often the number of experimental entities or subjects

sample standard deviation: an estimate of the spread of the probability distribution for the random variable; $s = \sqrt{[\sum(value-mean)^2 / (n-1)]}$

sampling error: differences between the features of a sample and a population due to the unrepresentativeness of the sample

scale model: a concrete physical object that is a downsized or enlarged representation of its target system

scatterplot: a graph in which the values of one variable are plotted against the values of the other variable

science: an inclusive social project of developing natural explanations for natural phenomena; these explanations are tested using empirical evidence and should be subject to additional open criticism, testing, refinement, or even rejection

scientific breakthrough: a radical shift in theory in some field of science

scientific knowledge: explanatory knowledge of why or how the world is the way it is

scientific model: constructed to represent phenomena of interest and investigated to learn about those phenomena

scientific revolution: a period of cultural, social, and technological changes that unfolded in Europe roughly between 1550–1750, which inaugurated the modern institution of science; also, a radical change of a reigning theory being overturned in favor of a new theory, often involving an alternative worldview

scientific theory: a comprehensive account of phenomena, broader and more explanatory than individual hypotheses and models and backed by more evidence

scientism: a derogatory term for an excessive belief in science as a solution to every problem

self-explanation effect: the phenomenon in which generating explanations to oneself or to others can facilitate the integration of new information into existing bodies of knowledge, leading to deeper understanding

set: any grouping of elements in no particular order

sexual selection: occurs when a trait is valuable because it appeals to potential mating partners, so increases reproductive success; this can lead the trait to evolve to be more pronounced over time

significance level: how improbable, given the null hypothesis, an experimental result must be to warrant rejecting the null hypothesis

Simpson's paradox: a correlation between two events that disappears, or is reversed, when data are grouped in a different way

social determinates of health: social factors relevant to disease susceptibility and severity, such as education, income and benefits, housing conditions and exposure to pollution, pervasive stress from poverty, access to healthy food and activities, and more

social values: group priorities and moral ideas accepted in a community

sound inference: a valid deductive argument with all true premises

spurious correlation: two events are correlated but aren't causally related in any way

standard deviation: a measure of the variability of a data set; the square root of the variance; for a population, $\int = \sqrt{[\sum(\text{value} - \text{mean})^2 / n]}$

standard error: estimate of how much the standard deviation of the sampling distribution of the mean varies from the true population mean; $SE = s/\sqrt{(\text{sample size})}$

statistically independent: two events for which the occurrence of one does not increase or decrease the probability of the other; that is, when $Pr(Y|X) = Pr(Y)$ and $Pr(X|Y) = Pr(X)$

statistically significant: in null hypothesis significance testing, the outcome is found to be unlikely enough if the null hypothesis is true that it provides grounds for rejecting the null hypothesis

statistics: a set of tools broadly used in science to systematically collect, curate, analyze, present, and interpret data

strength of correlation: how predictable the values of one variable are based on the values of the other variable

strength of inductive inference: the probability that the conclusion is true assuming that all the premises are true

subjects: humans, non-human animals, or inanimate objects in an experiment or non-experimental study; also called experimental entities or participants

subtraction rule: the probability that some outcome doesn't occur is the result of subtracting the probability of that outcome from the total probability ($Pr=1$)

sufficient cause: a cause that always brings about the effect

sufficient condition: a condition that, if met, guarantees the occurrence of the specified outcome

super-observational access: using tools to enhance our powers of observation beyond what they ordinarily include

target system: the real-world system that scientists want to study using a model

testimony: a spoken or written statement that something is the case

The Scientific Revolution: beginning with the work of Copernicus and ending with the work of Newton; a fundamental transformation in ideas about how knowledge claims ought to be justified, which led to the development of the scientific method

theorem: a statement deductively inferred from a set of axioms

theoretical claims: claims made about entities, properties, or occurrences that are not directly observable

thought experiment: an imagined intervention on an imagined system to learn about the role of the independent variable in the real world; may supplement or replace empirical evidence

total probability: the probability of the full outcome space for any random variable; always 1

tractability: the degree of ease in developing or using a model

triangulation: conducting an experiment or analyze data using different instruments or techniques to detect any variation depending on instruments or experimental design

type I error: a false positive; the erroneous rejection of a null hypothesis

type II error: a false negative; failing to reject the null hypothesis when it is actually false

underdetermination: when evidence is insufficient to determine which of multiple theories or hypotheses is true

understanding: grasping why or how something came about or is the way it is

uniform distribution: all values in a range are equally likely; a histogram shows a flat line

uniformity of nature: the idea that the natural world is sufficiently uniform, or unchanging, that we are justified in thinking our future experiences will resemble our past experiences

unimodal distribution: one value in a range is the most common; in a histogram, there is one peak

valid inference: a deductive inference in which the truth of the premises logically guarantees the truth of the conclusion

value of a variable: the particular state or quantity of a variable in some instance

value-free ideal: the thesis that scientific reasoning should proceed free from the influence of our values—such as moral, social, or political beliefs

variability: the distribution of values in a data set

variable: anything that can vary, change, or occur in different states

variable control: creating conditions such that no extraneous variable can change the value of a dependent variable during or as a result of an intervention on the independent variable

zero-risk bias: a preference for policies that entirely eliminate a risk instead of alternatives that more effectively reduce risk (by being cheaper, easier to implement, etc.)

Bibliography

CHAPTER 0

The discussion of HPV vaccination draws from Spayne, J., & Hesketh, T. (2021). Estimate of global human papillomavirus vaccination coverage: Analysis of country-level indicators. *BMJ Open*, *11*(9), e052016.

CHAPTER 1

On Manson's early climate change research see Manson, M. (1893). *Geological and solar climates— their causes and variations*. G. Spaulding & Co. The Arrhenius quote in section 1.1 comes from p. 54 of Arrhenius, S. (1908). *Worlds in the making: The evolution of the universe*. Harper & Brothers. The Callendar quote comes from p. 38 of Callendar, G. S. (1939). The composition of the atmosphere through the ages. *Meteorological Magazine*, *74*(878), 33–39. Historic comparisons of atmospheric carbon dioxide are supported by Ahmed, M. et al. (2013). Continental-scale temperature variability during the past two millennia. *Nature Geoscience*, 6, 339–346. Global mean temperature comparisons are drawn from NOAA National Centers for Environmental Information, Monthly Global Climate Report for Annual 2022, published online January 2023. www.ncei.noaa.gov/access/monitoring/monthly-report/global/202213/supplemental/page-3. For expert agreement about climate change see Anderegg, W. R. L., Prall, J. W., Harold, J., & Schneider, S. H. (2010). Expert credibility in climate change. *Proceedings of the National Academy of Sciences*, *107*, 12107–12110. On public acceptance of climate change see Lee, T. M., Markowitz, E. M., Howe, P. D., Ko, C. Y., & Leiserowitz, A. A. (2015). Predictors of public climate change awareness and risk perception around the world. *Nature Climate Change*, *5*(11), 1014–1020. On the illusion of explanatory depth see Keil, F. C. (2003). Folkscience: Coarse interpretations of a complex reality. *Trends in Cognitive Sciences*, *7*(8), 368–373. On falsification, see Popper, K. (1963). *Conjectures and refutations: The growth of scientific knowledge*. Routledge and Kegan Paul. The research on confirmation bias regarding views on the death penalty is Lord, C. G., Ross, L., & Lepper, M. R. (1979). Biased assimilation and attitude polarization: The effects of prior theories on subsequently considered evidence. *Journal of Personality and Social Psychology*, *37*(11), 2098–2109. https://doi.org/10.1037/0022-3514.37.11.2098. The discussion of Clever Hans draws from Pfungst, O. (1911). *Clever Hans (The horse of Mr. von Osten): A contribution to experimental animal and human psychology* (C. L. Rahn, Trans.). Henry Holt. (Originally published in German, 1907.) In the discussion of fabricated scientific results we refer to *Retraction Watch*. Tracking retractions as a window into the scientific process, http://retractionwatch.

com/. The discussion of the checklist approach to defining science is influenced by the Understanding Science website, https://undsci.berkeley.edu. In the discussion of pseudoscience, we use an example from Oreskes, N., & Conway, E. (2010). *Merchants of doubt.* Bloomsbury. Box 1.2 draws from Merton, R. K. (1942). A note on science and democracy. *Journal of Legal and Political Sociology, 1,* 115–126; the quote is from p. 117.

CHAPTER 2

The Covid-19 gender disparity research we consulted included Jin, J. M., Bai, P., He, W., Wu, F., Liu, X. F., Han, D. M., Liu, S., & Yang, J. K. (2020). Gender differences in patients with COVID-19: Focus on severity and mortality. *Frontiers in Public Health, 8,* 152. https://doi.org/10.3389/fpubh.2020.00152; Takahashi, T., Ellingson, M. K., Wong, P. et al. (2020). Sex differences in immune responses that underlie COVID-19 disease outcomes. *Nature, 588,* 315–320. https://doi.org/10.1038/s41586-020-2700-3; and Danielsen, A. C., Lee, K. M. N., Boulicault, M., Rushovich, T., Gompers, A., Tarrant, A., Reiches, M., Shattuck-Heidorn, H., Miratrix, L. W., & Richardson, S. S. (2022). Sex disparities in COVID-19 outcomes in the United States: Quantifying and contextualizing variation. *Social Science & Medicine, 294.* https://doi.org/10.1016/j.socscimed.2022.114716. The brief discussion of race in medicine draws from Tsai, J. (2018). What role should race play in medicine? *Scientific American.* https://blogs.scientificamerican.com/voices/what-role-should-race-play-in-medicine/. Discussion of social determinates of health was influenced by Yong, E. (2021). How public health took part in its own downfall. *The Atlantic.* www.theatlantic.com/health/archive/2021/10/how-public-health-took-part-its-own-downfall/620457/. Box 2.1 draws from Wallisch, P. (2020). How to read a scientific article: The QDAFI method of structured relevant gist. Critical reading across the curriculum: Volume 2. *Social and Natural Sciences,* 152–164.

CHAPTER 3

The case study of scientific research on light draws from Newton, I. (1704/1998). *Opticks: Or, a treatise of the reflexions, refractions, inflexions and colours of light.* Also two treatises of the species and magnitude of curvilinear figures. Commentary by Nicholas Humez (Octavo ed.). Octavo; Schaffer, S. (1989). Glass works: Newton's prisms and the uses of experiment. In D. Gooding, T. Pinch, & S. Schaffer (Eds.), *The uses of experiment: Studies in the natural sciences* (pp. 67–104). Cambridge University Press; Al-Khalili, J. (2015). In retrospect: Book of optics. *Nature, 518*(7538), 164–165; Herschel, W. (1801). Observations tending to investigate the nature of the Sun, in order to find the causes or symptoms of its variable emission of light and heat; With remarks on the use that may possibly be drawn from solar observations. *Philosophical Transactions of the Royal Society of London, 91,* 265–318. The urine-fertilizer example is inspired by Wald, C. (2022). The urine revolution: How recycling pee could help to save the world. *Nature, 602*(7896), 202–206. The discussion of population validity considers as an example Simon, V. (2005). Wanted: Women in clinical trials. *Science, 308*(5728), 1517–1517. Discussion of Eddington's experiment testing Einstein's relativity theories is based on Dyson, F. W., Eddington, A. S., & Davidson, C. R. (1920). A determination of the deflection of light by the sun's gravitational field, from observations made at the solar eclipse of May 29, 1919. *Philosophical Transactions of the Royal Society A, 220,* 571–581. The discussion of Planck's constant draws Haddad, D., Seifert, F., Chao, L. S., Possolo, A., Newell, D. B., Pratt, J. R., . . . Schlamminger, S. (2017). Measurement of the Planck constant at the National Institute of Standards and Technology from 2015 to 2017. *Metrologia, 54*(5), 633. The Herschel case study draws from Herschel, W. (1801). XIII. Observations tending to investigate the nature of the sun, in order to find the causes or symptoms of its variable emission of light and heat; with remarks on the use that may possibly be drawn from solar observations. *Philosophical Transactions of the Royal Society of London*, (91), 265–318. Box 3.2 draws from Yong, E. (2012). Nobel laureate challenges psychologists to clean up their act. *Nature.* https://doi.org/10.1038/nature.2012.11535; Romero, F. (2019). Philosophy of science and the replicability crisis. *Philosophy Compass, 14*(11), e12633.

CHAPTER 4

Section 4.1's case study is based on Wooller, M. J., Bataille, C., Druckenmiller, P., Erickson, G. M., Groves, P., Haubenstock, N., . . . Willis, A. D. (2021). Lifetime mobility of an Arctic woolly mammoth. *Science, 373*, 806–808; Miller, J. H., Fisher, D. C., Crowley, B. E., Secord, R., & Konomi, B. A. (2022). Male mastodon landscape use changed with maturation (late Pleistocene, North America). *Proceedings of the National Academy of Sciences, 119*(25), e2118329119. Examples of non-experimental studies in section 4.2 are drawn from Broca, P. (1861). Remarques sur le siège de la faculté du langage articulé, suivies d'une observation d'aphémie (perte de la parole). *Bulletins de la Société d'anatomie, 2e serie, 6*, 330–357; Chattopadhyay, R., & Duflo, E. (2004). Women as policy makers: Evidence from a randomized policy experiment in India. *Econometrica, 72*(5), 1409–1443; Dockery, D. W., Pope, C. A., Xu, X., Spengler, J. D., Ware, J. H., Fay, M. E., & Speizer, F. E. (1993). An association between air pollution and mortality in six US cities. *New England Journal of Medicine, 329*(24), 1753–1759; Harlow, J. M. (1848). Passage of an iron rod through the head. *Boston Medical and Surgical Journal, 39*, 389–393; Harlow, J. M. (1868). Recovery from the passage of an iron bar through the head. *Publications of the Massachusetts Medical Society, 2*, 327–347; Kogan, V., & Lavertu, S. (2021). *The COVID-19 pandemic and student achievement on Ohio's third-grade English language arts assessment* (p. 14). Early Childhood Longitudinal Study Program. The Ohio State University. https://nces.ed.gov/ecls/. The phenomenological analysis discussion in section 4.2 draws from Carel, H. (2011). Phenomenology and its application in medicine. *Theoretical Medicine and Bioethics, 32*(1), 33–46; Gallagher, S., & Sørensen, J. B. (2006). Experimenting with phenomenology. *Consciousness and Cognition, 15*(1), 119–134; Carel, H. (2011). Phenomenology and its application in medicine. *Theoretical Medicine and Bioethics, 32*, 33–46. The discussion of thought experiments draws from Brown, J. R., & Fehige, Y. (2022). Thought experiments. In E. N. Zalta & U. Nodelman (Eds.), *The Stanford encyclopedia of philosophy* (Winter 2022 ed.). https://plato.stanford.edu/archives/win2022/entries/thought-experiment/; Bokulich, A. (2001). Rethinking thought experiments. *Perspectives on Science, 9*, 285–307; (1638). *Discourses and mathematical demonstrations concerning two new sciences* [Discorsi e Dimostrazioni Matematiche intorno a Due Nuove Scienze, Elsevier, Leiden]; The System of the World, London, 1728. The original version of the third book of the Principia, retitled by the translator and reissued in reprint form, London: Dawsons of Pall Mall, 1969. Computer simulation discussion based on Adam, D. (2020). Special report: The simulations driving the world's response to COVID-19. *Nature, 580*(7802), 316–319. Big data discussion is influenced by Floridi, L. (2012). Big data and their epistemological challenge. *Philosophy and Technology, 25,*

435–437; Lazer, D., Kennedy, R., King, G., & Vespignani, A. (2014). The parable of Google Flu: Traps in big data analysis. *Science, 343*(6176), 1203–1205; Dastin (2018, October 10). Amazon scraps secret AI recruiting tool that showed bias against women. *Reuters.* Section 4.4 on meta-analysis is based on Kovaka, K. (2022). Meta-analysis and conservation science. *Philosophy of Science, 89*(5), 980–990; Wernsdorff, M. von et al. (2021). Effects of open-label placebos in clinical trials: A systematic review and meta-analysis. *Science, 11.*

CHAPTER 5

Section 5.1's case study of the Bay model and many other points in the chapter draw from Weisberg, M. (2013). *Simulation and similarity: Using models to understand the world.* Oxford University Press. Examples of models in sections 5.2 and 5.3 are drawn from (prisoner's dilemma) Axelrod, R. (1984). *The evolution of cooperation.* Basic Books; Rapoport, A., Seale, D. A., & Colman, A. M. (2015). Is tit-for-tat the answer? On the conclusions drawn from Axelrod's tournaments. *PLoS ONE, 10*(7), e0134128. (Lotka-Volterra model) Volterra, V. (1928). Variations and fluctuations of the number of individuals in animal species living together. *Journal du Conseil. Conseil Permanent International pour l'Exploration de la Mer, 3,* 3–51. (DNA) Watson, J. D. (1968). *The double helix.* Atheneum Press. (MONIAC) Morgan, M., & Boumans, M. J. (2004). Secrets hidden by two-dimensionality: The economy as a hydraulic machine. In S. de Chadarevian & N. Hopwood (Eds.), *Model: The third dimension of science* (pp. 369–401). Stanford University Press. The discussion of idealization draws from Potochnik, A. (2017). *Idealization and the aims of science.* University of Chicago Press; McMullin, E. (1985). Galilean idealization. *Studies in the History and Philosophy of Science, 16,* 247–273. Section 5.4 draws from Levins, R. (1966). The strategy of model building in population biology. *American Scientist, 54,* 421–431. Box 5.1 draws from Winsberg, E., & Harvard, S. (2022). Purposes and duties in scientific modelling. *Journal of Epidemiology and Community Health, 76,* 512–517.

CHAPTER 6

The discussion of Aristotle's views draws on John Philoponus's (1987). *Against Aristotle on the eternity of the world* (C. Wildberg, Trans.). Bristol Classical Press. Hubble's (1929) paper "A relation between distance and radial velocity among extra-galactic nebulae" was published in *Proceedings of the National Academy of Sciences, 15*(3), 168–173. See also Kragh, H., & Smith, R. W. (2003). Who discovered the expanding universe? *History of Science, 41*(2), 141–162. For more on Hubble's discovery from a nonspecialist's perspective, see Fox, K. C. (2002). *The big bang theory.* Wiley. The rehearsal of Semmelweis's case comes from his (1861/1983) book *The etiology, the concept and the prophylaxis of childbed fever,* and is also discussed in Hempel, C. (1966). *Philosophy of natural science.* Prentice-Hall. Box 6.3 draws from Wason, P. (1968). Reasoning about a rule. *Quarterly Journal of Experimental Psychology, 20,* 273–281; Cosmides, L. (1989). The logic of social exchange: Has natural selection shaped how humans reason? Studies with the Wason selection task. *Cognition, 31*(3), 187–276. Discussion of the JWST reflected in the extensive NASA site: https://webb.nasa.gov/.

CHAPTER 7

Discussion of the case of poisoned water in Flint, MI, derives from various journalistic sources, including reporting from *Politico*, *NPR*, and elsewhere. CNN has maintained a list of facts and timeline here: www.cnn.com/2016/03/04/us/flint-water-crisis-fast-facts/index.html. The URL for the US EPA report from (2015), "High Lead Levels in Flint, Michigan" is www.epa.gov/sites/production/files/2015-11/documents/transmittal_of_final_redacted_report_to_mdeq.pdf The locus classicus for the problem of induction comes from Hume, D. (1748/1999). *An enquiry concerning human understanding* (T. L. Beauchamp, Ed.). Oxford University Press. The discussion of abductive inference draws on Peirce, C. S. (1903/1904) (1931–1936). *The collected papers, volumes 1–6* (C. Hartshorne & P. Weiss, Eds.). Harvard University Press. See also Huygens, C. (1690/1962). *Treatise on light* (S. P. Thompson, Trans.). Dover Publications. The case of continental drift owes to Wegener, A. (1929/1966). *The origin of continents and oceans*. Dover. Sources for the Jebel Irhoud remains come from Callaway, E. (2017). Oldest *Homo sapiens* fossil claim rewrites our species' history. *Nature News*. https://doi.org/10.1038/nature.2017.22114; see also Hublin, J. J. et al. (2017). New fossils from Jebel Irhoud, Morocco and the pan-African origin of *Homo sapiens*. *Nature, 546*(7657), 289–292.

CHAPTER 8

There are many introductions to the theory of probability and its uses in science and everyday life. One enjoyable and accessible introduction is Olofsson, P. (2015). *Probabilities: The little numbers that rule our lives*. John Wiley & Sons. The rapid strep test example in section 8.1 is an example of the base-rate fallacy (cf., https://en.wikipedia.org/wiki/Base_rate_fallacy), which is discussed, among many others, by Gigerenzer, G., & Edwards, A. (2003). Simple tools for understanding risks: From innumeracy to insight. *BMJ, 327*(7417), 741–744. The estimates we refer to about the carriage rate of strep (*Streptococcus pharyngitis*) can be found in: Martin, J. (2016). The *Streptococcus pyogenes* carrier state. In J. J. Ferretti, D. L. Stevens, & V. A. Fischetti (Eds.), *Streptococcus pyogenes: Basic biology to clinical manifestations*. The University of Oklahoma Health Sciences Center; and Oliver, J., Malliya Wadu, E., Pierse, N., Moreland, N. J., Williamson, D. A., & Baker, M. G. (2018). Group A Streptococcus pharyngitis and pharyngeal carriage: A meta-analysis. *PLoS Neglected Tropical Diseases, 12*(3), e0006335. Exercise 8.12 is based on Tversky, A., & Kahneman, D. (1982). Judgments of and by representativeness. In D. Kahneman, P. Slovic, & A. Tversky (Eds.), *Judgment under uncertainty: Heuristics and biases* (pp. 84–98). Cambridge University Press. Exercise 8.16 is based on Gigerenzer, G. (2008). *Rationality for mortals: How people cope with uncertainty* (pp. 174ff). Oxford University Press. The three probabilistic biases described in Box 8.2, viz. the conjunction fallacy, neglect of probability, and zero-risk bias, have been widely studied in psychology, economics, law, and philosophy. Seminal studies of each include Tversky, A., & Kahneman, D. (1983). Extensional versus intuitive reasoning: The conjunction fallacy in probability judgment. *Psychological Review, 90*(4), 293–315. Sunstein, C. R. (2002). Probability neglect: Emotions, worst cases, and law. *Yale Law Journal, 112*(1), 61–107. And Allais, M. (1953). Le comportement de l'homme rationnel devant le risque: critique des postulats et axiomes de l'école américaine. *Econometrica, 21*(4), 503–546. There are many introductions to the interpretations of probability discussed in section 8.4. A book-length treatment is Gillies, D. (2000). *Philosophical theories of probability*. Routledge. A shorter, article-length introduction is Galavotti, M.

C. (2015). Probability theories and organization science: The nature and usefulness of different ways of treating uncertainty. *Journal of Management, 41*(2), 744–760.

CHAPTER 9

There are many introductions to statistical reasoning. One comprehensive resource relevant to the content of Chapters 9 and 10 is Freedman, D., Pisani, R., & Purves, R. (2007). *Statistics* (4th ed.). W. W. Norton & Company. A helpful, free, online textbook is Daniel Lakens's *Improving your statistical inferences.* https://lakens.github.io/statistical_inferences/. The statistics about world religions presented in section 9.1 come from recent reports by the PEW Research Center (www.pewresearch. org/topic/religion/): Pew Research Center, November 10, 2020, "In 2018, government restrictions on religion reach highest level globally in more than a decade"; Pew Research Center, April 11, 2017, "Global restrictions on religion rise modestly in 2015, reversing downward trend"; Pew Research Center, April 5, 2017, "The changing global religious landscape." The research about loud bars and alcohol consumption in section 9.2 is reported in Guéguen, N., Jacob, C., Le Guellec, H., Morineau, T., & Lourel, M. (2008). Sound level of environmental music and drinking behavior: A field experiment with beer drinkers. *Alcoholism: Clinical and Experimental Research, 32*(10), 1795–1798. Part of Galton's work on correlation can be found in Galton, F. (1889). *Natural inheritance.* Macmillan. Box 9.1 on the history of statistics and eugenics is based on Bodmer, W., Bailey, R. A., Charlesworth, B., Eyre-Walker, A., Farewell, V., Mead, A., . . . Senn, S. (2021). The outstanding scientist, R.A. Fisher: His views on eugenics and race. *Heredity, 126,* 565–576. https://doi.org/10.1038/s41437-020-00394-6; Gillham, N. W. (2001). Sir Francis Galton and the birth of eugenics. *Annual Review of Genetics, 35*(1), 83–101; Wijsen, L. D., Borsboom, D., & Alexandrova, A. (2022). Values in psychometrics. *Perspectives on Psychological Science, 17*(3), 788–804. The medical example in section 9.4 concerning the impact of relative vs. absolute risks on understanding risk comes from Kurz-Milcke, E., Gigerenzer, G., & Martignon, L. (2008). Transparency in risk communication: Graphical and analog tools. *Annals of the New York Academy of Sciences, 1128*(1), 18–28. The bar graphs in this section are based on their Figure 4. Another concise and helpful reference here is Spiegelhalter, D., Pearson, M., & Short, I. (2011). Visualizing uncertainty about the future. *Science, 333*(6048), 1393–1400. The study of correlations between cancer and food ingredients in section 9.4 is Schoenfeld, J. D., & Ioannidis, J. P. (2013). Is everything we eat associated with cancer? A systematic cookbook review. *The American Journal of Clinical Nutrition, 97*(1), 127–134. The case of IQ, education, and SAT scores in the US is based on Sackett, P. R., Kuncel, N. R., Arneson, J. J., Cooper, S. R., & Waters, S. D. (2009). Does socioeconomic status explain the relationship between admissions tests and post-secondary academic performance? *Psychological Bulletin, 135*(1), 1–22. Nisbett, R. E. (2013). Schooling makes you smarter: What teachers need to know about IQ. *American Educator, 37*(1), 10–39. Ritchie, S. J., & Tucker-Drob, E. M. (2018). How much does education improve intelligence? A meta-analysis. *Psychological Science, 29*(8), 1358–1369. On IQ and intelligence more generally, a helpful resource is Ritchie, S. (2015). *Intelligence: All that matters.* John Murray. Box 9.2 on panic headlines refers to "Panic Headlines: Alcohol Changes Face Shape, Induction Lowers IQ," ParentData by Emily Oster newsletter, e-mailed March 9, 2023. The GDP and austerity case in section 9.4 concerns research reported in Reinhart, C. M., & Rogoff, K. S. (2010). Growth in a time of debt. *American Economic Review, 100*(2), 573–578. Websites containing information about several real-life examples of bad data visualizations, which we discuss in section 9.4, include www.statisticshowto.com/misleading-graphs/ and https://venngage.com/blog/misleading-graphs/#Using-the-wrong-graph. On article-length advice about data visualizations, see Wolfe, J. (2015). Teaching students to focus on the data in data visualization. *Journal of Business and Technical Communication*, 29(3), 344–359. Chen, C. (2010). Information visualization. *Wiley Interdisciplinary Reviews: Computational Statistics*, 2(4), 387–403. The research on statistical reporting inconsistencies in psychology referenced in section 9.4 is by Nuijten, M. B., Hartgerink, C. H., Van Assen, M. A., Epskamp, S., & Wicherts, J. M. (2016). The prevalence of statistical reporting errors in psychology (1985–2013). *Behavior Research Methods*, 48, 1205–1226. Similar research showing a lower rate of misreporting in experimental philosophy compared to psychology is by Colombo, M., Duev, G., Nuijten, M. B., & Sprenger, J. (2018). Statistical reporting inconsistencies in experimental philosophy. *PLoS One*, 13(4), e0194360. Exercise 9.23 is based on https://marketinglaw.osborneclarke.com/retailing/colgates-80-of-dentists-recommend-claim-under-fire/ and Exercise 9.24 is based on www.washingtonpost.com/politics/2021/09/28/tucker-carlson-ties-his-vaccine-fearmongering-core-republican-insecurity/

CHAPTER 10

There are many introductions to statistical reasoning. Relevant to the content of this chapter and Chapter 9 is Freedman, D., Pisani, R., & Purves, R. (2007). *Statistics* (4th ed.). W. W. Norton & Company. A helpful, free, online textbook is Daniel Lakens's *Improving your statistical inferences.* https://lakens.github.io/statistical_inferences/ The discovery of the Higgs boson described in section 10.1 is by Chatrchyan, S., Khachatryan, V., Sirunyan, A. M., Tumasyan, A., Adam, W., Aguilo, E., . . . Friedl, M. (2012). Observation of a new boson at a mass of 125 GeV with the CMS experiment at the LHC. *Physics Letters B*, 716(1), 30–61. The 68–95–99.7 rule we cover in section 10.2 is explained in more detail in Pukelsheim, F. (1994). The three sigma rule. *The American Statistician*, 48(2), 88–91. Our discussion of confidence intervals on page is especially indebted to Hoekstra, R., Morey, R. D., Rouder, J. N., & Wagenmakers, E. J. (2014). Robust misinterpretation of confidence intervals. *Psychonomic Bulletin & Review*, 21(5), 1157–1164. The example of nonrandom sampling relies on Squire, P. (1988). Why the 1936 Literary Digest poll failed. *Public Opinion Quarterly*, 52(1), 125–133. The tea tasting experiment described in section 10.3 is from Fisher, R. A. (1956). Mathematics of a lady tasting tea. In J. R. Newman (Ed.), *The world of mathematics* (pp. 1512–1521). Simon & Schuster. (Original work published in Fisher, R. A. (1935). *The design of experiments*. Oliver and Boyd.) The discussion of statistical significance is indebted especially to Gelman, A., & Stern, H. (2006). The difference between "significant" and "not significant" is not itself statistically significant. *The American Statistician*, 60(4), 328–331. For the association between genes and smoking behavior, see Thorgeirsson, T. E., Gudbjartsson, D. F., Surakka, I., Vink, J. M., Amin, N., Geller, F., . . . Gieger, C. (2010). Sequence variants at CHRNB3-CHRNA6 and CYP2A6 affect smoking behavior. *Nature Genetics*, 42(5), 448–453. The criticism in section 10.4 that NHST says nothing about the probability different hypotheses are true is partly based on work by Cohen, J. (1994). The earth is round (p < .05). *American Psychologist*, 49(12), 997–1003. The second criticism of NHST is partly based on Lindley, D. V. (1993). The analysis of experimental data: The appreciation of tea and wine. *Teaching Statistics*, 15, 22–25. The problem of the prior we discuss is based on Gelman, A., & Hennig, C. (2017). Beyond subjective and objective in statistics (with discussion). *Journal of the Royal Statistical Society*, 180(4), 967–1033. https://doi.org/10.1111/rssa.12276.

CHAPTER 11

For the statistical information on poverty in section 11.1 we relied on United Nations.Department of Economic and Social Affairs. (2022). *The sustainable development goals: Report 2022*. UN. https://unstats.un.org/sdgs/report/2022. Causal evidence about the relationships between poverty and psychiatric conditions is discussed in Ridley, M., Rao, G., Schilbach, F., & Patel, V. (2020). Poverty, depression, and anxiety: Causal evidence and mechanisms. *Science, 370*(6522), eaay0214. Another key reference here is: Sen, A. (1983). Poor, relatively speaking. *Oxford Economic Papers, 35*(2), 153–169. The experiment on poverty in Niger is by Banerjee, A., Duflo, E., Goldberg, N., Karlan, D., Osei, R., Parienté, W., & Udry, C. (2015). A multifaceted program causes lasting progress for the very poor: Evidence from six countries. *Science, 348*(6236), 1260799. Evidence about the bearing of cash transfers on temptation goods is discussed in Evans, D. K., & Popova, A. (2017). Cash transfers and temptation goods. *Economic Development and Cultural Change, 65*(2), 189–221. One of the major works in philosophy, where Hume engaged with questions about causation, is Hume, D. (1738/2007). *A treatise of human nature* (D. F. Norton & M. J. Norton, Eds.). Clarendon Press. Evidence that "poverty in childhood can have effects much later in life, and poverty rates are influenced by geopolitical events that occur far away in time and space" is presented in Duncan, G. J., Magnuson, K., & Votruba-Drzal, E. (2017). Moving beyond correlations in assessing the consequences of poverty. *Annual Review of Psychology, 68*, 413–434. The Israeli case in Box 11.1 is described in Wadman, M. (2021). A grim warning from Israel: Vaccination blunts, but does not defeat Delta. *Science, 373*(6557), 838–839. For more on the Simpson's paradox, see Sprenger, J., & Weinberger, N. (2021). Simpson's paradox. In E. N. Zalta (Ed.), *The Stanford encyclopedia of philosophy* (Summer 2021 ed.). https://plato.stanford.edu/archives/sum2021/entries/paradox-simpson/. One key reference for the difference making account of causation in section 11.2 is Woodward, J. (2003). *Making things happen: A theory of causal explanation*. Oxford University Press. One reference on physical process theories of causation is Salmon, W. (1984). *Scientific explanation and the causal structure of the world*. Princeton University Press. The methods presented in Box 11.3 were originally developed in Mill, J. S. (1893). *A system of logic, ratiocinative and inductive: Being a connected view of the principles of evidence and the methods of scientific investigation*. Harper & brothers. The scenario described in section 11.4 is a direct quotation from page 30 of Korb, K., & Nicholson, A. (2010). *Bayesian artificial intelligence* (2nd ed.). Chapman & Hall/CRC. Scientific applications of Bayes nets are explained and discussed in more detail in e.g. Glymour, C. (2007). When is a brain like the planet? *Philosophy of Science, 74*(3), 330–347. The example of the two lamps and the principle of common cause is from Reichenbach, H. (1956). *The direction of time* (p. 157). University of California Press. Many cases of "spurious" correlations often explained by the presence of some common cause are discussed in Vigen, T. (2015). *Spurious correlations*. Hachette Books. Several fun examples of spurious correlations are also available at the website: www.tylervigen.com/spurious-correlations

CHAPTER 12

The case study of depression in section 12.1 draws from Sohn, E. (2022). Tackling the mental-health crisis in young people. *Nature, 608*, S39–S41. https://doi.org/10.1038/d41586-022-02206-9; also the WHO fact sheet on depression, www.who.int/news-room/fact-sheets/detail/depression. Section 12.2 on explanation draws from Gopnik, A. (1998). Explanation as orgasm. *Minds and*

Machines, 8(1), 101–118; Weisberg, D. S., Keil, F. C., Goodstein, J., Rawson, E., & Gray, J. R. (2008). The seductive allure of neuroscience explanations. *Journal of Cognitive Neuroscience, 20*(3), 470–477. The discussion of explanation is also influenced by Strevens, M. (2004). The causal and unification accounts of explanation unified—causally. *Noûs, 38,* 154–179. The mention of the illusion of explanatory depth is based on Rozenblit, L., & Keil, F. (2002). The misunderstood limits of folk science: An illusion of explanatory depth. *Cognitive Science, 26*(5), 521–562. The discussion of scientific revolutions in section 12.3 is based on Kuhn, T. (1962/1970). *The structure of scientific revolutions.* University of Chicago Press (1970, 2nd ed., with postscript). The discussion of the chemical revolution draws from Donovan, A. (1993). *Antoine Lavoisier: Science, administration, and revolution.* Blackwell; the Lavoisier quote is drawn from this source as well. The discussion of Merton's priority rule is based on Merton, R. K. (1957). Priorities in scientific discovery: A chapter in the sociology of science. *American Sociological Review, 22*(6), 635–659.

CHAPTER 13

The case study about sexual selection draws from Roughgarden, J. (2004). *Evolution's rainbow.* University of California Press and (2009). *The genial gene.* University of California Press; Hrdy, S. B. (1874). Empathy, polyandry, and the myth of the coy female. In E. Sober (Ed.), *Conceptual issues in evolutionary biology.* Darwin, C. (1874). *The descent of man, and selection in relation to sex* (2nd ed.). John Murray; Knight, J. (2002). Sexual stereotypes. *Nature, 415,* 254–256. The mention of helicopter research is influenced by Minasny, B., Fiantis, D., Mulyanto, B., Sulaeman, Y., & Widyatmanti, W. (2020). Global soil science research collaboration in the 21st century: Time to end helicopter research. *Geoderma, 373,* 114299. The discussion of diversity in science and participatory research are influenced by Angela Potochnik's forthcoming *Science and the public, elements in philosophy of science* (J. Stegenga, Ed.). Cambridge University Press. Section 13.3's discussion of how values influence science is guided by Elliott, K. C. (2017). *A tapestry of values: An introduction to values in science.* Oxford University Press. Section 13.4 on changes and challenges in science is influenced by Jumper, J., Evans, R., Pritzel, A., Green, T., Figurnov, M., Ronneberger, O., & Hassabis, D. (2021). Highly accurate protein structure prediction with AlphaFold. *Nature, 596*(7873), 583–589; Callaway, E. (2022). What's next for the AI protein-folding revolution. *Nature, 604,* 234–238; Lucier, P. (2019). Can marketplace science be trusted? *Nature, 574,* 481–485. The cancer-risk of food example comes from Schoenfeld, J. D., & Ioannidis, J. P. (2013). Is everything we eat associated with cancer? A systematic cookbook review. *TheAmerican Journal of Clinical Nutrition, 97*(1), 127–134.

Index

Note: Page numbers in *italic* indicate a figure and page numbers in **bold** indicate a table on the corresponding page.

T - #0107 - 031224 - C378 - 246/174/18 - PB - 9781032290966 - Matt Lamination